PRAISE FOR

FOOTBALL FOR A BUCK

"Pearlman's enthusiasm for his subject is infectious . . . He has channeled his youthful affection into a raucous, well-reported, supremely entertaining ripsaw of a story." — *NEW YORK TIMES BOOK REVIEW*

"Tells in delightful detail the story of a long-gone pro football league. To call it just a sports book, though, sells it way short. *Football for a Buck* far transcends other sports books, making it contemporary and fun for anybody to read . . . The book is nothing less than painstakingly thorough, with dozens of zany and colorful stories about players and executives in a renegade league that quickly made its mark, frightening the mighty NFL . . . It is a thorough, funny, fascinating, and entertaining sports book." — *FORBES.COM*

"Endlessly fascinating . . . Pearlman proves to be the perfect person to write [the USFL's] history. He approaches the USFL with both the critical eye of a sportswriter and the unbridled enthusiasm of a fan. He's also quite funny . . . Above all, Pearlman is a master storyteller — he draws the reader in with his vivid descriptions of the league's wild games and wilder players. He's clearly done his homework, drawing from media coverage of the league and interviews with players, who look back on their USFL days with varying degrees of fondness. *Football for a Buck* is a hilarious, engrossing roller coaster of a book." — *NPR.ORG*

"An amusing and entertaining tale of ego and excess . . . [a] wacky, fascinating narrative." — *WASHINGTON POST*

"Every bit as fantastic as you'd expect a Pearlman book to be, and you honestly don't need to have spent time as a devoted USFL fan (as Jeff and I both were) to enjoy." — MIKE VACCARO, *NEW YORK POST*

"A knockout: an affectionate, often funny, sometimes horrifying history of the madcap league." — *NEWSDAY*

"Masterfully chronicled . . . Even without the Trump subplot, *Football for a Buck* is a fascinating read on the short life of an upstart league that almost survived. But Trump's presence lends the book a depth and urgency. And in chronicling the ways Trump delivers less than he promises, then leaves others to suffer the fallout, Pearlman has also written a playbook to the current presidency." — *TORONTO STAR*

"A rollicking new book that captures the dream and the disaster that was the USFL." — *ARIZONA REPUBLIC*

"A wonderful, thorough, insightful look at a deeply weird moment in American football history. And it's also a primer of sorts for this current deeply weird moment in American political history." — *YAHOO! SPORTS*

"Engrossing, eerily relevant." — *TIME.COM*

"If you like football — and outrageous tales well told — *Football for a Buck* is for you." — *INC.COM*

"When I started this book, I wasn't interested in the USFL. By the time I finished it, I wanted a sequel, a USFL movie, and Tampa Bay Bandits action figures. Jeff Pearlman's joy for this wild old league is contagious as hell. Great read. I'm in. GO MAULERS!!" — KYLE BRANDT, COHOST OF NFL NETWORK'S *GOOD MORNING FOOTBALL*

"As we've come to expect from Jeff Pearlman's books, *Football for a Buck* is deeply reported, deftly told, smart, hilarious, sad, riveting — and prescient. Through the stories of USFL businessmen, coaches, benchwarmers, and stars who later became household names — Steve Young, Jim Kelly, Herschel Walker, and yes, Donald Trump — Pearlman captures the troubled lavishness of mid-1980s America and foreshadows so many of the issues that professional football — and the country — are experiencing today. Oh, and there's a lot of sex and drugs."

— SETH WICKERSHAM, SENIOR WRITER, *ESPN THE MAGAZINE*

"If foreshadowing was passing yards, this book would be Jim Kelly. A definitive history of a wild and woolly football league fondly remembered three decades after its demise, but also a definitive preview to the forty-fifth president."

— L. JON WERTHEIM, *SPORTS ILLUSTRATED*

"Scathing, action-packed account of the rise and fall of spring football in the 1980s, with a familiar villain to the piece . . . Pearlman's fluently told story provides context for why the sitting president holds the NFL in such contempt — and why the sentiment should be richly returned. Gridiron fans of all stripes will find this a fascinating exercise in the collision of money, entertainment, politics, and ego."

— *KIRKUS REVIEWS*

"Wonderful anecdotes . . . Fascinating and hilarious reading on a half-dozen levels. Just great for football fans who like to laugh."

— *BOOKLIST*, STARRED REVIEW

"Pearlman's hundreds of interviews with former players and coaches shine a light on this almost forgotten league. This is an excellent book for football junkies, but it's just as enthralling for a general audience."

— *PUBLISHERS WEEKLY*

FOOTBALL FOR A BUCK

FOOTBALL

AUGUST 16, 1984

MR. DONALD J. TRUMP
THE TRUMP ORGANIZATION
725 FIFTH AVENUE
26TH FLOOR
NEW YORK, NY 10022

DEAR DONALD:

ON A NUMBER OF OCCASIONS OVER THE PAST MEETINGS, I HAVE
LISTENED WITH ASTONISHMENT AT YOUR PERSONAL ABUSE OF THE
COMMISSIONER AND VARIOUS OF YOUR PARTNERS IF THEY DID NOT
HAPPEN TO ESPOUSE ONE OF YOUR CAUSES OR AGREE WITH ONE OF
YOUR ARGUMENTS.

IT IS OBVIOUS FROM THE RECORD THAT YOU ARE A TALENTED AND
SUCCESSFUL YOUNG MAN. IT IS ALSO A FACT THAT I REGARD YOU AS
A FRIEND AND AN OWNER WHO HAS MADE A CONTRIBUTION TO THE LEAGUE
IN GENERAL AND BEEN A SAVIOR TO NEW YORK/JERSEY IN PARTICULAR.

WHILE OTHERS MAY BE ABLE TO LET YOUR INSENSITIVE AND DENIGRATING
COMMENTS PASS, I NO LONGER WILL.

YOU ARE BIGGER, YOUNGER, AND STRONGER THAN I, WHICH MEANS I'LL
HAVE NO REGRETS WHATSOEVER PUNCHING YOU RIGHT IN THE MOUTH THE
NEXT TIME AN INSTANCE OCCURS WHERE YOU PERSONALLY SCORN ME, OR
ANYONE ELSE, WHO DOES NOT HAPPEN TO SALUTE AND DANCE TO YOUR TUNE.

I REALLY HOPE YOU DON'T KNOW THAT YOU ARE DOING IT, BUT YOU ARE
NOT ONLY DAMAGING YOURSELF WITH YOUR ASSOCIATES, BUT ALIENATING
THEM AS WELL.

THINK BEFORE YOU SHOOT AND WHEN YOU DO FIRE, STICK TO THE MESSAGE
WITHOUT KILLING THE MESSENGER.

KINDEST PERSONAL REGARDS,

JOHN F. BASSETT
/ST

4221 N. Himes, Suite 201, Tampa, Florida 33607/Telephone 813-875-2059
A Charter Member of the United States Football League

FOOTBALL FOR A BUCK

THE CRAZY RISE AND CRAZIER DEMISE OF THE USFL

JEFF PEARLMAN

Mariner Books
Houghton Mifflin Harcourt
Boston   New York

First Mariner Books edition 2019
Copyright © 2018 by Jeff Pearlman

For information about permission to reproduce selections from this book,
write to trade.permissions@hmhco.com or to Permissions, Houghton Mifflin Harcourt
Publishing Company, 3 Park Avenue, 19th Floor, New York, New York 10016.

hmhbooks.com

Library of Congress Cataloging-in-Publication Data
Names: Pearlman, Jeff, author.
Title: Football for a buck : the crazy rise and crazier demise of the USFL /
Jeff Pearlman.
Other titles: United States Football League | One dollar league.
Description: Boston, Massachusetts : Houghton Mifflin Harcourt, 2018. |
Includes bibliographical references and index.
Identifiers: LCCN 2018006360 (print) | LCCN 2018007170 (ebook) |
ISBN 9780544453685 (ebook) | ISBN 9780544454385 (hardcover) |
ISBN 9780358118114 (paperback)
Subjects: LCSH: USFL (Organization) — History. | Football —
United States — History — 20th century. | USFL (Organization) — Anecdotes.
Classification: LCC GV955.5.U8 (ebook) | LCC GV955.5.U8 P43 2018 (print) |
DDC 796.332/64 — dc23
LC record available at https://lccn.loc.gov/2018006360

Book design by Chloe Foster

Printed in the United States of America
DOC 10 9 8 7 6 5 4 3 2 1

Frontispiece courtesy of the author

To Catherine Mayhew,
who hired a young punk when no one wanted him — then saved his career.
(Sorry about the police scanner.)

I worked for the NFL for several years, so I knew people there. They called the USFL "the Useless." They thought that was funny.

— Tena Black, Oakland Invaders sales promotions manager

You always joke when you're scared.

— Kit Eavensham in *Married for His Convenience*

CONTENTS

PROLOGUE

ERIC HEIGHT WAS a large man.

I hate to open this book in such a shallow manner, but when you're 17 and green, and your radar focuses only upon the painfully obvious, size leaps from the page.

So, yes, Eric Height, my AP English teacher at Mahopac High School, was plump. I actually have an image of the guy permanently embedded in the medial temporal lobe of my brain. He is standing before the class, blathering on about writing, cheeks turning a fruit punch shade of bright red. Spittle soars from his lips, and — as a teenager damned to the dreaded front-row seat — I skillfully bob, weave, and watch as Jonathan Powell, my pal and the kid situated to my rear, absorbs the moist blow.

Along with the spittle, Mr. Height does this thing when he turns particularly excitable. He marches back and forth before us, arms waving, hands quivering. He desperately wants us to cherish writing in the way he cherishes writing, and insists, on a near-daily basis, that few things are more powerful than word applied to page. He even invents his own organizational system, the Constructive Schematic, that he swears will revolutionize the way people approach literary efforts.

On this particular day, in April 1990, Mr. Height is at his most bombastic. The end of the year is nigh, and the time to assign the final project has arrived. There are 22 of us in the class, and you can feel the vibrations of the collective shudder. Jon Kozak, editor of the student newspaper, sits to my left. Phil Mazzurco, bulky star wrestler, is on my right. I gaze longingly across the room toward Cathy Iannotta, every boy's crush, then spin to face Powell, whose face is awash in terror. "Pearl," he says, "what do you think this guy's gon—"

He is cut off. Mr. Height begins to speak . . .

"You will all be writing a paper of at least 20 pages in length," he says. "On the subject of your choosing . . ."

I'm pretty sure I don't hear another word.

Back in 2007, when my daughter, Casey, was three, we took her to Disney World for the first time.

It remains one of the greatest experiences of my life, and not because of the overpriced turkey legs or endless loop of brain-melting songs. No, what made the time so special was Casey's wide-eyed amazement over absolutely everything. *Is that Sleeping Beauty! Can we ride in the teacups! Do they paint faces! Daddy, the parade is starting! The parade is starting! The parade is starting!* Decked out in her special light-blue princess dress and matching tiara, Casey had a quickened pulse and a racing mind, and all I wanted was to bottle up the joy and place it in an eternal lockbox.

Alas, the magic faded. The magic *always* fades. We grow, we evolve, we experience, we touch, we taste, we feel. Our young minds become exposed to other phenomena, and what once evoked euphoria ultimately brings forth but a shrug.

Some 24 years earlier, in the spring of 1983, *I* was the giddy one, turned wide-eyed and ecstatic via an image resting before me on a counter inside the Mahopac Public Library. There, on the cover of the March 7 issue of *Sports Illustrated,* was Herschel Walker, the reigning Heisman Trophy winner from the University of Georgia. A few days earlier, Walker had nuked the news cycle by leaving college after his junior year to sign with a new entity, something called the United States Football League. Because I was but a young boy, not yet 11 and preoccupied with my bike and my basketball and my pals up the street, the intricacies of Walker's decision were beyond my scope. I had no idea there was a new spring football league on the horizon. I had no idea the NFL didn't allow juniors. I had no idea Walker was being compared to Joe Namath, the Alabama quarterback who shocked the nation in 1965 by joining the upstart New York Jets of the American Football League. I had no idea the NFL was terrified by the looming challenge of a newbie that had just added America's greatest football prodigy.

Nope, the only thing I knew was that the photograph of Walker blew my dome. There the running back stood, resplendent in the red-and-white uniform of the New Jersey Generals, a helmet — red, adorned by five gold stars — resting atop his right knee. The words HITTING PAY DIRT accompanied Walker's smiling face, as well as THE USFL'S COMMANDING GENERAL, HERSCHEL WALKER. I quickly scanned the index and turned to page 40, where I was greeted by nothing short of a visual sports orgasm. It was a photographic spread featuring 12 football helmets, the like of which I'd never seen. There were speeding horses and exploding stars and bright sunbursts and a clenched fist gripping a lightning bolt. The colors were silver and gold and rust and black and . . . and . . . and . . .

Holy shit!

Holy shit!

Holy shit!

From that moment forward, the USFL had me hooked. I would declare myself a loyalist of the new league, and arrive at Lakeview Elementary School armed with arguments why the NFL (home to my beloved New York Jets) was tired and stodgy. I would talk up the USFL's uniforms, its team names (Gold! Bandits! Wranglers!), its young stars (Walker, Kelvin Bryant, Craig James), its hot coaches (Philadelphia's Jim Mora, Chicago's George Allen, Washington's Ray Jauch). Touchdowns were celebrated with eclectic dances and explosive spikings of the ball. Offenses were inventive and high-flying. In Boston, Breakers coach Dick Coury would let a different fan design one play every week. In Tampa Bay, the Bandits were owned, in part, by Burt Reynolds, Hollywood's biggest star. There were players with funky names like Jo-Jo (Townsell) and Putt (Choate). The cheerleaders dressed (and moved) like dancers from a Mötley Crüe video. The mascots were offbeat and confusing. The 1983 season was promising, the 1984 season (highlighted by a pair of rookie quarterbacks named Steve Young and Jim Kelly) magical, the 1985 season transcendent. I had yet to kiss a girl, but I was attached at the hip to my first love.

Without exaggeration, I considered the USFL to be the absolute greatest thing the earth had hosted in its 4.543 billion years of existence. I couldn't wait for the next San Antonio Gunslingers–Arizona Outlaws clash. Or the next Los Angeles Express–Denver Gold clash. Or the next . . .

Then, without much warning, it vanished.

Forever.

I am back in Mr. Height's class. A couple of days have passed, and we are all debating what to write. The selections are generally obvious and lame. One kid picks an examination of Ronald Reagan's legacy. Another writes on the history of Lake Mahopac. We have a five-minute teacher-student powwow to suggest an idea and, hopefully, have it accepted. Thus far, no one has been turned down.

"Jeff," Mr. Height says, "what are you thinking?"

"Well," I reply, "it's a weird one."

To his credit, Mr. Height encourages weird.

"There was this football league, the USFL," I say. "I want to do a paper about what it was and why it went away."

Silence.

More silence.

"I don't know," he says.

More silence.

"Of all things, why that?" he asks.

I have an answer prepared. Because, I tell him, the USFL was awesome and colorful and explosive and amazing and killer and beautiful and brought me more happiness than I'd ever known. Because it was sports the way sports should be, and it caused me to pay attention and study everything and watch and learn. Because I could make the paper really great and maybe even get it published somewhere and . . .

"OK," he says. "But I'm a little worried."

I devote the next four weeks to living and dying with a league that lived and died. I find an old phone number for the USFL's Manhattan office, dial the digits, get a message. *Thank you for calling the United States Football League. No one is here to take your call right now, but if you leave your name after the tone . . .*

I leave messages. Lots of messages. But I never hear back. When the official submission date arrives, I present Mr. Height with the lamely titled "The Downfall of the United States Football League." The 21 other students submit papers that vary in length from 20 to 24 pages.

Mine is 40 — single-spaced. I have never felt better about a project, and tell Mr. Height as much. He nods and says, "Well, let's wait and see."

Three weeks later, the papers are returned. My palms are sweaty. My heart is racing. I wonder aloud whether "The Downfall of the United States Football League" will generate an A or an A+. Maybe an A++, if such a thing exists. Mr. Height walks from desk to desk, slapping down the thick packets like pancakes on a griddle. He slips past Powell, then comes to me. It's a slow-motion sort of thing — I can still see his hand delivering the goods with a solid *plop!* Mr. Height strolls past, and I turn the project over to spot, in bright red, a B+. The grade is bad enough, but it is the accompanying comment that haunts my soul for decades.

Solid job. But I feel like there's much more to this story.

I am angry.

But, ultimately, I am awakened.

Mr. Height is 100 percent correct.

There is much more to this story.

FOOTBALL FOR A BUCK

1

ORIGINS

Early on everyone was into the idea of having a budget and sticking to it. But sports is a dizzying thing. It's the only business to have its own daily section in the newspaper.

— Randy Vataha, co-owner, Boston Breakers

THE IDEA WAS a good one.

That's the thing lost long ago, well before men like Steve Young and Jim Kelly and Reggie White became Hall of Famers; well before Donald Trump became a reality-TV-personality turned 45th president of the United States; well before the NFL owned Sundays and Monday nights and Thursday nights and Thanksgivings and Christmases; well before people were paying tens of thousands of dollars not for a seat to a football game, but for the *right to buy a seat to a football game.*

Wait.

Hold on.

Let's start again.

The idea was not a good one.

The idea was a *great* one.

It was hatched way back in 1961, when a little-known New Orleans antiques and art dealer named David Dixon wondered why the National Football League — an entity he both loved and resented — seemed steadfast in its refusal to move into new cities. "It dawned on me that the NFL never in its history expanded willingly," he said. "They always expanded to head off competition." As a Tulane University graduate and son of the Crescent City, the 37-year-old Dixon's first priority was landing his hometown a franchise in the 14-team NFL. So he thought, *I should threaten the NFL*. Not with a lawsuit, however, or taunts and insults. No, Dixon decided he would provoke the NFL . . . with the idea of a spring football league. "I did reason correctly that football is known as a fall sport simply because Rutgers and Princeton played their famous first game of football in the fall," he once recalled. "And the sport took off from there. If those two great institutions had played their first game in the spring, the chances are football would have become known for its spring tradition." Dixon broached the concept with friends and colleagues. Everyone seemed to like it. Spring wasn't merely the season of love and renewal. It was also a vast sports wasteland. The heart of Major League Baseball season could be painfully dull. March Madness had yet to become March Madness. Hockey? Yawn. Horse racing? Double yawn. It all made too much sense. At the time there were the NFL and its rival, the American Football League. Both played in the fall, both fought for players and attention, both were transforming the United States from a baseball country to a gridiron country. In particular, Dixon was inspired by the AFL, which commenced in 1960 with little fanfare and minimal attendance, but — thanks to a generous television deal with ABC — quickly established itself as a thrill-per-minute profitable alternative to the NFL. With its colorful uniforms and wide-open offenses, the AFL was electric viewing, the kind of thing that kept Dixon's eyeballs glued to the screen.

So, with nothing to lose, in 1963 he took a trip to La Jolla, California, to visit Paul Brown, the legendary former Cleveland Browns coach who had recently been fired from the team he founded over a power struggle with Art Modell, the owner. What was supposed to be a quick meeting spanned more than nine hours. Brown was mesmerized by Dixon's vision, and as they shook hands and parted ways the old man looked the entrepreneur in the eyes and said, bluntly, "Never, ever let anyone talk you out of doing this. This will work." Dixon was giddy, especially when — powered by Brown's endorsement — he began to receive commitments from a bevy of Fortune 500 mainstays. Kemmons Wilson, founder of Holiday Inn, wanted a team. So did

Jerry O'Neil, General Tire and Rubber Company's president. And Nelson Bunker Hunt, oil tycoon and one of the world's richest men. And Gussie Busch, of the Anheuser-Busch beer empire. More and more powerful, ambitious men with means called Dixon to express their interest. "By this time," Dixon said, "even a slow-witted southerner knew he had a tiger by the tail."

The headline was splashed across newspapers all around the country — NEW FOOTBALL LEAGUE ANNOUNCED. The date was June 25, 1966, and the Associated Press story spoke of a league — the United States Football League, to be precise — with "a pooled $24 million" and a group of potential owners who would be "willing and adequately prepared to spend money for outstanding players — and can go to half a million dollars to get any one of them." Frank Leahy, the former Notre Dame coach, would serve as chairman of the board. Bruce Erlhof, an attorney from Santa Ana, California, was executive vice president. "I always liked the name United States Steel Company," Dixon said. "I had a share of U.S. Steel company stock when I was a young boy and I liked the name. I thought U.S. Steel . . . you know, that was great. So I went with USFL." It was terrific. There would be a draft and an opening day and . . .

No.

Earlier that month the NFL merged with the AFL, then expanded into (among other cities) New Orleans. In the name of civic pride, Dixon was thrilled when the Saints were born, but knew his idea was dead. None of the television networks expressed an interest, largely out of fear of crossing the established league and its bulldog young commissioner, Pete Rozelle. The bushel of wealthy owners quickly dissolved. Dixon dropped out, too. Before long, talk of the USFL petered out. There would be no more headlines and no grand plans.

Spring football?

Who needed spring football?

The years passed. The NFL and AFL merger created a juggernaut of unstoppable professional-sports dominance. By 1979 the league was home to 28 franchises, with an average value of $30 million. Both NBC and CBS televised games, as well as ABC every Monday night. Whereas once the famed American athletes emerged primarily from the world of Major League Baseball, we now had household names wearing helmets and shoulder pads. Terry Bradshaw. Roger Staubach. Walter Payton. Tony Dorsett. Though baseball loyalists continued to refer to their sport as the national pastime, the sentiment was nonsense. The NFL was king.

Dixon was no longer merely an art and antiques collector with some wealthy friends. He had helped convince the NFL to add the Saints and urged John McKeithen, Louisiana's governor, to endorse the financing of a domed stadium, ultimately to be known as the Louisiana Superdome. Then he joined the effort to create World Championship Tennis (WCT), and persuaded many of the planet's greatest players to enlist in the professional ranks. "Dave was all about shaking things up," said Tim Brando, the longtime sportscaster and native Louisianan. "He was an incredible promoter always looking for the next big thing."

By the late 1970s Dixon really wasn't looking. In 1979 he underwent open-heart surgery and a triple bypass. His doctors insisted he rest and relax and take up something soothing, like knitting or bird-watching. He was a content Saints fan, no longer infatuated with bringing forth a challenger. "The way I figured it," he said, "I needed to start a new league like I needed a hole in the head." But as much as he tried to focus on peace and tranquility, Dixon couldn't stop thinking about this crazy phenomenon called cable television. Thanks to coaxial cables and radio frequency, there was suddenly a way to broadcast sports on channels other than the three major networks. He started hearing about something called ESPN, which launched on September 7, 1979, as the world's first 24/7 all-sports, all-the-time network. Dixon pondered back to the 1960s, when ABC, NBC, and CBS wanted nothing to do with his plan. He pondered back to the now-defunct World Football League (WFL), a 1974–75 fall rival to the NFL that lasted only one and a half seasons, in large part because the games appeared only on the TVS Television Network, a little-seen syndicator of sports programming. "Dave was a visionary, and visionaries don't rest easily," said Jerry Argovitz, a players' agent and future USFL team owner. "He had this idea in his head and he wouldn't let it go. Cable TV changed everything to Dave."

Just as he had once paid a visit to Paul Brown, Dixon now made a trip to

Palos Verdes Estates, California, and the home of George Allen. A revered NFL coach who guided the 1972 Redskins to the Super Bowl, Allen's last taste of the sidelines had been an ugly one: in 1978 he was hired to coach the Los Angeles Rams, then dismissed six months later after players revolted over his tyrannical rules and demands. For all his flaws (and there were many), Allen's love of football was never questioned. So Dixon knocked on his door, sat on his couch, sipped from a glass of orange juice, and explained his idea.

"My God, this will work!" Allen said, echoing Paul Brown's words from 15 years earlier. "Don't let anyone talk you out of this!"

Dixon floated out the door, convinced his concept finally had legs. His next step was to hire a public opinion firm, Frank N. Magid Associates, out of Cedar Rapids, Iowa, to conduct a national survey on the subject of spring football. The results, presented in a 126-page document, were, technically, mixed. But not through the prism of Dixon's rose-tinted glasses. Magid made 800 telephone calls, and 600 people identified themselves as having "an interest" in professional football. Wrote Jim Byrne in his book about the USFL, *The $1 League*: "As far as Dixon was concerned this meant that three out of four adult Americans were professional football fans. This was 'incredible' in Dixon's view. It was also a considerable exaggeration. Respondents who expressed an 'interest' in watching a USFL game 'would watch' in Dixon's analysis. This led him to conclude that 67 percent of pro football fans would watch a USFL game and 82 percent of NFL fans would watch the USFL on television. He described the former as a 'powerful figure — extremely encouraging under any circumstances'; the latter as a 'bonanza for the USFL by any measure.'"

In other words — it was on.

In November 1980 the USFL made its first-ever hire, naming former Stanford and Denver Broncos coach John Ralston as the league's chairman. This was one of the most important moves of Dixon's business career, and not because Ralston was particularly bright or articulate. No, what the 54-year-old ex-coach added was gravitas. He was a man who could tell stories about fourth and 3 against USC at the Los Angeles Coliseum; about staring down Kenny Stabler and the Oakland Raiders. He knew both the college and professional worlds, and offered in-the-flesh proof this new league would be no joke. Before long, Dixon and Ralston were traveling the country, seeking out potential owners with two primary qualifications:

1. Deep pockets
2. Large egos

"We just started virtually calling people on the Forbes Top 400 richest people in the world," Ralston recalled. "We called on people all over the country." Dixon and Ralston were a dreamy marketing tandem. The old coach regaled with sideline tales that played to the egos of people itching for access and glory. The businessman followed with statistics, figures, details, information. Dixon was a mixture of P. T. Barnum and Dale Carnegie. The USFL, he raved, would be like no other league. "We are going to play football, the most wildly popular sport in the history of this country, at a time of year when there is no football competition whatsoever!" he raved to potential investors. "No NFL, no college, no high school!"

Dixon's closing line was his most powerful. "I've never believed that Americans buy chewing gum only in the spring, make love only in the fall, go to movies only in the summer," he said. "We will play spring football. And people will watch."

Would they watch? Who the hell could say? But they *might* watch. And *might* wasn't so far from *will*. One by one, Dixon and Ralston went on the recruiting warpath. This wasn't merely a football league, they insisted, but a revolution. The USFL would be the NFL, only better and smarter. Each team could boast a couple of superstars, but the majority of rosters would be regionally stocked via a specific player-allocation system. They explained how a franchise in Southern California, for example, would consist primarily of standouts from schools like UCLA, USC, and Stanford; how a Chicago operation would pluck from such colleges as Illinois and Notre Dame. Most important, this wouldn't be a league where the filthy rich held competitive advantage over the somewhat filthy rich. Every team, wrote Byrne, "would operate under strict budgetary controls with a limitation on player salaries dictated by anticipated income from broadcast, attendance and marketing/licensing revenues." Franchises sold for just $50,000, but an irrevocable $1.5

million letter of credit was also required, as well as a pledge of at least $6 million for operating expenses for the first season.

Again — the idea was a *great* one.

"Revolutionary," said Argovitz. "I don't see how there was much fault to it."

Dixon arranged multiple meetings for potential owners, a conga line of the rich and (aspiring) famous. Many came once or twice, then vanished. Herb Kohl, future Wisconsin senator, repeatedly promised to build a team in Chicago — then stopped returning calls. Others stuck. On March 7, 1981, Dixon and 14 strangers gathered in a Chicago hotel suite. Two of the attendees were Bill Daniels and Alan Harmon, Denver-based cable TV executives. They wound up with the San Diego team. John Bassett, a Canadian film producer, once owned the Memphis Southmen of the WFL. He had little desire to sink more of his dough into professional football — until he met Dixon and fell in love with the plan. He committed a team to Florida. "I remember when I was first approached about this league, I said, 'No way,'" Bassett said. But upon hearing of the other owners, "I thought they'd really be astute, budget-conscious guys." One by one, people signed up. Allen, the longtime coach and early USFL enthusiast, partnered with a noted cardiovascular surgeon, Dr. Ted Diethrich, to be the leaders of the Chicago club (Allen was named team chairman of the board). Tad Taube, a San Francisco real estate developer, put his foot down on the Bay Area. Marvin Warner, a banker, former U.S. ambassador to Switzerland, and onetime owner of 48 percent of the Tampa Bay Buccaneers, claimed Washington, D.C. — but then opted for Birmingham. George Matthews, a Massachusetts businessman, snagged Boston along with his partner, former Patriots wide receiver Randy Vataha. Ron Blanding, another real estate developer, secured Denver. A. Alfred Taubman, one of the United States' wealthiest men and a shopping mall developer, would go with his hometown of Detroit.

Before long, the USFL's 12 first-year franchises were all claimed.

In October 1981 many of the owners gathered at the Fairmont Hotel in San Francisco for an organizational meeting. It was a who's who of new ownership, as well as Dixon. Peter Spivak, a former judge of the Third Judicial Circuit in Michigan and Taubman's co-owner, was named the USFL's acting chairman. Leslie Schupak, the cofounder and senior managing partner of KCSA Worldwide, a highly regarded New York City–based marketing firm, was flown in to make a presentation along with his partners, Herb Corbin and Les Aronow. "It was a big deal," Schupak said. "And we're all sitting around a large table, and there's a speaker phone in the middle. Everyone's chatting, time is passing, and we're waiting for the [prospective] owner of the New York franchise. Food comes—he's not there yet. The time goes from 12:30 to 12:45 to 1:00. Where is this guy?" Finally, the phone rang. Only no one in the room knew how to answer. After fumbling with the buttons, Daniels mastered the technology.

"Hello! Hello!"

The voice emerging from the speaker was loud and young and brash.

"Hey, guys! How are you? Hello? Hello? It's Donald."

Donald—as in Donald Trump, the 35-year-old New York City real estate developer.

"Donald," said Dixon, "we're all waiting for you. Where are you?"

A pause.

"Right, well, I've gotta apologize. I'm not gonna be able to make it. In fact, I can't be a member of your project. Things are just going unbelievably for me. I have this casino project, and it's gonna be big. Really big. Very exciting. Lots of money in it. OK, so anyway, sorry. Bye."

Click.

Dixon's face turned ashen. He asked Schupak, Corbin, and Aronow to leave the room for a few moments. "We went to the lobby," Schupak recalled. "And Herb looks at me and says, 'We came all the way out here for nothing. They're going to disband.'" Schupak wasn't having it. He dashed out to a nearby convenience store and came back with a small bag filled to the brim. When the trio returned to the meeting, the owners were speechless. Without a New York franchise, the USFL was dead. There was no debating the point.

"So I gave my presentation," Schupak said. "And then I got original." He pulled a bottle of aspirin from his bag and said, "We know you guys have just had a really big headache. But I have an antidote for that. This won't be the first time you have these headaches. It might get worse. You might have real indigestion — but I have an antidote for that." Once again he reached into the sack, this time retrieving a pink bottle of Pepto-Bismol.

Schupak paused for dramatic impact.

"Look," he said, "the one thing we are confident of is that the USFL will be a success. With or without the people who are supposed to be here. And when we make the official announcement of the league and have the first ball kicked off a tee, we're going to do this . . ." He stuck his hand into the bag for a final time and removed a bottle of champagne.

George Allen, standing in the rear of the room, sprinted forward. "I love these guys!" he screamed. "I love them! We're gonna make it!"

The other owners joined in, forming the first victory dance in league history.

Seven months later, theory became reality.

On May 11, 1982, Dixon and a roster of owners and co-owners booked a room inside New York City's famed 21 Club to officially announce what had been rumored for some time: come spring 1983 the NFL would no longer be the only professional football league around. Trump was a distant memory, replaced in the New York/New Jersey market by a well-financed Oklahoma oilman named J. Walter Duncan (he initially craved Chicago, but when Trump bailed, Duncan agreed to New Jersey). Berl Bernhard, a senior partner in a law firm, was owner of a Washington, D.C., franchise. Behind closed doors, there were significantly more questions than answers. The league still lacked a commissioner, as well as a television deal. There were no team names, no rosters, no stadium deals, no concessions or publishing or legitimate financial figures to disclose.

Schupak's firm was once again brought in, this time to pull off the event. About an hour before the press conference it occurred to the marketing wizard that the USFL had neither a football nor helmets. "So I ran out to the Modell's [Sporting Goods] on 42nd Street and bought two footballs and two helmets," he said. "We came up with this nice red, white, and blue USFL logo, and we placed decals on the ball and the helmets. You'd think these things are all thought of well in advance. That wasn't exactly the case."

With two dozen reporters in attendance, Spivak stood in the middle of a long metal table, surrounded by the 12 owners, a helmet, and a ball with a crudely placed sticker. Flashes flashed. Cameras rolled. Notes were taken. The press were offered tiny bottles of Coca-Cola and 7 Up, and some forgettable finger food. Spivak cleared his throat and leaned into the microphone. "Our league believes that the sports fan in the United States wants to see more than the current 16-game professional football season," he said. "After all, his favorite baseball team plays a 162-game schedule and the basketball season runs 82 games."

He explained, in broad terms, that the men before them pledged to commit more than $100 million in capital over the ensuing two seasons. "[Our owners] have the financial strength, personal reputation, and accomplishments that are equal of any group of professional league owners today," he said — and a lie this was not. These were *wealthy* men. The details were intriguing. The 12 markets would be: Birmingham, Boston, Chicago, Denver, Detroit, Los Angeles, New York/New Jersey, Philadelphia, San Francisco, San Diego, Tampa, and Washington. The emphasis would be on fast, fun football. The message was clear. The NFL was fine, but stilted and uninteresting. The USFL would be energetic, powerful, thrilling.

The following day's reaction hardly inspired confidence. None of America's major newspapers covered the press conference as a particularly big event, and the *New York Times* seemed to go out of its way to minimize the news. In a small article on the third page of the sports section, Michael Strauss wrote that Spivak was "vague in response to several questions as to how his group could gain credibility and strength in an era when the efforts of other fledgling sports leagues, such as the World Football League, had ended in failure."

And yet, the league's owners were enthusiastic. Mike Trager, the USFL's media consultant, had debated whether to hold the press conference before or after securing a television contract. "I told everyone, 'Look, we might be able to get a TV deal first, but we'd have no leverage,'" he said. "We needed a dynamo press conference that said, 'We're playing and it's going to be amazing.' That's exactly what we did, and it was the best thing we could have done.

All of a sudden we were legit. It turned out to be the biggest move in those early years — announcing first. It gave us a lot of leverage."

Behind the scenes, the USFL was working on two important moves: finding a full-time commissioner and landing a major broadcast deal. The tasks were not mutually exclusive. Back in the late months of 1981, when Dixon was deep into pursuing his dream, he placed a call to Rozelle, the legendary NFL commissioner, and asked what would need to be offered for him to jump leagues. It was, all later acknowledged, a pipe dream — with a $400,000 salary plus myriad perks, Rozelle was one of the three or four most powerful people in sports. But Dixon knew the weight the NFL commissioner carried. Were he to come to the USFL, the networks would follow.

With Rozelle's rejection, Dixon and Co. went after choice number 2. More than two years earlier, when the USFL was little more than a thought in his head, Dixon reached out to Chet Simmons, president and chief operating officer of ESPN, to solicit his opinions on spring football and television. The conversation, via phone, was purely informational — *Here's what we're doing, keep your eyes open for us, maybe we can work together in the future.* Now, however, the league sought out a leader who could see beyond the playing field. Simmons was, at age 53, on the verge of becoming a television legend; a visionary who, while at ABC, helped create *Wide World of Sports* before spending 15 years at NBC Sports and the past three years at ESPN, which went from fringe to mainstream at breakneck speed. "I was at ESPN at the time, and I kept hearing about this new football league by David Dixon and I said, 'Who the hell's David Dixon?'" Simmons recalled. "I had enough problems and challenges at ESPN to worry about a new football league, but like everything else in sports I think that if somebody had an idea, and something was new, I'll keep watching; keep my eye on it." In the absence of a commissioner, Trager had been doing some good work in the TV department. The USFL secured lucrative multiyear television deals with ABC *and* ESPN (ABC paid $9 million to televise a Sunday national game, ESPN $4 million for games on Saturday and Monday nights). The contract required the USFL to schedule a minimum of three games on Sunday, with ABC guaranteed to broadcast one game nationally as its "Game of the Week" or two or more regionally. The deal included no clauses regarding "blackouts" (local games are not televised unless they're sold out) — something the league would soon regret. ABC was obligated to televise 21 games in 1983, and ESPN committed to two prime-time games per week. "It's instant credibility," said Steve Ehrhart, the league's soon-to-be-hired executive director. "Having your games televised across the country . . . that says you're real. You're legitimate and authentic."

On June 14, barely a month after the press conference, Simmons was hired to become the USFL's first commissioner. He was a unanimous choice by the owners and — wrote Jim Byrne — confirmed "a widely held view among the media that the USFL was a 'made-for-television' league."

He was also a bit of an odd fit. Chet Simmons didn't *love* football the way Rozelle loved football. Really, his first passion was televised sports. He had no hobbies or genuine outside interests, and would spend his free time at home, sitting before three side-by-side-by-side TVs, watching various sporting events and paying close attention to the commentary, the camera angles, the zoom lenses. First downs? Ho-hum. Deep bombs down the sideline? Interesting, but mainly because the arc of a Steve Bartkowski bomb filled the screen like a glorious rainbow. Also, while he had been calling the shots for much of the past 20-plus years, it was always with network underlings who knew their places. The USFL — with 12 wealthy ownership groups looking out for their own wealthy interests — would be a different beast. Simmons believed he could handle it, but did he know, with 100 percent certainty, this was the job for him? No. "Chet wasn't an Xs and Os guy," said Doug Kelly, the league's coordinator of information. "But that's not what we wanted him to be. We knew sports were increasingly TV driven. And *he was* sports television."

Before long, everything seemed to be flowing beautifully. The season would last 18 games (as opposed to the NFL's 16 games), and the players and coaches would play central roles in selling the product. Unlike the NFL, which seemed happy to hide its gladiators beneath helmets and shoulder pads and strict media guidelines, the USFL wanted its participants front and center. So there would be midgame sideline interviews, extra media days, endless promotional appearances and fan engagements. "It felt like the people's league," said Jed Simmons, Chet's son. "They wanted to be more out there than the NFL." The powers that be were also willing to mess with well-established pro football rules. For example, the USFL would break with the NFL and allow teams to go for a two-point conversion after touchdowns. Also, in other departures, the USFL clock would stop after first downs in the final minutes of both halves and overtime, and wide receivers had to have just one foot down — not two — for a catch. Kickers could use one-inch tees on field goals and extra points. "All in the name of excitement," Ehrhart said.

Although it was far too early to sign players, the league committed itself to the hard pursuit of veteran coaches, and it paid dividends. Allen was officially named the head man in Chicago and Ralston in Oakland. (The franchise initially planned on rotating between San Francisco and Oakland. But because of lease issues with Candlestick Park, San Francisco fell through.) The New York/New Jersey organization hired Chuck Fairbanks, the legendary University of Oklahoma coach, and Denver swooped up the wildly popular Red Miller, who guided the Broncos to Super Bowl XII in 1978. Though they would play nary a down, the coaches added immediate credibility. This wasn't a semipro league coached by Biff, the part-time truck repairman who once played a little Division I-AA ball at Delaware. No, these were titans of the sport, here to place the USFL on the map.

With the hirings came positive headlines. BY GEORGE, IT'S ALLEN crowed the *Chicago Tribune,* and MILLER STRIKES GOLD AS DENVER COACH raved the *Philadelphia Daily News.* Yet, beneath the surface, there were concerns. The concept of football in the spring seemed a confusing one for some to grasp. In a brief *New York Times* editorial headlined UNSEASON-ABLE, the newspaper's editorial board made its case against the league:

> Doesn't the U.S.F.L. season run straight into the basketball, hockey and baseball seasons? Of course, but so what? None of them get great national TV ratings in those months, says the new league's president. "The second quarter of the year is traditionally very good for male demographics. With the weather bad, you might get women to watch football, too."
>
> Somehow, we doubt it, and find ourselves hoping that men won't

either. You don't have to be a quiche eater to want to see the seasons change. You can like football and still be glad there's an off-season to whet the appetite for its reappearance. What the male, and female, demographics around here suggest is that there can be way too much of a good thing. When football season ends, we'll be content to wait till next year.

While some regions expressed tremendous glee in having the USFL come to town, others were less enamored. In Boston, the NFL's New England Patriots owned Schaefer Stadium, and weren't about to lease it to a rival league. San Diego's Jack Murphy Stadium was owned by the city, but the three tenants (football's Chargers, baseball's Padres, soccer's Sockers) repeatedly objected until Daniels and Harmon — holders of the USFL's Southern California territorial rights — abandoned the municipality for Los Angeles. As a result, Jim Joseph, who first planned on having a team in his hometown of San Francisco, then in L.A., shifted to Phoenix.

Ah, spring football in Phoenix (average high temperature: 94 degrees) . . .

Geography, though, would largely work itself out. What kept Dixon up nights was a lingering fear that belied his peppy chatter: ego. Actually, two lingering fears: ego and greed. Having watched the World Football League crumble, John Bassett shared Dixon's concerns. It was all well and good to hear owners talk unity and restraint and moderation. But enacting such discipline was an altogether different exercise. What would happen once fans demanded results? What would happen when an expensive NFL quarterback was available? What would happen if a rival club began throwing around big dollars? During an owners' dinner around this time, Bassett was asked what it would take for the USFL to make it. Sipping from a martini, he stood, took a drag from a cigarette, and said dryly, "We'll have 12 teams. Every week six are going to win and six are going to lose." He returned to his chair. "John was dead on," said Randy Vataha, a co-owner of the Boston franchise. "It was perfectly stated."

Not long after the league's introductory press conference, Dixon issued a memo to the owners that concerned what they could expect, financially, from the first season. He wrote that each team would likely generate $4.1

million in revenues (based upon $2.5 million in gate receipts, $1.5 million in television revenues, and $100,000 in radio revenues). There was, however, a caveat. In a follow-up memorandum, he wrote, "40 percent of revenues for player salaries is not unreasonable; 45 percent would be the basis for a maximum if one was to use a maximum. On 45 percent, the salary budget would be $1.845 million, which allows plenty of money to sign one or two glamorous players, particularly with area notoriety." Ultimately, the league's owners embraced the idea of setting a spending limit of $1.8 million per team, covering the costs of the 40-man roster and 10-man developmental squad. An additional $500,000 was allotted to sign the two "star" players who did not count against the cap. "That way you could sign a No. 1 pick and not break the bank," said Tom Brand, the Tampa Bay Bandits' director of player personnel. "It was a smart plan."

It was never formally voted upon.

"We had a gentleman's agreement," said Tad Taube, owner of the Invaders. "Of course, that's only OK as long as you have gentlemen agreeing."

This is where the fun begins.

First, with the names. You can't have football teams if you don't have names, so the USFL's new owners jumped right in. Some seemed to use the unique branding opportunity to bore the regional populace. The Philadelphia franchise, owned by Myles Tanenbaum, held a naming contest via TV and print advertising, then narrowed the thousands of suggestions down to Fireworks, Colonials, Franklins, Spirits, Militia, Sentinels, and Stars. Tanenbaum chose Stars, because, well, *zzzz*. In Denver, team vice president Bill Roth said the name "Gold" was selected because the malleable, ductile metal "represents first place or first class. Also, the cities around here were formed during the gold rush." Again, *zzzz*.

Washington brought forth the Federals ("the essence of our government structure," raved the USFL's official magazine), Chicago the Blitz (suggested in a contest by a man from Naperville, Illinois, named Mike Wehrli), Birmingham the Stallions (owner Marvin Warner liked horses), New Jersey the Generals (Fairbanks dug military evocations), Boston the Breakers (George Matthews was an avid sailor), Michigan the Panthers (a "screw you — we're here, too" nod to Detroit's Lions), Arizona the Wranglers, Los Angeles the Express. (Daniels and Harmon initially selected "Olympics," but trademark complications nixed the idea. Then they were going to go with "Satellites," a nod to their work in television. But they agreed the name stunk. "Express" worked well enough.)

Of the 12 monikers, two leapt from the page. In Oakland, the NFL's Raiders had recently announced their relocation 371 miles south to Los Angeles, thereby crushing one of America's most passionate fan bases. Howard Friedman, the USFL franchise's vice president, explained to writer Dan Herbst that "Invaders" was purely coincidental — as was John Wilkes Booth's appearance at Ford's Theatre the night a president was in attendance. "A lot of people have read into that," said a smirking Friedman. "The fact is, we liked the idea of the strength of an Invader."

Tampa Bay, meanwhile, settled upon "Bandits," which was presumed to be influenced by the presence of Burt Reynolds, the actor/minority investor whose biggest-grossing film was the 1977 non-classic *Smokey and the Bandit*. Truth be told, Bassett's daughter, a professional tennis player named Carling, owned a German Shepherd named Bandit.

So, naturally, the Bandits mascot was *a guy riding a horse*.

2

ALBERT C. LYNCH

You came to the USFL either as a marquee guy brought in to bring credibility to the league, or as a washed-up NFL reject. That was me. The reject.

— Matt Braswell, center, Michigan Panthers

THEY STARTED WITH nothing.

No desks.

No staplers.

No stacks of paper.

Not even a name.

When the franchise that would ultimately be known as the Chicago Blitz first formed in June 1982, it was little more than an owner (Ted Diethrich), a coach (George Allen), and a vacant office suite on the 18th floor of the Merchandise Mart, a downtown building situated alongside the Chicago River. The space had been rented by Diethrich because, located 1,453 miles away in Phoenix, he needed a presence in the city where his team would play. So he dispatched Bruce Allen, the coach's 25-year-old son and newly appointed general manager, to handle the organizational intricacies of kicking off a franchise lacking organizational intricacies. "We had no money," said Bruce Allen. "Nothing really to work with."

The early Blitz staff featured the young Allen; a business manager named Jim Banahan; Mike McCarthy, the director of player personnel; and Kay

Fred Schultz, the pipe-smoking media relations director. Every morning at six o'clock the men would congregate in Suite 1889, then spend the next 16 hours compiling lists, devising ticket plans, thinking up ways to birth a team in a city already dominated by the NFL's Chicago Bears. "We were staying in the downtown Marriott for a while," said McCarthy. "It got too expensive, so I moved to Banahan's couch. The Playboy Club was still around, and I got a membership so we could go. And Bennigan's had Tuesday-night Heineken Night, where every bottle was $1. That's how we survived — on football and drinking."

From the get-go, the Blitz were unlike any other USFL franchise. While the Washington Federals and New Jersey Generals and Los Angeles Express and the other eight clubs had yet to even look at office space or ponder the future, the Blitz — boosted by George Allen's directive and Diethrich's meaty wallet — charged ahead. Eleven years earlier, when George Allen was in his first season as coach of the Washington Redskins, Edward Bennett Williams, the team's president, famously said, "I gave George an unlimited expense account, and he's already exceeded it." Those words affixed themselves to Allen, but he didn't mind. Yes, George Allen spent excessively. But for a man who compiled a .712 winning percentage in 12 years as a head coach, it was all in the name of victory. Even as he stayed behind at his home in Palos Verdes Estates, California, his drive to succeed extended eastward. "The guys in the office in those early days, we'd go out most nights," said Bruce Allen. "We'd go to a bar or restaurant, and every night we had to sell two tickets to a future Blitz game. Just two. Usually to someone in the bar. And whoever didn't sell had to pay for the next dinner. All we wanted to do was build something big."

By August 1982, as the rest of the league was largely asleep, the Blitz were arranging a series of free-agent tryout camps across the nation. They held nine in a three-week span, highlighted by two in Chicago and a scattered handful in, among other cities, Milwaukee, Dallas, and Indianapolis. The first, scheduled for August 21 at Soldier Field (future home to the Blitz), brought a whopping 518 men into the stadium. United Airlines served as sponsor, so every participant received a red-and-white T-shirt with the Blitz logo screen-printed on the front and the airline logo on the rear. Thanks to George Allen's connections, the Blitz were able to use the field for free, but on the condition that workouts were held while the turf was being painted. "I asked, 'How the hell do we try out 500 guys while they're applying paint?'" said McCarthy. "Somehow we did."

The system was simple: a dozen temporary coaches worked out the players, and those who impressed were sent toward the stands, where Bruce Allen awaited with a contract and an invitation to a future mini-camp. Only 13 were signed, and none wound up making the Blitz. Yet one reigns eternal . . .

"Albert C. Lynch!" McCarthy said some 30 years later.

"Ah, Albert C. Lynch!" bellowed Bruce Allen.

Indeed, Albert C. Lynch was a 24-year-old Chicago native with no football experience. He stood five foot eight and weighed 185 pounds, and clocked a turtle-like 5.7 in the 40. His vertical jump measured 23 inches (last among all signees), and he was the only man who failed to break 3 seconds in the 20-yard sprint. He was listed as a linebacker, though no one is quite sure how or why. According to a file labeled BLITZ MINI-CAMP I and dated November 13, 1982, Lynch's measurables were closer to those of the office receptionist than a passable football player. "Terrible," said McCarthy. "Couldn't play a lick. But give the guy credit. He figured us out." Indeed, Lynch noticed that the players who did well were summoned to Allen. So, after his dreadful workout, Lynch looked left, looked right, and approached the general manager, who robotically signed him to a contract.

Later that day, while sorting through paperwork of the 13 successes, McCarthy turned toward Allen, dumbfounded.

"We signed Albert C. Lynch?" he said. "The slow guy with the Afro?"

"You sent him to me," Allen replied.

"I certainly did not," said McCarthy. "No way."

"Yes you did," said Allen.

"No," McCarthy replied. "I didn't."

Silence — until the appearance of a simultaneous we've-been-duped lightbulb. The two laughed and laughed. Lynch would never appear in a Blitz uniform. But on the early listing of the team's linebackers, he's right there — wedged alphabetically between Northern Illinois's Frank Lewandowski and Wade Mauland of Illinois State.

Over the course of a few months the team added more employees, including a scout from the University of Evansville named John Butler and Joe Haering, a former Canadian Football League assistant coach. One day Bruce Allen, McCarthy, Haering, and Butler woke at dawn to drive a Chevy Impala 356 miles from Chicago to Columbus, Ohio, for a tryout camp. "We get there," said Allen, "and we realize we forgot to bring footballs." It happened again two weeks later in Los Angeles.

Camp attendees were a buffet of the fat, the skinny, the minuscule, the enormous. With no experience requirements, there were plumbers and doctors and lawyers and lab technicians, taxi drivers, fish tank cleaners. "I went to Scranton, Pennsylvania, for a tryout and a bunch of steelworkers came out," said John Teerlinck, the Blitz's future defensive line coach. "I signed none of them, and I thought they were gonna kick my ass." A handful of applicants had played Division I football. The majority were veterans of high school gridiron magic, looking to reclaim that long-ago taste of Friday-night glory. Some wore shorts. A couple wore jeans. A few stripped down to underwear. "There was a guy with one kidney," said Teerlinck. "Didn't make it."

Across America, coaches, administrators, and players from small colleges and independent leagues caught wind of the Blitz's all-points-bulletin manhunt. "We had someone's sister write us," said McCarthy. "Her brother was a football player who'd been stabbed, and she would do anything to get him a tryout. By anything, she meant *anything*. She was willing to pull down her pants if it meant he'd get a shot."

Notes to the team's headquarters poured in — dozens by the day, featuring the multicolored letterheads of America's myriad colleges and universities. Todd Ratliff, a six-foot, 180-pound cornerback and punter from Georgetown College in Kentucky, raved of his "reasonable statistics" and "eight *ent*erceptions." An attorney named J. Carlton Cherry took up the cause of Chowan College fullback Delbert Melvin, writing, "I have never seen this young man play football and although he was not a regular player on some outstanding teams at Chowan . . . he is very persistent." Letters praised the intelligence of Humboldt State wide receiver Eddie Pate (he had a 3.1 GPA as a wildlife-management major), the military-squatting skills of Azusa Pacific punter Jay C. Mellberg (600 pounds), the toughness of East Carolina nose guard Gerry Rogers (he played with two broken hands), the weight loss of Utah fullback Hilria Johnson (from 210 to 192 pounds!), and the "strong safety zone reading" talents of Bruce Buchla, who spent five years quarterbacking the New England Crusaders of something called the Atlantic Football League (his coach, Frank Fucci, wrote that Buchla's 34 touchdown passes warranted an "honest" shot. The Blitz were "honest"—thanks but no thanks). Even the man who wound up playing the team's mascot, Captain Blitz, was a United Airlines flight attendant discovered on a plane. "No stones went unturned," said Teerlinck.

All told, a preposterous 3,148 people received looks from the Blitz, and George Allen and his staff logged more than 20,000 air miles seeking out players. "I genuinely believe we started what would become the modern combine system," said Bruce Allen. "You ran a 40, did a vertical jump, a version of the short shuttle. We'd break off into groups—quarterbacks here, linebackers there. And if they ran well, more often than not we'd invite them to a regional mini-camp." The Blitz even traveled to the Logan Correctional Center in Lincoln, Illinois, to offer a tryout to Michael Sifford, an inmate who wrote a letter to the team, saying, "I love football. I can play." (Just a third of the way through a 12-year sentence for armed robbery, Sifford was eligible via a work-release program. He failed to impress.) In a December 1 press release, Kay Fred Schultz, the club's media relations head, wrote that the Blitz signed 258 players to contracts—"an unofficial Guinness Book world record."*

Before long, the USFL's other organizations saw what the Blitz were up to and panicked. The Denver Gold ordered one of its first hires, a defensive backs coach named Deek Pollard, to covertly attend the Chicago tryouts and try to sign those who seemed worthy. In Los Angeles, the Express's new coach was Hugh Campbell, straight off of five consecutive Grey Cup titles with the Canadian Football League's Edmonton Eskimos. Tom Fears, the re-

---

* Two sidenotes: One player the Blitz did not sign was Karl Mecklenburg, a linebacker out of the University of Minnesota who attended a tryout camp. Mecklenburg went on to appear in six Pro Bowls with the Denver Broncos. One player the team *did* sign was former Northwestern cornerback Rick Telander, who was 11 years removed from his final game and working as a writer at *Sports Illustrated*. Telander chronicled his experience in an *SI* piece headlined A FINAL FAREWELL TO FOOTBALL.

cently hired director of player personnel with the club, suggested they hold an open tryout on the Long Beach State University football field. "I figured some people would come," Campbell said. "But . . . *holy cow*." More than 500 aspiring superstars arrived for the chance to run a 40-yard dash. One man — African American, slender, in his mid-20s — hollered at Campbell from behind a fence.

"Hey, they won't let me in!

"Hey, they won't let me in!

"Hey, they won't let me in!"

An exasperated Campbell walked over. "Well, we're filled up," he said.

"Coach," he replied, "I can run a 4.4 40."

Yeah, right. "I'll tell you what," he said, "if I let you in here to run the 40, will you stop screaming?"

"Yes!" he said. "Absolutely."

The man hopped over the fence and — wearing Levi jeans and tennis shoes — burst across the line in exactly 4.4 seconds. "We let him come to training camp," said Campbell.

How was he?

"Well," the coach said, "fast."

Along with the anonymous speedster, 62 punters and placekickers showed their faces, forcing the Express to hold a kicker-only audition later in the week. One kicked with both legs. Another smoked a corncob pipe and had no teeth. "Old kickers, young kickers," said Campbell. "Just not many good kickers."

The Boston Breakers didn't have such a problem. At the same time USFL franchises were locating players, Tim Mazzetti was getting antsy. From 1978 to 1980 Mazzetti had been a solid NFL kicker, nailing 45 of 68 field goals for the Atlanta Falcons. Yet when Cleveland claimed him off waivers, the University of Pennsylvania graduate angrily retired. "I was visibly shaken by the way the NFL treated me like a slab of meat," he said. "I'd had enough." Mazzetti worked as a television reporter at Atlanta's ABC affiliate, but missed the football life. When he learned of the USFL, he called the organizations one by one. The first franchise was Birmingham. Mazzetti dialed the main office number and introduced himself. "Fuck you," a man replied. "We're not a joke and we don't appreciate the prank call." *Click!*

The next team was Boston. Tom Marino, the director of player personnel, answered. "Wait!" he said. "You're Mazzetti! *The* Tim Mazzetti?"

"That's me," he said.

Marino handed the phone to Dick Coury, Boston's coach. "Tim," he said, "don't call anyone else. We're signing you, and you're our kicker."

On the campus of San Mateo Junior College, 35 miles outside Oakland, the Invaders held a tryout that drew 700 players. *One* was signed. "John Ralston told me it was just for publicity," said Geno Effler, the team's media relations director. "A way to get some early press." At a Bandits camp at Jesuit High School in Tampa, the team was thrilled to have former Miami Dolphin Garo Yepremian audition for kicking duties. It was only after seeing Yepremian boot a ball 12 yards on a bend that coaches realized the man before them wasn't actually Yepremian, but a 36-year-old civil engineer named Sam Salman. "I thought I'd let them believe what they want to believe," Salman later explained to the *Fort Lauderdale Sun-Sentinel.* "I was asked if I was Mr. Yepremian and I said that some people call me that."* The open try-outs generated fun sports-section headlines (TAMPA KICKER NOT GARO) but little-to-no talent. From afar, the NFL watched with indifferent bemusement. It was a cute little league doing cute little league stuff. *Ha ha, hee hee, ho ho . . .*

Everything changed.

The headline appeared in the August 7, 1982, *Washington Post,* and while it was bold and big, it was neither bold nor big enough. ALLEN'S USFL TEAM SIGNS TIGHT END DRAFTED BY BEARS topped an article about the Blitz acquiring Tim Wrightman, the recently graduated UCLA star who had been selected in the third round by the Bears just four months earlier. Because a contract between the NFL and the NFL Players Association had recently

---

* When asked about Salman, the real Yepremian was amused, telling the Associated Press, "All I can say is that maybe somebody who was bald-headed, very virile and masculine showed up and said he was me. I can understand why somebody would want to do that."

expired, the Bears were unable to negotiate with players, and Mike Ditka, the team's meatheaded coach, ordered from afar that Wrightman report to training camp . . . *or else*. "They offered me less money than they paid their third-round pick from a year earlier," said Wrightman. "So when I was approached by George Allen, it felt fresh and exciting and new. It was different than the standard 'You work for us, now get here and do what we tell you.' It felt like becoming part of a family. That was the atmosphere and the energy the Blitz had going." The Blitz gave Wrightman a two-year, $400,000 deal that required he also serve as a team spokesman at local dinners and events. "It was awesome," he said. "I felt like a salesman and a football player."

Ditka was horrified. So was Jim Finks, the Bears' general manager. Within the next nine days, George Allen's club added two other noteworthy gems, Jim Fahnhorst, an All–Big Ten linebacker from the University of Minnesota, and Greg Landry, a 14-year NFL quarterback who, though approaching his 36th birthday, was a name familiar to football fans. "Greg was the biggie for us, because he gave us immediate legitimacy," Bruce Allen said. "It was 'Wait, these guys mean business.' It was instant credibility."

Still, the NFL refused to take the USFL seriously. Yes, Wrightman was a minor blow, and Landry signified . . . *something*. But outside of their coaching hires and goofy tryouts, most of the franchises concluded the 1982 calendar year making little-to-no noise.

That seemed destined to continue into 1983, even with the USFL's inaugural collegiate draft scheduled for January 4–5 inside a ballroom at the Grand Hyatt in Midtown Manhattan. Because the league needed to fill its rosters, the two-day bonanza would begin with a *twenty-four-round* regular draft (where teams could pluck the college seniors of their liking — geography be damned), followed by the territorial portion, where each franchise added 26 more players from five colleges designated their regional property. This was a big part of the Dixon plan — give local fans an emotional attachment to the roster. Before the draft began, each organization selected five colleges and gained the protected signing rights to the players from those schools. The colleges were, in most cases, within the geographical region of the USFL team. For example, the Tampa Bay Bandits were tied to Florida, Florida State, Florida A&M, Albany State, and Bethune-Cookman. The whole exercise was, from the NFL's vantage point, useless. There was no way America's top collegiate players would choose to join an experiment over the fulfillment of the lifelong NFL dream. So let the USFL have its dinky party. It was, for the older league, mere entertainment.

What the NFL didn't realize — *couldn't realize* — was that the USFL's franchises were about to shake up football. Inside the Blitz offices, an enormous chalkboard listed dozens of desired college players from across the United States. Kathy MacLachlan, an administrative assistant with the club, found herself shocked to see Anthony Carter, Michigan's All-American wide receiver, near the top. "Why do we have guys like him on the board?" she asked.

"Because," Bruce Allen replied, "we plan on getting them."

The Blitz were well organized. The Philadelphia Stars were meticulous. The team's general manager was Carl Peterson, who had spent the previous six years as an assistant coach and director of player personnel with the Philadelphia Eagles. He took the USFL job because Myles Tanenbaum, the owner, promised complete authority and an emphasis on winning. Although George Allen's name recognition brought much attention to the Blitz, Peterson was the Jedi Master of landing talent from the most unusual of places. Seven years earlier, Peterson helped generate much attention for the lowly Eagles when the team signed Vince Papale, a local kid who had never played college football but went on to excel as an NFL special teams standout. Now the Stars, like the Blitz, were inviting in plumbers and nurse technicians for looks. "I knew the NFL waiver wire really well," Peterson said. "I'd watch them, see where guys were, who was available. It was thrilling, because it was a new adventure."

The USFL Draft did its all to mimic the NFL Draft, what with each team given a table inside a ballroom; what with decorative helmets and footballs; what with a telephone for each organization; what with Simmons approaching the microphone to announce picks. The league prepared well, going so far as to rank the top 250 available players and distributing the information to all 12 franchises. Yet unlike the NFL's event, the USFL Draft failed to garner much attention. The *New York Times* sent a reporter, William N. Wallace, but mainly because it cost the newspaper a mere 75-cent subway fare. The Associated Press and United Press International also had journalists in attendance. Those outlets, too, were local.

When, at 10:00 a.m., the Los Angeles Express (given the first pick based on a lottery) led things off with the selection of Dan Marino, the University of Pittsburgh's splendid quarterback, the sporting world shrugged. The same reaction occurred when Ohio State's Tim Spencer went second to Chicago, then Southern Miss quarterback Reggie Collier third to Birmingham and Craig James, the Southern Methodist halfback, fourth to Washington. Over an exhausting 16-hour span, the USFL's teams grabbed 288 players, the vast majority of whom no more than a handful of humanoids had ever heard of. Afterward, the NFL's 28 franchises pretended the event was never held. There was no public acknowledgment, no nod toward the other league. The USFL Draft? What USFL Draft?

The faux indifference lasted less than 24 hours.

On the day after the event Marino, a Pittsburgh native, flew to Los Angeles and attended a press conference held by the Express. Decked out in a dark suit and tie and standing alongside Alan Harmon, the team's president, Marino insisted he was open to the USFL. "The opportunity with a new league is very exciting," he said. "I'll have an opportunity to play early. It'll be a very unique situation because it's the beginning of a new league and I'll be a part of it."

Marino was posturing. He *had* to be posturing.* Pete Rozelle, the NFL's commissioner, assured his league's owners not to worry. There was no way marquee players like Marino would simply sign with the USFL without waiting for the NFL Draft. "All I know," said Tex Schramm, the Dallas Cowboys' general manager, "is the great players and the great competitors are going to want to play in the NFL." BLESTO, the scouting combine that worked with a large number of NFL franchises, sent telegrams to the top-rated players that read, in part, "You have earned the opportunity to play for the best. You can look forward to a long and rewarding career in the NFL."

The established league enjoyed a good laugh when, immediately after the

---

* Indeed, he was posturing. Marino wound up a Miami Dolphin.

draft, the Boston Breakers announced they had agreed to terms with Dave Rimington, Nebraska's senior center. The reigning Outland Trophy winner was widely regarded as the nation's best offensive-line prospect, so the Breakers sent a team of representatives to roll out the red carpet and greet him at Logan International Airport. When Rimington didn't exit the plane, the Breakers placed a few calls. Lo and behold, the Rimington they had been negotiating with via phone was not the All-American, but someone playing a cruel joke. "We've been had," said Gary Gillis, the team's publicity director. "I guess we were so enthusiastic about the way things were going that we forgot to make sure we were talking to Rimington. I think little gremlins are making phone calls."

The NFL happiness was short-lived. No, Marino would not become the Express quarterback, and Dave Rimington wasn't Dave Rimington. But George Allen's Blitz offered Spencer, a second-team UPI All-American halfback who ran for 1,538 yards as a senior, a four-year, $800,000 deal, and he accepted. The news was stunning. Wrightman was a nice piece, but Spencer would have been a top-15 selection in the upcoming NFL Draft. "Signing Spencer gave us so much credibility, because there are all those Ohio State fans in the area," said Bruce Allen. "It put us on the map."

Within a 10-day span, four other first-round picks joined USFL franchises. Collier, the first quarterback in NCAA history to rush and pass for 1,000 yards in a single season, inked a five-year, $1 million deal with the Stallions; safety David Greenwood, a University of Wisconsin star picked 10th by the Michigan Panthers, received a multiyear deal; and Trumaine Johnson, a gifted Grambling wide receiver, snagged 11th by the Blitz (via a trade with the Breakers), went for a smidge less than $1 million over four years. Those were all troubling to the NFL, but absolute devastation came on January 12, when Craig James and the Washington Federals agreed to a four-year, $2 million contract.

In their four years together at SMU, James and Dickerson gained fame as the Pony Express, an explosive two-back tandem that resulted in All-American nods for both players. Not only was James a likely top-10 NFL pick, but he was a marketing director's dream — handsome, engaging, white. In Washington, the biggest gridiron star was John Riggins, the Redskins running back, who was old (33) and frumpy. James, by comparison, was Robert Redford. He also understood precisely what the USFL needed. "Everybody laughed at [Joe Namath and Billy Cannon] when they went to the AFL," he said. "But I tell you what, they're not laughing at them anymore and we all know the end of the story of that."

It was all smiles in the USFL offices, all terror in the NFL offices. The James addition, however, was also a harbinger of things to come for a new league with much to learn.

The Federals' general manager was Dick Myers, who came to the USFL after four years as an assistant GM with the Redskins. Just 36 at the time of his hiring, Myers was uniquely unprepared and unqualified. Bobby Beathard, the Redskins' general manager, relied on Myers as a number cruncher, but later recalled, "Dick was not a football guy by any means. So I was a little surprised when the Federals gave him that position." Berl Bernhard, the Federals' owner, wanted to make a splash, and he instructed Myers to travel to Dallas and bring back a signed James. So that's what he did. *Sort of.*

"Dick went to Texas, and he calls me in the evening and says, 'We're coming home tomorrow! We've got Craig James signed, so get ready for a press conference!'" said Rick Vaughn, the Federals' director of media relations. "I didn't sleep all night. It was *that* exciting."

The next morning Vaughn arrived at the stadium at 8 o'clock. He received a call from Myers, confirming the press conference should be slated for 2:00 p.m. Vaughn alerted the local media and prepared a room. At noon Myers, James, and a man named Sherwood Blount entered the stadium and retreated to a private office. Vaughn later learned that Blount served simultaneously as a millionaire SMU booster (whose actions would result in the school receiving the NCAA's death penalty) and James's agent — something the Federals were previously unaware of.

At 1 o'clock, Vaughn poked his head into Myers's office and was told to go away. He returned a half hour later — same. At 1:50, a panicked Vaughn knocked again. Much of the Washington sports press corps was inside the conference room, waiting. "We're running into a few snags," Myers told him. "Tell them we might be a few minutes late." At 2 o'clock, Vaughn apologized to the masses and suggested they enjoy a soda. The clock continued to tick — 2:15, then 2:30, then 2:40. Finally, at 2:45, the door opened and a smiling Craig James, a smiling Sherwood Blount, and a battered Dick Myers emerged. In front of the press, everything was happy-happy. Behind the scenes a catastrophe was narrowly averted. In Texas, James and Blount had agreed to terms. In Washington, with the media looming, Blount asked for $500,000 more than the initial deal. "Here's what I want and here's what Craig wants," Blount told Myers. "You give it to us or we're leaving. You decide what to tell the press . . ."

Myers buckled. "I don't know what choice he had," said Vaughn. "This was our real introduction to Washington."

By the beginning of February, all 12 USFL teams seemed reasonably prepared for the tasks at hand. The Blitz, for example, left the Merchandise Mart for an abandoned school, Maine North High School, in Des Plaines, Illinois. George Allen renamed the facility Blitz Park and outfitted it with a weight room. The Wranglers, featuring the league's coolest logo (a fiery red-and-gold *W* branding iron), built its headquarters at Phoenix's East High School, while the Express turned a dilapidated Manhattan Beach elementary school into a dilapidated football complex. For players and coaches arriving from Division II and III colleges, it would be, facility-wise, a neutral leap. For players and coaches out of schools like Penn State and Alabama, or those via the NFL, it was a plunge not unlike the one from the St. Regis to a Motel 6.

Between Wrightman's breakthrough signing and the opening of training camps across Florida, California, and Arizona, teams were busy hiring coaches and filling rosters and acquiring equipment and renting space. It was a crazy time for general managers, none of whom had ever been charged with constructing franchises from scratch.

One of the first things many organizations sought to do was add familiar names with potential headline appeal. The USFL was not going to compete, quality-wise, with the NFL, and no one was fooled into thinking otherwise. But that didn't mean the teams couldn't try to drum up some hype. That's why the Federals signed Coy Bacon, a 40-year-old chain-smoking defensive lineman who last played for the Redskins in 1981. ("He was football's first 1,000-year-old man," said Brett Friedlander, a writer for the *Annapolis Eve-*

ning Capital.) It's why the Philadelphia Stars brought in Lydell Mitchell, the former Colts halfback whose last NFL snap occurred in 1980. It's why the Denver Gold added quarterback Joe Gilliam, an on-again, off-again drug addict who once started over Terry Bradshaw with the Pittsburgh Steelers.

The early USFL transactions page became a sports nostalgist's almanac and no one excelled quite like the Los Angeles Express, where the roster featured the boldfaced names of Anthony Davis *and* Chuck Foreman, a couple of former halfbacks itching for a last taste of glory. Better known for his long-ago stardom at Southern Cal than his two-year NFL flameout, Davis pulled up to camp in a Rolls-Royce, and entered the locker room wearing a full-length mink coat, with no clothes beneath it. "I was a shell of my old self," said Davis, who had last played in 1978. "But I wanted to make an impression." Foreman, too, stood out, in that he was a five-time Pro Bowler who ran for 5,950 yards over eight seasons before retiring in 1980. In his Minnesota Vikings heyday, Foreman was a purple-and-white blur of physicality, a hard-charging, forearm-to-the-chin ode to power running. "When I heard about the USFL, I thought, *Why not?*" Foreman said. "I wasn't ready to let the game go." Many of the Express aspirants were in awe of the ex-Viking. Then, during a drill, he took a handoff, stepped forward, and was laid out by Jerome Franey, a nondescript linebacker from San Diego State. "Franey ran Foreman on his ass," said Danny Rich, a linebacker. "Foreman gets up and says, 'OK, motherfucker, let's see you do that again.' Franey ran over his ass again. That was the last we'd see of Chuck Foreman."

Another of Los Angeles's recognizable figures was Mike Rae, who 11 years earlier quarterbacked USC to a national championship before embarking on a journeyman run through the NFL and Canadian Football League. The team considered building its early marketing campaign around Rae, which was an exciting proposition for a faded collegiate headliner. Then the quarterback reported to camp. "I walked up to get my helmet," he said, "and it had a cage on it. I explained to the guy that I was a quarterback and he said he wouldn't make an adjustment. I literally went to my car and pulled out a screwdriver.

"A different time I was throwing in the rain, and the balls got heavy. I asked for a new football. Nope. They only had two."

With the exception of the handful of famous high draft picks and faded big-name NFL veterans, the men initially signed by the league formed a riveting and unparalleled mishmash of football players from all walks of life, from all corners of the gridiron globe, of all sizes, shapes, backgrounds, injury histories, astrological signs, birthmarks. Danny Buggs, a five-year NFL wide receiver, had spent the past three seasons in the Canadian League, winning a Grey Cup title with Edmonton but freezing his butt off. "It just wasn't good for me, weather-wise," he said. "The Tampa Bay Bandits called. I thought, *Tampa — shorts and T-shirts. I'm in.*" Andy Cannavino played linebacker at the University of Michigan, then was cut in back-to-back preseasons by the Lions (1981) and Eagles (1982). The Michigan Panthers made a $36,000 offer — $10,000 more than Alvin Hall, the Lions' starting free safety, was making in the NFL. "All I wanted to do was play pro football," he said. "I honestly would have done it for free." Joe Bock was a defensive lineman from the University of Virginia. He was cut by the Buffalo Bills in 1981. He failed a physical with the New York Jets in 1982. He spent half a season with the CFL's Hamilton Tiger Cats. The Washington Federals asked that he switch from defensive tackle to center. "Sure," Bock told the team. Phil Denfeld had been a tight end at Wake Forest. He was signed by the Wranglers, reported to camp, and was asked whether he had ever kicked. "Sometimes in college," he replied. "But I'm not really a kicker." *Would you kick for an extra $15,000 this year?* he was asked. "I'm a kicker!" Denfeld replied.

Perhaps the best training-camp narratives belonged to a pair of Boston Breakers. First, there was Jeff Gaylord, a nose tackle and former fourth-round draft pick of the Los Angeles Rams who, while attending the University of Missouri from 1978 to 1982:

- Went on a five-month cocaine binge that, he recalled, "fried my brain."
- Agreed to fill in for an absent male stripper at the nightclub where he worked as a bouncer. Gaylord purchased a tin of green paint, coated his body head to toe, and nicknamed himself "the Incredible Hulk." That night he was arrested for indecent exposure and fined $50.
- Filled his body with every imaginable steroid and steroid knockoff.

- Served as a certified chiropractor who arrived early to the stadium to crack his teammates' backs before games.

The other Breaker of note was Billy Don Jackson, a onetime standout defensive end at UCLA who arrived in camp less than four months after wrapping an eight-month stay at the L.A. County Jail and Wayside Honor Rancho for manslaughter. On October 27, 1980, Jackson allegedly killed 28-year-old Mark Bernolak in an argument over marijuana. "I remember being told that Billy Don beat the man to death with a tennis racket during a drug deal gone bad," said David Cataneo, who covered the Breakers for the *Boston Herald*. "Well, the day I interviewed him they set us up in a private room, and I turn around and notice a pile of tennis rackets stacked up against a wall near his chair . . ."

"Billy Don would carry around a two-by-four in the locker room," said Marcus Marek, a Breakers linebacker. "I was never sure why."

Training camps officially opened on January 31, which meant 12 variations of *North Dallas Forty* springing to life. The two best-prepared teams, Chicago and Philadelphia, were located 1,845 miles apart — the Blitz in Glendale, Arizona, the Stars at Stetson University in DeLand, Florida. George Allen insisted on the finest for his club, which meant players stayed at the luxurious Phoenix Westcourt Hotel and ate like kings. Long known for his love of veterans, Allen and the Blitz signed a handful of NFL elders, including former Jets halfback Kevin Long and a pair of ex-Colt stars, linebacker Stan White (who was the first starter to leave an NFL team for the newborn league) and defensive lineman Joe Ehrmann. "It was great, like a reunion of hardened NFL players," said Ehrmann. "Our linemen were older than our coaches."

The Stars, meanwhile, suffered through a messy start. Peterson's first choice for the coaching position was George Perles, the longtime Pittsburgh Steelers assistant who accepted the job in the summer of 1982, then resigned on December 3, 1982, to take the same gig at Michigan State, his alma mater. Peterson immediately turned to an obscure New England Patriots defensive coordinator named Jim Mora, whose no-nonsense approach appealed to the Stars' GM. Peterson and Mora had briefly coached together at UCLA in 1974. "I visited and I liked what they were trying to do," said Mora. "So I accepted just two weeks before training camp."

Mora's coaching staff was composed of men hired mostly via phone conversations. "I introduced myself to some of the guys when I arrived in DeLand," he said. "We came together as a staff that first night before the first practice, and we put together an offensive and defensive playbook. In one night! That first day I knew none of the players, so they all wore names on their helmets. I had zero idea what type of team we had."

"We had a week or two to accomplish what usually takes six months," said John Rosenberg, the defensive backs coach. "It was craziness."

Peterson's talent cobbling was nothing short of breathtaking. Philadelphia's first-round pick, UCLA offensive tackle Irv Eatman, signed a four-year, $1 million contract, and its second-round selection, BYU center Bart Oates, signed a three-year, $300,000 contract. He brought in a slew of former Penn State standouts, including quarterback Chuck Fusina, wide receiver Scott Fitzkee, and defensive lineman Dave Opfar, and added Kelvin Bryant, a gifted halfback out of North Carolina, in the regional college draft. As some of the other franchises reached out to any stiff with a pulse and a résumé, Peterson factored in temperament, decency, commitment. He wanted players who didn't merely view the league as a place to earn a paycheck, but as a place to quench a burning desire to compete. Because of his NFL ties, Peterson often had friends from the other league suggest this safety, that quarterback, this kicker. "Frankly, I had an inside source, Lynn Stiles, who replaced me as the director of player personnel with the Eagles," Peterson said. "Lynn and I kind of had an agreement that when he was getting ready to release a player he'd let me know first. So we'd be kind of standing in the parking lot waiting for that player as he was released to say, 'Hey, you didn't make it here in the NFL, but you have a great chance to start playing right away.'"

Out of the blue, Peterson received a phone call from Sam Rutigliano, head coach of the Cleveland Browns. "Carl," he said, "we're cutting a guy, and the only reason we're doing so is because he's not 6 feet. He might not even be 5 foot 10, and he's a middle linebacker. But I think the Stars should sign him."

Peterson was befuddled. Why would he want a dwarf middle linebacker? Middle linebackers are the heart of a defense. They need to see over offensive linemen. They need to be able to scan the entirety of a field. "Look, whatever you do, get him signed," Rutigliano said. "And don't cut him until you see him play in full pads."

The next day, Peterson offered a horn-rim-glasses-wearing, Volvo-se-dan-driving former Division III linebacker from Montclair State a two-year deal for $42,000, plus a $500 bonus. It was the first time he had heard of Sam Mills, who at the time was working as a photography teacher at East Orange High School in New Jersey.

"So we start training camp, and nobody stands out more than Sam," Peterson said. "He just lit it up. On the snap of the ball, it didn't matter if the lineman was six foot five, Sam would slip underneath and nail people. They called him the Field Mouse, because he was a mouse running around a field of elephants, but the elephants wanted nothing to do with him."

"I don't think they even expected Sam to make the team," said Antonio Gibson, a safety and the team's fourth-round pick, from Cincinnati. "Nobody knew who he was. Nobody. Then you'd be in the huddle with him, and his eyes were wide open, like a wild animal ready to kill. That intensity, that motor — he has to be the greatest find in USFL history."

With Mills leading the way, the Stars were sharp and, under Mora's iron fist, disciplined.* The team's first-ever meeting was called for 7:00 p.m. Players arrived on time, but the door was locked. Mora finally let everyone in while announcing, "Seven means be here by six fifty!"

"We called it Vince Lombardi Time," said Allen Harvin, a halfback. "Jim Mora was no joke."

The same could not be said for the Denver Gold, which held its camp at the woeful Francisco Grande Resort complex in Casa Grande, Arizona. The location had once been the spring-training home of the San Francisco Giants, who abandoned ship in 1979. "It was like a juvenile prison with bars everywhere," said Matt Braswell, a center who spent time with the Gold. "The bathrooms had no doors, so one of my first impressions with the Gold was

---

* "Our nickname for Mora on the team was 'Dick,'" said one Stars player. Why? "Because he was a dick."

walking in and seeing this black linebacker, built like Zeus and covered with tattoos and scars, sitting naked on the toilet taking a dump. I looked at the guy standing next to me and said, 'We're not in Kansas anymore.'"

The Gold were coached by Red Miller (good) and owned by Ron Blanding (iffy). The former had guided the Denver Broncos to Super Bowl XII. The latter was a Denver native identified by *Sports Illustrated* as "an entrepreneur in construction, real estate and recreational ventures." Blanding was the poorest of the USFL's 12 primary owners. He only agreed to buy the Gold when John Ralston, his friend and racquetball partner, pleaded. "They needed a guy in Denver," Blanding said. "It was a city they considered extremely important."

Like the other 11 franchises, the Gold outfitted their players with training-camp automobiles. Unlike the other 11 franchises, who went with either Hertz or Avis, the Gold contracted the services of Rent-A-Wreck. "That became our early legacy," said Jim Van Someren, Denver's media relations director. "Our guys were driving around in these old clunkers that would fall apart without warning." Indeed, the vehicles were dented, rusted. Some were missing windows. A handful lacked operational trunks. Oil slicks pocked the Francisco Grande parking lot. "Five of us pooled money and upgraded to a Chevy Chevette," said Braswell. "Top of the line for the Gold."

Members of the team were supplied with $10 food per diems. The meals at the facility were bare-bones and limited. "We stayed at the Rodeway Inn, a motel across from the airport," said Van Someren. "It wasn't the Hyatt." Every morning, Gold players boarded a pair of out-of-circulation school buses to travel from lodging to work. The vehicles had neither seat belts nor air-conditioning. The practice fields were overgrown with weeds and marred by deep holes and scattered rocks. Denver invited approximately 120 men to camp, but owned only 100 uniforms. So, between shifts, players would remove their outfits and hand them to those waiting along the sideline as Carter Tate, the equipment manager, screamed instructions. "It was the most violent football ever played by mankind," said Harry Sydney, a free-agent halfback. "You didn't have to be a great athlete. Everybody who thought they were a tough guy tried out."

Whereas the Stars signed four of their first five draft picks, Blanding and the Gold seemed to view the draft as a useless exercise. The team not only failed to ink any of its first seven draft picks — it didn't even bother to try. Too much money. "The first argument I ever saw between an owner and a coach in pro football was between Ron and Red Miller," said Sydney. "Ron wanted to send the cut players home by bus . . ."

Denver's most famous employee was Miller. His assistant coach, former Broncos quarterback Craig Morton, was a close second. Gilliam, the 32-year-old quarterback, would arrive for workouts either drunk, high, or drunk and high. He chain-smoked cigarettes along the sidelines. The roster was an obscure wasteland populated by random nobodies and nobodies who, at one time or another, had played for the Denver Broncos. Miller called as many of his former NFLers as possible, and wound up with ex-Bronco quarterbacks (Ken Johnson, Craig Penrose), an ex-Bronco halfback (Larry Canada), and an assorted buffet of ex-Bronco linemen. Some were passable. Most were not. "We were in the USFL for a reason," said Johnson. "That was pretty clear."

As camp progressed, and Miller cut down the roster, the relationship with Blanding turned increasingly hostile. Miller was a tough old boot who learned the game under Lou Saban at Western Illinois in the late 1950s. He snarled and barked and refused to let outsiders interfere with his business. Blanding, on the other hand, bristled at every unnecessarily spent dime. One morning, Blanding and Miller could be heard screaming at each other. The coach named Red turned, literally, red. He grabbed Blanding by the throat, picked him up, and pinned him against the wall. "You either act like a fucking owner of a professional football team," he growled, "or you get someone else to coach!"

"Red tried to kill the owner," said Don Frease, the wide receivers coach. "He came close."

Blanding stormed out of the room. Miller wouldn't last the season.

Of all the USFL training camps, the spot that drew the least interest seemed to be the campus of the University of Central Florida in Orlando, where the Boston Breakers and New Jersey Generals shared a facility.

When Chuck Fairbanks, the Generals' coach, initially joined the USFL on June 2, 1982, it was considered a major coup for the new league. He had been a standout coach for six years at Oklahoma, then led the woebegone New England Patriots to a pair of 11-win seasons. Fairbanks, though, was historically dull. "He looked like the Marlboro Man, and he talked really slow," said Charley Steiner, the team's play-by-play voice. "He was an old football coach from Oklahoma who was very reluctant to join the 20th century. There wasn't much color to be found."

"If you asked him who was starting at receiver, he'd just say, 'Well, you know, we're looking at everybody . . . blah, blah,'" said Barry Stanton, the Generals beat writer for the *Journal News.* "There was nothing there."

New Jersey's roster was remarkable solely for its lack of remarkability. The USFL hoped that the Generals would serve as its marquee franchise, what with its grand media market and the presence of two NFL teams. If you could make it in New York, you could make it anywhere. At least that's what Simmons was prone to remind people. Yet the Generals were a mess. "We had a lot of holes," said Tom Woodland, the starting nose guard and a Cleveland Browns castoff. "Too many guys in too many spots who didn't know the game." The team's presumptive starting quarterback was Bobby Scott, the longtime New Orleans Saints backup whose claim to fame was injuring his knee by tripping over a TV cable. "He couldn't throw the ball 15 yards," said Dom Camera, the league's director of marketing. "Bobby Scott was a good player in his day, but his arm was shot." The halfback, Larry Coffey, was West Virginia Wesleyan's all-time leading rusher. Its best training-camp linebacker, Sherdeill Breathett, found himself cut after being observed in the company of a white woman. ("My room was right near the coaches' room," he said. "They saw that and I was gone ASAP.") One of the defensive linemen, Steve Williams, would go on to a lengthy professional wrestling career as "Dr. Death." He was significantly better at the top-rope elbow drop than tackling halfbacks. In the *Sporting News'* 30-page USFL preview issue, Paul Domowitch could barely string together two positive sentences about Fairbanks's gang. Wrote Domowitch: "They seem well set in the pass catching department."

Get yer tickets! Get yer Generals season tickets!

The Generals were flat.

The Generals were dull.

"Then," said Steiner, "a nuclear bomb went off."

The sporting universe would never be quite the same.

3

HERSCHEL

We've acquired a guy named Herschel Walker.

— Chuck Fairbanks, New Jersey Generals coach,
to his players, February 26, 1983

THE PHONE CALL came on the afternoon of January 17, 1983. This was well before the technology that brought the world caller ID, so when Steve Ehrhart, the USFL's executive director, picked up the receiver in the league's new digs on Manhattan's Vanderbilt Avenue, he had no idea his world was about to change.

"Steve, this is Jack Manton . . ."

Pause.

". . . and I'm Herschel Walker's representative."

Um . . .

This was odd for two major reasons:

1. Herschel Walker, the University of Georgia tailback, was only a junior.
2. College juniors didn't have "representatives."

"Herschel wants to come out, and he doesn't want to go back," Manton said. "He's ready to make a living, and the USFL could provide that."

Ehrhart was speechless. A former players' agent who represented a handful of Denver Broncos in the 1970s, Ehrhart was well known for his decency and integrity. That's one of the reasons Chet Simmons, the USFL commissioner, hired him to help start the league. That's also why the call from Manton, an agent whose clients included such NBA players as Sidney Moncrief and Mike Mitchell, proved shocking. As long as Ehrhart could remember, the NFL had a strict policy when it came to college underclassmen. Namely, they were not permitted. Although it had never actually been discussed in early meetings, he presumed the USFL would follow the same rule. "Look," Ehrhart told Manton, "this could cost him his eligibility—just us talking and just you serving as a representative. You know the rules, Jack. I don't even want to talk with you."

Click.

In Ehrhart's mind, that was that. Yes, Herschel Walker was a transcendent football star, the sort of iconic figure whose presence could take the USFL from here to there. He was not merely the Heisman Trophy winner, but the greatest all-around player college football had seen in decades. Perhaps ever. Walker was a power runner with Olympic burst and shoulders the size of mountains. He was Jim Thorpe after Jim Thorpe; Bo Jackson before Bo Jackson. "[Walker is] a 220-pound running back with Jim Brown's strength and O. J. Simpson's speed," wrote Dave Anderson of the *New York Times.* "He is the epitome of strength—that thick neck, those huge sloping shoulders, those powerful but swift legs." Walker ran for 5,259 yards over three seasons, starred on the Bulldogs track and field team, and even wrote poetry (*I wish they could see / The real person in me*). He refused to lift weights, and was famous for his regimen of never drinking, never smoking, and doing hundreds of sit-ups and push-ups every night before going to bed. He aspired to join the United States Marine Corps because, he said, "I wanted to kill people." In short, he was mythology brought to life. "Ridiculously gifted," said Jim Mora, the Stars coach. "There's no getting around that."

But Herschel Walker was not coming to the USFL.

No way.

No chance.

Well . . .

What Ehrhart and the other men and women embedded in the USFL office claimed not to realize was that the Herschel-Walker-to-the-league mechanisms had been in place for weeks. First, well before any of the rosters had been filled, the Chicago Blitz placed some calls to people who knew Walker, wondering if — hint, hint — he might want to skip out on school and make some good money. The team even sent Walker a USFL contract, just on the off chance he would sign it. The move was typical George Allen, who once, as coach and general manager of the Washington Redskins, traded a draft choice his team didn't possess. "Allen," wrote the *New York Times*, "has never been mistaken for an ethics professor."

The Blitz's efforts didn't go far. But they did place a bug in some ears. In fact, nearly a month before Manton officially reached out, a short Jewish man with a Texas drawl and a game show host's charisma traveled to Wrightsville, Georgia, home to 1,600 people and the Old Fashioned Fourth of July Festival, to meet with Willis and Christine Walker, Herschel's parents. His name was Jerry Argovitz, and although he was a players' agent whose client list included such noteworthy NFL stars as Detroit halfback Billy Sims and Tampa Bay linebacker Hugh Green, his trip to Wrightsville was, he insists, not an effort to snag a new client. No, it was to propel a new league.

"I had heard the Walkers were thinking about having Herschel sign in Canada, because that league had no eligibility restrictions," said Argovitz (who was correct. Two years earlier Walker rejected a $1.5 million contract offer from the CFL). "They were not wealthy people [Herschel's father was a tenant farmer who raised soybeans, peas, and cotton and earned approximately $30 per week], and they were rightly afraid that if Herschel got hurt playing another season of college ball, he would never be able to play professionally. As an agent, I desperately wanted the USFL to work, because it would give players an alternative to the NFL, and that would create a bidding war and open up everything, financially. I wasn't about to sign a junior and ruin my name. No way. But I at least wanted to go and let them know what was coming . . ."

According to Argovitz, a follow-up trip to Wrightsville included a yellow legal pad, a pen, a hug from Christine, and serious talk. The Sugar Bowl was approaching, and the Walkers were united in the belief that Herschel should play the game against Penn State and then turn pro. Argovitz said he spoke with some USFL owners, and that the idea of adding Walker was Christmas and Easter combined. "If I can get you serious money from the USFL, would you go?" Argovitz asked. "I'm talking about a guaranteed contract, and you can pick which team you play for . . ."

Herschel, he said, nodded in agreement.

The Georgia football team arrived in New Orleans for the Sugar Bowl on the afternoon of December 28, 1982. Walker—the game's marquee attraction—was given his own room in the Bourbon Orleans Hotel, outfitted with a king-size bed fit for a king. The contest featured America's No. 1 team (Georgia) against No. 2 (Penn State), and the throng of media in town befitted such a matchup. Yet for all the attention lavished upon Walker, no one in the press seemed aware of what, behind the scenes, was happening.

Two nights before the game, Argovitz, who rented a suite in the Bourbon Orleans, said he invited Walker and Ted Diethrich, owner of the Blitz, to his room for a chat. Chicago had been the USFL's most aggressive franchise thus far, as well as one located in a major market that would suit Walker's inevitable status as the face of the league. A few days earlier, Argovitz had prepared Diethrich for the encounter, and now the famed surgeon arrived with a suit, a tie, a handshake, and an envelope.

Following some brief pleasantries, Argovitz looked at Diethrich and said, "Ted, do you have something for Herschel?"

Diethrich reached into the inside pocket of his suit jacket, pulled out the envelope, and passed it to Argovitz. "Herschel," he said, "this is for you."

Walker took hold of the envelope, methodically ripped it open, and gazed emotionless at the content. He stared and stared and stared—"probably for 15 seconds, without making a sound," recalled Argovitz.

It was a cashier's check, payable to HERSCHEL WALKER, for $1 million.

"I can't accept this now," Walker said.

"Herschel, it isn't meant for you to accept now," Argovitz said. "It's meant to show you that these people are serious. This is a cashier's check. You take it to any bank in America and cash it, you'll walk out with $1 million in your pocket. This is just to show how serious the USFL is."

Walker was both dazzled and dumbfounded. There would be no more talk of the Canadian Football League. It was now either the USFL or a return to Georgia. On the evening of January 1, 1983, Walker ran for 103 yards in his team's 27–23 loss to the Nittany Lions. The game was televised by ABC, and the announcers, Keith Jackson and Frank Broyles, reassured despondent Bulldog fans that they need not worry—there was always next year with Herschel carrying the ball.

Manton placed the January 17 call to Ehrhart, which, a few days later, was followed by a direct call to the USFL from Walker himself. "Mr. Ehrhart, it's Herschel," he said. "I just want to tell you directly that I'm ready to come out and I'm ready to go and I definitely want this."

What to do? Ehrhart could make no promises, or even reassurances. However, he knew, by having an agent reach out on his behalf, Walker was already technically ineligible to return to college; that were anyone to find out about Manton and the conversations, his time at Georgia was done. Ehrhart and Simmons engaged in many long, detailed discussions about the implications of adding a college junior to the league. On the one hand, it might have the groundbreaking impact of Joe Namath shunning the NFL to go to the American Football League's New York Jets in 1965. Walker wasn't just another great player, or even just another Heisman winner. He was a generational icon. This wasn't rinky-dink nonsense, or a silly goof. The USFL would be screaming, "We're for real!"

And yet, the negatives weighed heavily on both men. First, Walker would break the USFL's bank. Salary restrictions? What salary restrictions? He would probably wind up being the highest-paid player in football—something Simmons was unsure the USFL could handle. There was also the impact signing Walker would have on USFL-NCAA relations. Would college coaches even allow USFL scouts on campus, knowing they would be evaluating seniors *and* juniors? Were Walker to join the professional ranks, the floodgate could never be closed. How would the NCAA react? Hell, how would sports fans react? The line between groundbreaking signing and vulture-like carcass pecking was thin.

After enough back-and-forth, Simmons and Ehrhart decided the rewards outweighed the negatives. The two men instructed Mike Trager, the USFL's media consultant, to contact ABC's senior vice president, Jim Spence. "We asked if they would contribute to Herschel's salary, because it would add a great deal of value to the telecasts," said Ehrhart. "They said they didn't want to pay any extra money, but they encouraged us strongly to do it." Simmons and Ehrhart then reached out to the league's 12 owners, none of whom objected to Walker. "We agreed the only place for Herschel to go was New York, because of the Namath-Jets parallel and the market size," Ehrhart said. "It just made sense."

J. Walter Duncan, the Oklahoma oilman who owned the Generals, was one of the league's wealthiest men. On the evening of February 8, he met in Orlando with Manton, Chuck Fairbanks, and Jim Valek, New Jersey's general manager. The rendezvous was kept hush-hush, but for the first time terms of a Herschel Walker professional football contract were discussed. Manton wanted a three-year deal at $1.2 million per year, as well as a $1 million signing bonus, interest in one of Duncan's Oklahoma oil wells, and a share in the Generals' home-gate receipts. The next morning, Duncan called Manton — he could do everything but the gate receipts. That sounded fine to the agent.

In the meantime, the Dallas Cowboys secretly reached out to Walker. Yes, the NFL had made clear that no underclassmen were allowed in the league. But, well . . . eh . . . that was before the USFL stepped up. "The Dallas people had an intermediary that offered him more money to sign with the Cowboys because Dallas always coveted him," said Ehrhart. "They were very careful to have cutouts in between, because [Pete] Rozelle would have never allowed it . . . despite the NFL rule, the Cowboys made an offer . . . I'm sure Jack Manton, if he was under a polygraph, would say that he got an offer from the Cowboys for more money than we signed him for in the USFL. I think the Cowboys had probably been in contact with Rozelle, and they were worried about letting a great player like that get out of the league." Walker, to his credit, wasn't listening. The USFL had struck first, and that was enough to gain his loyalty.

On February 17, Duncan, along with Jim Valek, flew to Athens, Georgia, and sat down with Walker and his fiancée, Cindy DeAngelis, in their apartment. He wanted to personally connect with the man who might be taking so much of his money. Walker had never heard of the New Jersey owner, and when he mentioned the name to Cindy, she said, "Isn't he the guy on the Monopoly game? The one with the fancy suit and fistfuls of cash?" The initial discussion took place at a small table. Duncan wore slacks, a cowboy shirt, and a bolo tie. Herschel and Cindy had been playing Space Invaders on an Atari 2600 when he arrived, and Duncan seemed intrigued by the unfamiliar sounds and images. They chatted for several hours, then again over breakfast the following morning. Walker liked Duncan, and insisted this was what he wanted to do; that there was no other choice; that he was done with college. The agreed-upon contract — signed by Walker — included the 25 percent interest in an oil well. "The oil well — that really did it," said Ehrhart. "Everything was good. Herschel was coming to the United States Football League."

Immediately after Walker inked his deal, Duncan — as honorable as he was wealthy — told the running back to take 24 hours to make certain he wanted to go pro. "I told Herschel that if he came to me by 7 a.m. the next morning, I would give him the contract back if he didn't want it," Duncan said. "He drove me back to my room and I told him to go see Dooley."

"Dooley" was Vince Dooley, Georgia's 19-year football coach. Upon hearing the words "I'm going pro" from Walker, he had what can best be described as a legendary freak-out. "Herschel, do not do this!" he told the halfback. "It's an enormous mistake." Walker was no more or less naive or impressionable than any other 20-year-old kid from the backwoods of Georgia. He first lied and said there was no contract. Then he told Dooley that, in actuality, there was a contract, but it included an out clause, and he wanted to remain a Bulldog, and he was *so, so, so* sorry. The following morning he contacted Duncan and said he had decided to return to Georgia, and that evening he was honored during the Georgia Sports Hall of Fame banquet. When asked about USFL rumors, Walker made stuff up. "Like I said, I haven't seen a contract," he said. "I've heard a lot of rumors."

"He was confused," Duncan said.

When Walker changed his mind, Duncan and Valek flew back empty-handed to New Jersey. The team employed a low-level gopher to handle tasks, and on this day he was charged with picking up the men from the airport. En route he overheard what had happened, and upon reaching the team complex shared the news with Rick Buffington, a friend and New Jer-

sey scout. Later that day Buffington received a call from an old pal, Will Mc-
Donough of the *Boston Globe*. "Rick," he said, "I know something happened
in Georgia. What was it?"

"I can't say," Buffington replied.

Then he reconsidered. Walker had, indeed, signed a contract, thereby ruling him ineligible to play college football. The news would almost certainly
get out at some point, possibly midway through Walker's senior year at Georgia. "A lot of people would get in trouble if that happened," said Buffington.
"So I decided to anonymously tell Will the whole story."

The *Globe* broke the news, and Buffington began to worry Fairbanks
would find out, thirdhand, that he was the Herschel Walker Deep Throat. So
he knocked on the head coach's door, expecting to be terminated.

"Coach," Buffington said, "can I talk to you for a second . . ."

"Rick," Fairbanks replied, "you don't need to do this. I know why you're
here, and you can go back to your office. Without you we don't have Herschel
Walker. So thank you."

The narrative was partially true. As powerful a sway as Dooley and the
Bulldogs held over Herschel Walker, it paled in comparison to that of Christine Walker, his mother. The family lived in a white frame house on a dirt
road in a poor town. Were Herschel to suffer a permanent injury his senior
year, there would be no money, no guarantees of a secure future. Though
possessing only a high school education, Christine Walker was no dummy.
She understood a system that used athletes while providing them with very
little. "His mother was begging him, imploring him to keep his word [on the
contract]," Duncan said.

Which, in the end, thanks to the *Globe* and the mother, Herschel Walker
did. He was the newest member of the New Jersey Generals.

The local media sold the story as the lecherous USFL robbing Georgia's
cradle — and it caused Simmons, a mild-mannered man, to explode. The
USFL dispatched an attorney to the University of Georgia campus, armed
with the signed contract. Farewell, national championship. Farewell, Zeus.
"Herschel played with fire and got burned," Dooley said. "I know my children
have lied to me, too. I still love them." (He later curtly told Walker to take his
millions and "do the best you can with the rest of your life.") Shortly there-

after, state senators in Georgia wore black-and-red armbands in mourning and college coaches across the nation ripped the USFL for violating a code. An overwhelmed Walker apologized for deceit. "No one realizes more than I that I am a human being," he said. "I ask for your forgiveness and ask God for his forgiveness."

Wrote Tom Callahan in *Time* magazine: "That seems like a lot of apology just for acting like a human being, a 20-year-old one at that."

A few days later, the *Dallas Morning News* ran an official ranking of the highest-paid players in professional football.

The list included established superstars like Saints quarterback Archie Manning, Cowboys halfback Tony Dorsett, and Bears running back Walter Payton. Topping the chart, however, was Herschel Walker — whose deal paid $2 million up front, plus $1 million bonuses in 1983 and 1984, a $1.25 million bonus in 1985, and a $750,000 loan for his investment portfolio.

It was craziness.

He arrived on Saturday, February 26, and the fanfare at the University of Central Florida was — no exaggeration — explosive.

There were television cameras everywhere. Reporters from near and far. All the national morning shows sent correspondents, seeking out that first giant interview, that initial glimpse of a big-time player in an unfamiliar uniform.

When he entered the locker room, the newest USFL star took in the scene. He had played at a large college, and knew not whether this league with its funky colors and unfamiliar rosters would be a step down or a step up. His coach gathered the entire team around, requested silence, patted the newcomer on the shoulders, and said, loudly, "Fellas, we've got someone here who a lot of you probably have heard of . . ."

Then Dick Coury introduced Marcus Marek — and laughed.

The former Ohio State linebacker, coming off of an All-American career with the Buckeyes but projected to be no higher than a fourth-round NFL Draft selection, had agreed to terms with the Boston Breakers just a day earlier, so when he pulled up to training camp along with all the hubbub, he momentarily believed maybe, just maybe, the USFL press was appreciating his potential. "But I'm Marcus Marek," he said years later with an exaggerated sigh. "And the other guy was *Herschel friggin' Walker* . . ."

As Marek and the Breakers began practicing on their side of the facility, the rest of the campus was abuzz. Earlier that morning Duncan sent his private plane to Wrightsville to pick up Walker, Manton, and Bob Newsome, an automobile dealer and family friend. The three flew to Orlando Airport, then took a helicopter to the Generals' practice field. While the chopper was en route, Chuck Fairbanks, New Jersey's coach, summoned everyone for a meeting. "As some of you have probably heard," he said, "we've acquired a guy named Herschel Walker . . ."

Amazingly, most of the Generals had been in the dark. According to Tom Woodland, a nose guard, two running backs immediately packed their bags. Another halfback on the roster was Terry Miller, the presumed starter who, five years earlier, had rushed for 1,060 yards as a Buffalo Bill. He knew his No. 34 jersey would soon belong to another. Indeed, by day's end Miller was wearing No. 40. By the weekend, Miller was traded to Denver.

The Generals were on the field practicing when the sound of the chopper disrupted drills. The craft came closer and closer and closer to the turf when — *plop*. There it was. And there *he* was. Walker exited the aircraft, his herculean frame neatly encased in a brown-and-beige Adidas warm-up suit. Spring football's Christ-like figure had arrived. He followed Duncan inside to a large dining room, where 200 credentialed media members and nine TV cameras awaited. Though five days shy of his 21st birthday, Walker spoke with poise and maturity. "The money is great, but I love the enjoyment of playing the game," he said. "I think a lot of college athletes, football players like myself, really don't have a chance to think. You really don't have an option to do what you want. The option I made was that I needed to do something else."

Walker changed into a white-and-red Generals practice uniform and jogged out to meet his teammates. It was nothing but high fives and butt slaps — few of them authentic. Truth be told, most of New Jersey's players were either resentful or stressed. The average USFL grunt was making $36,000. Walker was making $1.2 million. The average USFL (skill) player ran the 40 somewhere around 4.9 seconds. Walker ran it in 4.3. There were

12 running backs in camp, and all but 1 was rendered invisible. "Herschel's arrival was a big distraction," said Greg Murtha, an offensive lineman who had played with the Baltimore Colts in 1982. "But then we saw Herschel in action. Oh, man."

"We ran 40s that day, and the field was not dry," said Tom McConnaughey, a wide receiver. "Herschel ran a 4.28 on grass. A 4.28! That was unheard of. I was the guy unlucky enough to run right after him. It was as if I were invisible. A 4.28!"

In a 12-hour span, the team sold more than 1,000 season-ticket packages.

As one day turned to two, and two to three, Walker's teammates began to get a feel for their meal ticket. Walker would only have nine days to prepare for the March 6 season opener at Los Angeles, and while his athletic gifts leapt from the page, there were flaws. Walker was not used to catching balls out of the backfield. Or blocking. Or reading defenses. "He was fast, but not quick," said Larry Coffey, a running back. "That said, he picked things up very quickly. It was hard on me, being a running back. But you had to respect Herschel. He didn't big-league guys. He was there to be a part of the team."

Walker was assigned to room with Maurice Carthon, the rookie fullback out of Arkansas State, and a bond formed. Walker was in camp long enough to collect two end-of-week $30 per diem checks, and he was meticulous about depositing the loot. "He could have been a big shot who tossed the money in a desk," said Carthon. "But that wasn't Herschel. He valued what he had. He appreciated it.

"He knew how lucky he was to be getting paid to play football."

ACT ONE

We're about to fly back from Oakland after a game and we have this guy on our team — Ray Costict, a linebacker from Mississippi State. A big motherfucker. And Ray starts having an anxiety attack. He's walking up and down the aisle of the plane screaming, "No! No! No! No! No!" And he gets off the plane. We're on the tarmac. Chuck Fairbanks, our coach, tries to coerce him back on, but he won't budge. He was not getting on the plane. So Ray Costict takes a bus from Oakland to New Jersey. That was the absurdity of the USFL.

— Charley Steiner, radio voice of the New Jersey Generals

4

SEASONAL

I was at a party in Beverly Hills, and everyone in sports was there. Magic Johnson, Wayne Gretzky, Kareem. And I spotted my idol — Hank Aaron! I walked up to him and introduced myself. He said, "I know who you are. You made a great decision going to the USFL." I was shocked. Then he leaned in and whispered, "Tom, you always have to get the money. Get the fucking money. Because they don't care about you."

— Tom Ramsey, quarterback, Los Angeles Express

THE INAUGURAL SEASON of the United States Football League was to begin on March 6, and the 12 teams were evenly divided into three divisions — the Atlantic (Boston, New Jersey, Philadelphia, and Washington), the Central (Birmingham, Chicago, Michigan, Tampa Bay), and the Pacific (Arizona, Denver, Los Angeles, Oakland). Three division winners and one wild card would reach the playoffs.

As the big opening weekend approached, however, there was nothing but panic and terror inside the league offices.

Yes, the addition of Herschel Walker added heft and oomph. But as training camps came to a close (there were no preseason games), Chet Simmons, the league's commissioner, grew increasingly aware of the disparity between the USFL's upper-echelon organizations and its sinkholes.

On the bright side, Walker made the Generals immediately marketable, and both the Chicago Blitz and Philadelphia Stars seemed bolstered by a combination of high-level rookies and viable NFL veterans. The Michigan Panthers featured three rookies (safety David Greenwood, quarterback Bobby Hebert, and receiver Anthony Carter) who would have been certain NFL Draft selections, and the Tampa Bay Bandits and Birmingham Stallions cobbled together units that were, worst-case scenario, passable.

Those were the good teams.

On the downside, the Denver Gold and Boston Breakers were both painfully cheap, and it showed in uninspired, unrecognizable rosters. The Los Angeles Express's two most famous players were their quarterbacks, Tom Ramsey from UCLA and Mike Rae from USC (neither of whom was particularly famous).

The Oakland Invaders, meanwhile, were downright bewildering. Early on, John Ralston, the coach and general manager, brought in two former Oakland Raiders standouts, tight end Raymond Chester and defensive end Cedrick Hardman, and the league was encouraged. The NFL's Raiders had recently bolted to Los Angeles, so the rhyming name (Raiders-Invaders) and the familiar faces made perfect sense. The Invaders, however, were a troubling combination of kooky and stingy. Ralston's choice for the important position of equipment manager was John Daggett, a blind Mississippian. "Legally blind, couldn't see unless something was within two inches of his face," said Geno Effler, the club's public relations director. "The first time I met John was when the team sent me to pick him up at the airport. 'Why doesn't he just rent a car?' 'Because he can't see.' 'Oh.'"

The Invaders training-camp facility featured high school goalposts. Their fences were rusted and their paint was peeling. "They put us up in a semi-scuzzy hotel across the street from a flea market," said Ron Lynn, the defensive coordinator. "Streetwalkers everywhere."

"I broke my thumb in one of the drills," said Kevin Graffis, an Invaders guard. "That was on a Thursday, but we had to wait until Saturday to get it treated because there was a free clinic nearby that only opened on weekends." Tad Taube, the owner, purchased a run-down school bus that read ASSEMBLY OF GOD along the side to ferry the players from the motel to the training-camp facility in Arizona. The bus broke down every fifth or sixth trip, and one day the players were ordered to exit, choose a side, and push the vehicle up a steep hill. Mike Livingston, a longtime Kansas City Chiefs quarterback in camp with Oakland, turned to Lynn and said, "Welcome to the goddamn USFL."

He wasn't laughing.

"We were cheap, and we flew on a cheap charter," said Gary Plummer, the Invaders middle linebacker. "So we're on this flight to Denver, and suddenly the plane just drops for what feels like a minute. People are screaming, holding on to their seats, and a flight attendant flew, like, three feet in the air. I look over at [running back] Arthur Whittington. He has his head sticking out the middle of the aisle, and I couldn't figure out why. I said, 'Art, what are you doing?'" I swear to God — he looked at me and said, 'If I'm going to die, the last thing I want to see is pussy.'"*

With the exception of the ex-Raiders, the Invaders' roster was a desert. In particular, Simmons and Co. were alarmed by the team's choice of a starting quarterback. Around the league, most franchises seemed to have, in one form or another, a familiar name to situate behind center. In Michigan, Los Angeles, and Birmingham, three promising rookies (Hebert, from Northwestern State–Louisiana, Ramsey of UCLA, and Reggie Collier of Southern Miss) headlined the offenses. New Jersey (Bobby Scott, the ex-Saint who smoked cigarettes in the huddle), Boston (Johnnie Walton, three-year Philadelphia Eagle and a long-ago star of the World Football League's San Antonio Wings), Chicago (former Lion Greg Landry), Tampa Bay (John Reaves,

---

* It's debatable whether this is the best USFL plane story. During the 1983 season, the Wranglers were stuck at Newark (N.J.) Airport on a delay, and members of the team convinced the terminal bar to remain open an extra hour. Players drank to historic levels of excess. "One of our linemen was walking onto the plane, through first class where all the coaches sit," recalled Phil Denfeld, a tight end. "He puked all over the coaches, and they smelled of his vomit the whole ride home. He was fined $1,000." One week earlier, the Wranglers had flown home from Denver when the pilot asked players to lie down, single file, in the aisle. "They had to balance out the plane," said Todd Krueger, a quarterback. "That was terrifying." There was also the time a Chicago Blitz quarterback named Tom Porras was overheard laughing as the team flew home from Boston after a loss. George Allen, the coach, stormed to the rear of the plane, jabbed Porras in the chest, and screamed, "If I could, I'd throw you off this plane at 40,000 feet!"

the former Eagles first-round draft pick who battled substance-abuse issues), and Philadelphia (Chuck Fusina, Doug Williams's backup with the Buccaneers) all employed NFL journeymen of one form or another. It was, Simmons acknowledged, an early USFL weakness. In the NFL, quarterbacks like Pittsburgh's Terry Bradshaw, San Francisco's Joe Montana, and San Diego's Dan Fouts were the reasons fans tuned in. Why, the upcoming NFL Draft would feature a record five quarterbacks selected in the first round. "We were not strong at the position," said Steve Ehrhart, the league's executive director. "It was a real concern."

No place was more worrisome than Oakland. Simmons impatiently waited throughout camp for Ralston to find a legitimate starter, but the name that kept being mentioned was Fred Besana. Which led to the $1 million question: "Who in God's name is Fred Besana?"

He was, in fact, a former University of California backup who, after having gone undrafted by the NFL, was released in training camp by the Buffalo Bills in 1978, then the New York Giants a year later. He spent the next three seasons with the Twin City Cougars of the California Football League, where he earned $50 a week while also teaching at Marysville High, selling insurance, and working as a beer distributor. "My first reaction when the Cougars called was, 'I don't wanna play in a beer league,'" Besana said. "But I ended up throwing the ball a lot." Ralston had coached against Cal when he was at Stanford, and he offered Besana an Invaders tryout. "Honestly, Coach, I think I'm over it," Besana said. "Football isn't my thing anymore."

"C'mon, Freddie," Ralston said. "What do you have to lose?"

Besana was assured he would be walking into professional football, but arrived to the battered bus, the $5 per diem, the mediocre facilities, the goofy lightning-bolt-in-a-fist logo, the utter weirdness. During camp the Invaders stayed at a Red Lion Inn, and one night one of the leaders of the team's regular Bible-study group was caught having anal sex with a prostitute against the railing outside his room. "We were a beautiful mess," said Plummer. "We'd be on that church bus, mooning people out the back windows. We had guys like Ray and Cedrick who were actually listed as player-coaches. What they'd do is, if they got too drunk the night before practice, they wouldn't practice the next day and say, 'Hell, I'm just a coach today.' That was a saying for 'I'm hung over, man.'"

Besana loved it. His contract paid $40,000, and to the shock of most everyone in Invaders camp (few of whom had heard of the Twin City Cougars), he possessed a strong arm and pinpoint accuracy. The football exploded from his hand, sliced through the air, hit his receivers between the numbers. "He was a legitimate NFL quarterback," said Chester. "Freddie could sling the ball."

While the Invaders' situation was unsettling to the league, it paled in comparison to the monstrosity that was quickly developing in the nation's capital. The Federals and the Blitz were slated to open for business on March 6 with a 1:30 p.m. kickoff at RFK Stadium, and odds were 50-50 whether Washington would field a team with people who could capably catch and throw footballs.

The Federals were supposed to be a benchmark USFL franchise (big city, established football fan base, Craig James as a marquee star), and all looked good when Marvin Warner, a wealthy businessman, had been awarded the team. At the last moment, however, he opted for a club in Birmingham, and Washington was owned by a limited partnership that was in turn owned by a joint venture that was operated by three different corporations. The majority owner was Berl Bernhard, a founding partner in a law firm who was also involved in Democratic politics (Bernhard managed Edmund Muskie's failed presidential campaign in 1972). A well-liked and respected man, Bernhard thought it might be fun to own a team — until he realized the expense. "We didn't come in with enough capital," he later said. "We needed more lead time, and we gave up too much ground to everyone else."

Were there one cow pie in the field, the Federals found a way to step in it. The team's general manager, former Redskins' assistant GM Dick Myers, had no personnel experience. ("The Redskins wanted us to be a four-star disaster," said Gordon Davenport, the director of marketing. "They gave us Dick.") The choice for head coach was Ray Jauch, a Canadian Football League refugee who had once led the Edmonton Eskimos to three straight Grey Cup appearances, but came to America unprepared for what awaited. Though only 45, Jauch's gray hair and methodical shuffle made him seem closer to 80. "In Canada everything is very relaxed," said Willie Holley, a Federals defensive back. "So guys come to Washington and maybe they show up for film, maybe they don't. There was no punishment."

The Federals held part of their training camp in Jacksonville, Florida, at the Samuel W. Wolfson Baseball Park, a foul-smelling sinkhole surrounded by a crooked fence with billboards for places like Sonny's Real Pit Bar-B-Q and Nimnicht Chevy. "It felt like a junior college," said Randy Burke, a wide receiver. "A crappy one." The organization's first-ever cut was, appropriately, a lineman out of the University of Maryland named Les Boring. The bus ferrying all the Federals' equipment from Florida to Washington broke down en route. When camp ended, Federals ownership told the players they had to find their own transportation to the nation's capital. "It was their test to see how mature we were," said Mike Hohensee, the rookie quarterback from the University of Minnesota. "They made it clear we had to drive back, find an apartment, find the practice facility, learn our way around the city—all in four days."

One player, a defensive end named Dallas Hickman, returned as instructed to Washington after camp. Having spent six seasons in the NFL, he was eager to continue his career. "So I report to RFK Stadium, enter the locker room, and there's no locker for me," he said. "I asked and the equipment manager said they wanted to see me upstairs."

Hickman entered Myers's office, only to be told he was cut.

"Dick, you must be fucking kidding me!" he screamed. "Shit, I had to come back from Florida, then walk into the fucking locker room to find out there's no locker! How about a little respect? Go fuck yourself!"

With the exception of James, Hohensee, and Joey Walters, a fleet wide receiver who had been a CFL star, Washington acquired people who were too old, too fat, too uncoordinated. "Ray brought in *so* many Canadian castoffs," said Doug Greene, a safety. "They were in Canada for a reason—they sucked."

On more than one occasion newspapers mistakenly referred to the Washington Federals as the Washington Generals—the basketball team that lost to the Harlem Globetrotters more than 14,000 times. It wasn't a leap. When the Federals signed a lease to practice and play at RFK Stadium, Bernhard assumed that meant free access to all the facilities. Wrong. The Redskins refused to give an inch, and the Federals were forced to practice on a small patch of rocks and gravel that wasn't even the size of a proper football field. It sat directly across from a prison.

Entering the season opener, the Federals led the USFL in three unofficial categories:

1. Football players no one had ever heard of
2. Cigarette smokers
3. Coke addicts

The third was a by-product of the drug's emergence as a 1980s American phenomenon. At the start of the decade, cocaine was pouring in through the Bahamas and the Dominican Republic, and was largely available on the streets of United States cities. Most professional sports teams made their players aware of the dangers and temptations of cocaine. The Washington Federals did not. Such was the result of a clueless personnel department that pretty much signed anyone with operational hands. In the days leading up to the season opener, running back Buddy Hardeman was arrested for attacking a police officer. "We had a lot of hoodlums," said Hohensee. "We outdrank everyone, we outsnorted everyone, we outsmoked everyone. We probably had more women than any other team. We had lots of fun off the field, but we were miserable on it."

Who was the most preposterous member of the Federals? Tough call. After he was released by the Denver Gold, Joe Gilliam was signed by Washington. The former Pittsburgh Steelers quarterback had not played in a professional game since 1975. He was a drug addict who long ago pawned off his two Super Bowl rings. Rick Vaughn, the Federals' media relations director, met Gilliam at the airport and drove him to a TV station for a local introduction. "He said we needed to stop at a liquor store," said Vaughn. "I was 24, he was a Steeler. What choice did I have? He got a bottle in a brown bag, and he drank on the ride." Gilliam lived in his car (a 1960 Ford Falcon with bald tires) until a spot opened inside a Washington-based halfway house for drug addicts. "I look back and I realize the truth about Joe," said Victor Jackson, a defensive back. "He didn't stay late after practices to work out. He did so because he needed a ride to the rehab center."

"One time we were flying back from a game, and Joe—who wasn't supposed to drink—had a drink on the plane," said Billy Taylor, a Federals halfback. "Everyone was sleeping, and he told me he was going to talk to the coaches right then and there because he wanted to be the offensive coordinator and he knew he could do a better job. I stopped him. He was drunk, and he would have been cut right there."

Another addict was Bob Cobb, a defensive end out of the University of Arizona who had been the Los Angeles Rams' third-round pick in 1981. Cobb began snorting cocaine in college, and by the time he reached the NFL he was a full-fledged junkie. During his rookie year with the Rams he failed to show up for a game, and his wife, Myca, drove the streets of Los Angeles before finding him strung out on a corner. He was suspended twice, then bounced from Tampa Bay to Chicago before joining the Federals. Because they failed to run background checks, Cobb's size (six foot four, 248 pounds) and speed won Myers over. "Cobb looked like Tarzan," said Dana Moore, the punter. "All muscle." With a $100,000 contract, Cobb was Washington's second-highest-paid player. But he lived in the projects in a one-bedroom apartment with a refrigerator and a couch. The Federals soon learned their top pass rusher was a drug fiend. He failed to last the season, and later died of AIDS.

The most unique backstory belonged to Doug Greene, a safety out of Texas A&M–Kingsville who spent 1978 with the Cardinals and 1979 and 1980 with the Bills. During his year in St. Louis, Greene had befriended Leon Spinks, the former world heavyweight champion. When Greene joined the Bills, Spinks asked whether he knew anyone in Buffalo.

"Nope," Greene said. "Never even been there."

"Well, I'm gonna hook you up," Spinks promised.

Fast-forward to one of the Bills' first practices during the 1979 season. Greene was working out when an Excalibur parked near the fence, and two men unfolded from the vehicle. One was a six-foot-nine bodyguard holding a 9 mm pistol. The other was Rick James, the music superstar. He had long dreadlocks, a crooked smile, and wore a full black-leather bodysuit.

The bodyguard swaggered over to Greene and said, "Rick wants to speak to you."

Greene introduced himself to James. "I know who you are," the singer said. "I promised Leon I'd take care of you." James handed Greene an envelope containing $500 in cash, and insisted he check out of his hotel after practice and relocate to the James household in East Aurora. "It was the most buff, the baddest mansion ever," Greene said. "Indoor tennis court, indoor pool. And the women — oh, man. Black, white — but mostly white and blonde. They came and went nonstop." Before long, Greene was devoting his spare off-field time to work as one of James's bodyguards on the road. It was first-class travel and sexual and financial perks galore. "I wasn't the best player in football," Greene said. "But I was the only one who could say he was Rick James's bodyguard."

With the swirling negativity and gross ineptitude, no team needed a successful opener more than the Federals. Despite the thin roster and mediocre coaching staff and drug-infested locker room, an impressive 38,010 fans bought tickets for Blitz-Federals at RFK Stadium. The day felt historic. Two stud running backs, Washington's Craig James and Chicago's Tim Spencer, would be making their professional debuts.* George Allen, the Redskins' longtime coach, was returning to the scene of his greatest triumphs. Lou Rawls, a man Frank Sinatra once said had "the classiest singing and silkiest chops in the singing game," brought forth an A-level a cappella national anthem. Groundskeepers dyed the grass green so it would look presentable on television. ABC televised the matchup, with a baby-faced Jim Lampley sitting alongside Lee Corso in the booth. "Hello, I'm Jim Lampley," the broad-

---

* Unlike Spencer, who was embraced in Chicago, many of James's Federals teammates were turned off by the rookie. "He was conceited as hell," said Billy Taylor, his backup at halfback. "The first time I met him he told me that, were it not for [college teammate Eric Dickerson], he would have won the Heisman. I was shocked when he said that. Craig was good — but not *that* good."

cast began, "and I'll be honest with you. I will wait to see how many of you are that anxious to see football in the spring . . ."

It was, meteorology-wise, not a good day for football. When David Dixon hatched the USFL idea, his vision featured sunny skies and temperatures in the 60s. That was a big slice of the appeal — fans wouldn't have to drive through sludge and sleet to sit through sludge and sleet. Yet here, in Washington, the temperature was 54, the skies were pewter, and it was raining. The field, not fully recovered from the Redskins, featured as many divots as strands of grass. "It was horrible," said Hohensee. "Everything about that day was horrible."

Throughout his storied coaching career, Allen was infamous for seeking out every possible advantage — legality and morality nonfactors. In his days leading the Redskins, he became a PhD-level practitioner of football espionage. Among other things, Allen wiretapped visiting locker rooms. He installed tinted one-way glass in the facility so he could look in on opposing coaches unobserved. He ordered his defensive and offensive linemen to lather the front of their uniforms with Vaseline, thereby making them impossible to grip. He messed with the temperature controls of the visiting locker rooms — on a hot day, the furnaces blasted; on icy afternoons, the heat magically failed to operate. All USFL teams were required to submit player contracts to the league. The Blitz had two sets of contracts for most of their veteran standouts — the ones with lower figures that were presented to league offices to appease the USFL's financial concerns, and the real contracts, with eye-popping bonuses. "George Allen did what he felt he had to do," said George Heddleston, future general manager of the Pittsburgh Maulers. "And no one would dare stop him."

In the week leading up to the opener, Allen reached new heights (or depths). He supplied a couple of low-level Blitz employees with yellow USFL STAFF windbreakers and sent them to Washington. The two informed (a.k.a. lied to) Myers and Jauch that they were part of the league's film crew, and needed to tape the Federals' practices.

"We knew every play they were running," said a Blitz assistant coach. "There were no surprises."

In a game that was lopsided, wet, and ugly, the Blitz decimated the Federals, 28–7. Washington's first play from scrimmage was a sweep by James, who was smothered for a 3-yard loss. If that didn't tell the afternoon's story, Chicago's opening score did. Quarterback Greg Landry dropped back and threw a short pass to Trumaine Johnson, who ran 33 yards before the wet football slipped from his hands. Another Blitz receiver, Wamon Buggs, fell on the ball in the end zone. "I remember people bitching, 'This isn't anything like the Redskins,'" said Brett Friedlander, who covered the game for the *Annapolis Evening Capital*. "They were right about that."

By midway through the third quarter, RFK Stadium was a ghost town. "I can't express this quite adamantly enough, but the Washington Federals just sucked," said David Remnick, the *Washington Post*'s beat writer. "They were just unnecessary, and unnecessarily bad."

"We got our clock cleaned by Chicago," said Gordon Davenport, the Federals' director of marketing. "Washington never forgave us for that."

The Federals went on to finish 4-14, playing largely before pigeons and billboards. In an effort to generate some interest, the organization once held a T-shirt giveaway for the first 10,000 people to enter the stadium.

Less than 7,000 attended.

Despite the Federals' awfulness, the USFL was pleased by the attendance figures from the first four games of the opening day. In Florida, the Bandits drew 42,437 to Tampa Stadium for a riveting 21–17 triumph over Boston. An inspired 45,102 came to Mile High Stadium to see the hometown Gold lose to Philadelphia, 13–7. And in Arizona, the Wranglers — bad in a Washington Federals sort of way — were pummeled by Besana and the Invaders, 24–0. "The thing I remember most is the statistics were all wrong," said Bob Hieronymus, the Wranglers' assistant media relations director. "We went cheap and hired a crew that didn't know what the hell it was doing."

Still, 45,167 filled the stadium. And if it was mainly to catch the one-hour pregame Beach Boys concert, hey, who was counting?* Save for the Oakland-Arizona and Boston–Tampa Bay clashes, the games were televised re-

---

* "Yes, we were awful," said Alan Risher, Arizona's quarterback. "But we had a defensive back whose name was Admiral Dewey Larry. That's amazing."

gionally by ABC, and ratings were impressive. The Stars, for example, drew a 19.5 rating in Philadelphia — which means nearly 20 percent of the region's television sets were tuned to the game. Other regions followed suit. "It was good stuff," said Gordon Banks, an Oakland wide receiver. "Not amazing, but good. If the NFL was a 100, we were about a 40."

The contest that mattered most was slated for 4:00 p.m. Eastern, and featured Walker's Generals traveling to Los Angeles for a nationally televised meeting with the Express. This was, for the league, both thrilling and terrifying — thrilling because it was akin to unwrapping a highly anticipated birthday present (Walker) and being allowed to play with it; terrifying because only 12,471 things could possibly go wrong.

Although the Express were not nearly as dysfunctional as the Federals or Wranglers, the franchise had its issues. To begin with, one of the owners, a cable television executive named Alan Harmon, acted as team president and often arrived at the office coked up after a night of drug use. Behind his back employees nicknamed him Arthur after the eccentric, alcoholic movie character played by Dudley Moore. "He'd come in, either strung out or drunk or both, and start swearing at the girls who worked in the office," said Paul Sandrock, the team's controller. "And I'd have to send him home."

There was also a problem with the facility. The Express signed a lease with the mammoth (and dilapidated) Los Angeles Coliseum, which meant either: (a) the team would play to weekly sellout crowds of 92,516, or (b) the USFL's largest stadium would feel (and look) like one of its emptiest.

As for coach Hugh Campbell's 40-man roster, it was largely a collection of anonymous grade-C USC and UCLA products, as well as a funky cobbling of weirdos and castoffs. Linebacker John Barefield used to arrive most days with a black cowboy hat atop his head and a nickel wedged into his right ear ("I still have no idea why," said Campbell years later). The starting nose guard, a kid out of Georgia named Eddie Weaver, was nicknamed Meat Cleaver and refused to wear a jockstrap. "Meat just let it hang," Campbell said. "Free." Charles Philyaw had been a productive defensive end with the Raiders in the late 1970s before a knee injury ruined his career. Back in Oakland, he gained fame for trapping his head in the sunroof of a car.

When the Express called, Philyaw eagerly signed. Now 29 years old, he played with a pronounced limp, and after every practice and game applied WD-40 motor oil to his legs. "I'd seen many things in football," Campbell said. "But never that."

The clash against the Generals was important not only to both organizations, but also to the league as an entity. ABC sent its A-team of play-by-play man Keith Jackson and color commentator Lynn Swann. Simmons attended, gazing down from above in a sealed box. Before Walker's signing, less than 14,000 tickets had been sold. By the time the ink dried on his contract, however, that number doubled, and as kickoff approached, 34,002 fans entered the stadium, paying between $7.50 and $11 for tickets. There were even a handful of celebrities, including Wilt Chamberlain, Dionne Warwick, and Lee Majors, a future Express minority partner and the actor best known as Steve Austin from *The Six Million Dollar Man*. With television in mind, the league planned certain details beforehand, first and foremost packing the majority of the spectators on one side of the coliseum so the stands would appear full. "Here's the thing—34,000 people in a 100,000-seat stadium is not good," said Mike Hope, the Express's vice president of marketing. "So we sold every *other* row of seats to make it look more like 60,000 people."

The organization wasn't entirely prepared. "Our full uniforms never arrived for the game," said Pennie Orcutt, an Express cheerleader. "We wore silver boots, white T-shirts . . . and our own underwear." The largely male fan base seemed unperturbed.

On the strength of Walker and hype, the Generals were merely one-point underdogs on the road. But the running back had been with the team for little more than a week, and head coach Chuck Fairbanks was a stubborn old coot who didn't worry much about perception or headlines. Throughout his three years at Georgia, Walker ran almost exclusively out of the I-formation, directly behind a blocking back. When New Jersey's offense opened the game, however, Walker situated himself alongside—not behind—fullback Dwight Sullivan. It was uncomfortable and unfamiliar, and after rolling for a 9-yard gain on his first career carry, then appearing in 15 of the team's 17 offensive plays, Walker spent the afternoon doing pretty much nothing.

Fairbanks committed himself to not overusing the youngster, and it made for both a flat game and a flat television-viewing experience. "Herschel's biggest problem was he wasn't ready yet," said Mike Stock, the New Jersey running backs coach. "We put in all this stuff for him, but he wasn't familiar with pass protection, checking linebackers, making route decisions, whether to stay in and block. He had a few days to learn. It wasn't enough." Fans pining for Herschel Walker were forced to watch a lot of quarterback Bobby Scott. Walker carried 11 times in the first half, but only 5 in the second.

"It was strange, because I've run into softer walls than Herschel . . . the guy was all muscle," said Danny Rich, an Express linebacker. "There was one play I remember most. The Generals handed the ball to Herschel, and I reached in and grabbed his crotch and twisted it as hard as I could. And he puts this kind of kung fu grip on my wrist. I'm grabbing his wiener, he's grabbing my wrist. I'm now trying to yank my hand away, and he stands up and just points at me. Doesn't say a word. He was a *baaaad* man. They needed to use him."

In the fourth quarter, after the Generals scored a touchdown that brought them within five points, 20–15, Fairbanks sent his offense onto the field for a two-point conversion from the 3-yard line. It was the perfect time to go with a new professional football rule, and the stadium buzzed. This was quintessential Georgia Herschel Walker turf — tough runner also able to soar over the defense. Only Fairbanks refused to insert him. Instead, the unremarkable Larry Coffey lined up at halfback, and Scott's wobbly duckling of a pass was intercepted. The coach defended himself afterward, meekly noting that "three yards is a lot to get with the running game." (Walker averaged 5.3 yards per carry in three collegiate seasons.)

"I don't want to question Coach Fairbanks' decision," Anthony Davis, the Los Angeles halfback, said later, "but most people anywhere would say Herschel's got to be in there."

The Generals mustered a final drive, and with a third and 3 at the Express 5, Fairbanks again kept his superstar on the sideline. Coffey — three inches shorter, 24 pounds lighter, and significantly slower than Walker — came out of the backfield but did nothing as Scott was sacked for a 9-yard loss. With 57 seconds left, the Generals now faced fourth and 12 from the Express 14. Scott stood behind center. Fullback Maurice Carthon stood behind Scott.

And there, at the top of the I, resplendent in his white-and-red uniform, was . . . Coffey.

Coffey!?

Scott faked the handoff to his halfback, who missed a block on defensive end Greg Fields. The quarterback rolled right, twisted, and rushed a throw to wide receiver Larry Brodsky. He juggled the ball, caught it, but appeared to step out of bounds two inches before the first-down marker. From his faraway perch in the broadcast booth, Charley Steiner, the Generals' play-by-play man, was lost. "As the football gods would have it, I'm up here and the play is all the way over there on the far side," he said. "As distant as we were, you couldn't tell what number he was wearing or whether he caught the ball. It's the biggest play of the opening game, and I'm not just sweating — I had the Mississippi River flowing down my back. Somehow I didn't fuck up the call, but the Generals fucked up the play. They didn't make it."

Game over. New Jersey loses.

On the sideline, Walker lowered his head. He expressed neither anger nor agitation, and refused to blame Fairbanks. His final statistical line (16 carries, 65 yards) was forgettable. "I ran the ball a little better than I expected and I caught a pass," Walker said. "It's tougher than I thought it would be. A lot of the guys had more speed than I expected to see. And the execution was better. I guess that's the biggest adjustment I have to make.

"I'm just sorry we didn't win."

Lost in the headlines (The *Journal News* sports section led with HERSCHEL WALKER A FORGOTTEN GENERAL) was the unspoken beauty of the entire USFL experiment, where football dreams turned real. While Walker spent too much time watching from the sideline, an Express halfback named Tony Boddie stole the spotlight, running for 77 yards on 13 carries, catching five passes for 49 yards, and scoring the go-ahead touchdown on an 11-yard completion from Tom Ramsey. Boddie was the team's 12th-round pick out of Montana State, a 5-foot-11, 195-pound honorable mention Division I-AA All-American. "Our crowds averaged 8,000 in college," Boddie said. "It was the only school in the country that offered me a scholarship."

Now, thanks to the new league, Tony Boddie was standing in the Los Angeles sunshine, basking in the glow of having outperformed, in one writer's words, "the best running back on God's earth." Herschel Walker would go on to great accomplishments, while Boddie faded into the mist. But for one glorious afternoon, with much of the nation watching, it was his time to shine.

"That was the coolest thing," Ramsey said. "In the USFL, nobodies became somebodies."

5

WEATHERING A STORM

Nate Newton was one of our offensive linemen. He was enormous. Nate came to practice one day on a motorcycle. He rode it every day for two weeks. Then there came a day when he arrived with a huge dent in his head. We never saw the motorcycle again.

— Zenon Andrusyshyn, punter, Tampa Bay Bandits

WHEN DAVID DIXON initially went about selling the idea of a spring football league, he raved at length about potential.

The potential to draw fans.

The potential for TV deals.

The potential for exciting games.

The potential for marketing.

The potential for player development.

Dixon was deemed a visionary by nearly all who knew him, but, as the novelist Salman Rushdie once noted, "Not even the visionary or mystical experience ever lasts very long."

Or, in the context of the USFL, spring came with its unanticipated complications.

If the league's opening weekend, what with its 14.2 television ratings and 40,511 average attendance, had Dixon and Co. crowing, by the end of the fourth week of the 18-game schedule, moods — and optimism — had shifted.

First, there was the underwhelming Herschel Walker, who took the league by (non)storm with games of 65, 60, 39, and 97 rushing yards for the 0-4 Generals and their increasingly agitated roster of ensemble players. "You had a bunch of veterans on the team who walked around with stripes drawn on their sleeves," said Marc May, a tight end. "They felt underpaid, and liked to say, 'Herschel's the only General — we're noncommissioned officers.'"

Second, there was the dreadful quarterback play, which resulted in some of the wobbliest ducks to ever travel above a football field.

Third, there was the weather. All of the Week 4 contests were played in rain, and four of the six drew less than 20,000 spectators. Some of the stadiums endured only light mists. For others it was borderline torrential. The worst conditions came to Birmingham, a city whose average spring temperature is 67 degrees and where precipitation is generally a nonissue. One week earlier, in a Monday home loss to the Stars at Legion Field, the Stallions played in front of 12,850 fans during rain and hurricane-force winds. That was thought to be a mere aberration. "If anything, Birmingham was too damn hot," said Herbert Harris, a Stars wide receiver. "There were games there where, come third quarter, you'd have blisters on the bottoms of your feet."

Now, on the night of March 26, the precipitation was happening again — only worse than ever. The surprising 2-1 Arizona Wranglers came to Legion Field, and with a threat of freezing rain, the local police department warned residents to stay off the roads. Chet Simmons, however, refused to cancel or postpone. ESPN was in town, and the league needed the exposure. "I would've liked to not play," moaned Doug Shively, Arizona's coach, "but we had to."

The conditions — and game — could not have been uglier.* The announced attendance was 5,000. The real attendance was closer to 1,500.

---

* Well, perhaps they could have been uglier. Later in the season the Wranglers traveled to Detroit for a July 3 game against the Michigan Panthers. The air-conditioning inside the Pontiac Silverdome stopped working, and 31,905 fans slow-roasted. Recalled Arizona quarterback Dan Manucci: "It was like playing in your garage with the door shut."

That night showers bombarded the city. If the slop and monsoon-like winds weren't enough of a deterrent, there was the unfortunate fact that Reggie Collier, Birmingham's fabulous young quarterback, was out with a hip pointer. His replacement, Bob Lane, shared a name with the Detroit Lions Hall of Fame quarterback and a skill set with Bob Lane, one of the finest marketers of meat products in all of Decatur, Illinois.*

The Stallions' Lane threw 16 passes, and completed 5 for 47 yards.

With 50 mph winds howling and precipitation arriving from all angles, the inept Stallions and inept Lane somehow held off the inept Wranglers, 16–7. It was the USFL's worst nightmare — ESPN scored awful ratings, and the following morning's headlines were all along the lines of TINY CROWD SEES BIRMINGHAM VICTORY (*Nashville Tennessean*). When asked about the catastrophe, Peter Hadhazy, the USFL's director of operations, sighed. "If I didn't work for the league," he said, "I wouldn't have gone to a game, either."

The attendance and weather were troublesome, but the television numbers were devastating. The national ratings for the fourth week were a 6.4, with a 15 share of the audience. In other words, within a month's time viewership of USFL games dropped by more than 50 percent. The franchise owners were wealthy, egotistical men used to getting what they demanded, and this was not meeting their expectations. Simmons had been hired for his TV acumen, but most of the teams were furious that the deal with ABC did not include a blackout clause (whereby a team's game would not be televised locally if certain attendance thresholds were not met). Without that stipulation, it was rightly argued, what would inspire fans to attend when they could watch from home for free? "The NFL had a blackout policy, so our guys were screaming about it, wondering why we lacked one," said Steve Ehrhart, the USFL's executive director. "That issue stuck in the throats of a lot of owners. They blamed Chet, but it wasn't his doing."

The season's fifth week wasn't quite as bad as the fourth, but it brought little to celebrate. The attendance totals at the site of ABC's two home games,

---

* Indeed, there was a Bob Lane who sold meat products in Decatur in 1983. Odds are he was quite good at his job.

Philadelphia and Los Angeles, were 10,802 and 17,139, respectively. Clearly, the nonblackout policy wasn't helping the situation. "It was hard," said Ehrhart. "I think we'd all admit it was a mistake."

In the absence of better television policy, and with the dual plagues of weather and a general apathy among the public, many of the USFL's franchises turned toward X-factors that might boost marketability. In Boston, for example, the Breakers were hamstrung by a limited budget, a mediocre roster, and a home stadium (Nickerson Field, on the campus of Boston University) that held a USFL-low 21,000-capacity crowd and provided negligible space for parking (the Breakers led the nation in attendees returning to their vehicles with $20 citations stuck to the windshield). Hence, Dick Coury, the head coach, came up with an idea. In the week leading up to every home game, he invited fans to send trick plays to the Breaker offices. The coaching staff would dig through the submissions and pick a winner. Not only would that play be used in the upcoming game, but its designer would stand along the sideline and hold the wires extending from Coury's headset. "We had a lot of fun with it," Coury said years later. "We'd get some crazy plays." One, in particular, involved a spread formation where the center and quarterback stood on one side of the field and the other nine players stood on the other. Johnnie Walton, the Boston quarterback, threw a screen to the tight end that gained 10 yards, and the sideline went berserk. "Some might think it high-schoolish," Coury said. "I just thought it was really fun."

"Coach once put 12 guys in the huddle, lined up in a formation, and then, at the last second, motioned a guy out of bounds," said J. D. Shaw, a Breakers ball boy. "They outlawed that. But it was brilliant fun."

Coury was the right man for zaniness, in that he kept football in perspective. On November 13, 1971, a plane chartered by three of his assistant coaches at Cal State–Fullerton crashed en route to San Luis Obispo. All three coaches, as well as the pilot, died when the single-engine Piper plunged into the Santa Ynez Mountains. Coury had missed the flight, and it changed him. Football remained important, but he no longer viewed the game through a more-important-than-anything-in-the-world prism. "I gave serious thought about giving it all up," he said. "It's stayed with me all those years."

There were times, after particularly rough Breakers practices, where Coury would steer his players toward enormous buckets filled with soda and beer. Or host the entire team on a movie night or bowling trip. "He was the best," said Walton, the quarterback. "I played for a lot of coaches who took football far too seriously. Dick knew it wasn't life or death."

The undermanned Breakers opened the season 4-1 and emerged as one of the USFL's hot stories. The other team that somehow overcame the ills weighing down myriad franchises was the Tampa Bay Bandits, also 4-1. Unlike the Generals, who relied on Walker for hype, or the Chicago Blitz and their merry (and heavily promoted) band of geriatric NFL refugees, the Bandits drew some of the USFL's largest crowds to Tampa Stadium on the strength of three factors:

1. The rote dreadfulness of the NFL's Tampa Bay Buccaneers, who were preparing to embark on a 2-14 season.
2. An exciting, up-tempo playing style that featured lots of vertical passing.
3. An uber-emphasis on fun.

John Bassett, the Bandits' managing general partner, was unlike any other USFL owner, in that while he certainly enjoyed making money, his guiding life principles primarily concerned entertaining, engaging, embracing. Bassett's main sources of income involved real estate development and film production, but sports were his passion. At age eight, while attending Bishop's College School in Quebec, young John invented a dice baseball game, created a league, charged a 10-cent franchise fee, and typed up a weekly one-page league newsletter that sold for a penny. A former Davis Cup tennis player, Bassett was a squash champion in his home of Ontario, Canada, as well as a quarterback at the University of Western Ontario. His paternal grandfather, John Bassett, had been publisher of the *Montreal Gazette*. Another grandparent was the mayor of Sherbrooke, Quebec. His daughter, Carling, was a professional tennis player who ultimately rose to No. 8 in world rankings, and he had once been an owner of the World Hockey Association's Birmingham Bulls, as well as the Memphis Southmen of the short-lived World Football League.

One of the lessons Bassett learned from the failed WFL experience was that football alone was rarely enough of a sell. An organization needed to place as much emphasis on image and fan friendliness as it did on first downs and interceptions. A game had to be a scene. An event. Pomp and circumstance and oomph and pizzazz. That's why, shortly after he agreed to form the Bandits, Bassett enlisted Burt Reynolds, America's hottest actor, to serve as a part owner. "My dad understood the value of celebrity," said John Bassett Jr. "He knew the appeal." Though Reynolds was only given 5 percent of the team (he put no money into the operation; it was a tradeoff for exposure and hype), he went all out. Back on August 4, 1982, for example, Bassett held a press conference to announce the franchise name, and Reynolds not only arrived, but did so accompanied by bottles of champagne and a sexy young woman on each arm. He was brought onto the field via stagecoach, as female fans threw their bras in the air. "I'll be around a lot, involved," he said — and it was no lie. Reynolds's mug shot graced the team's 1983 media guide. His girlfriend, the buxom, blonde actress Loni Anderson, was featured on billboards, her bare midriff peeking out from beneath a white Bandits T-shirt alongside the words ALL THE FUN THE LAW ALLOWS. At the conclusion of the season, Reynolds gifted every player with a western-style gold-dipped Bandits belt buckle, as well as team jackets with WITH LOVE, BURT REYNOLDS embroidered on the inside. "John's mantra was 'Community first!'" said D. J. Mackovets, the team's media relations director. "If the fan is happy, the team is happy."

That helps explain what was, at the time, a curious head-coaching decision. Bassett compiled a lengthy list of candidates to lead the Bandits, topped by John Rauch, the man who coached the Oakland Raiders to Super Bowl II, and Jack Pardee, the former Chicago Bears coach. Before making a hire, however, Bugsy Engelberg, the club's director of football operations, reached out to Steve Spurrier, Duke University's offensive coordinator, about holding the same position with the Bandits. The appeal was as much about PR as football; 17 years earlier, Spurrier had won the Heisman Trophy quarter-backing the University of Florida Gators. He wasn't merely a Sunshine State hero, but royalty. Engelberg, a portly character who referred to everyone as Bubba, got Spurrier on the phone. "Bubba," he said, "we want you to be our offensive coordinator."

"I'm already an offensive coordinator," he replied. "I want to be a head coach."

Bassett decided to fly to Durham, North Carolina, to interview Spurrier in person. He went to the coach's house. A broken TV was flickering in a corner. On top of the set was the 1966 Heisman Trophy. Spurrier owned a little white poodle named, of all things, Bandit.

"There was no way out," Bassett said.

Spurrier came back to Florida as a new head coach. At age 37, he was the USFL's youngest; with a $60,000 salary, he was also the USFL's cheapest. "John and those guys had no idea whether I could even do the job," Spurrier said. "But John was a maverick. He wasn't afraid to take chances."

Spurrier's return to Florida was enormous news. George Steinbrenner, Tampa native and owner of the New York Yankees, attended the introductory press conference. The Bandits proceeded to overload their roster with regional players from Florida colleges, then send them off to every super-market opening, Rotary meeting, and 4-H fair. The starting quarterback, John Reaves of Florida, was handsome and approachable. His backup, Jimmy Jordan of Florida State, was handsome and approachable. The team sold 22,600 season tickets and drew crowds that made the other owners salivate. A magical 42,437 fans packed Tampa Stadium for the opening-day victory over Boston, then 38,789 came a week later for another win, this over Michigan. Tampa was demolished, 42–3, by visiting Chicago in the fifth week, but 46,585 people watched. "You talk about putting it together with a marketing blueprint — we had all the elements lined up," said Jim McVay, the Bandits' director of marketing. "You have to have the right product, you have to have the right promotion, you have to have the right packaging, and you have to have the right pricing. It was a wonderful combination, and that's why the Bandits did so well. It's the other teams in the USFL that struggled a little. Not everybody was as well managed as the Bandits." One never knew when

Reynolds and Anderson might appear on the sidelines alongside such stars as Robert Urich, Ernest Borgnine, Ricardo Montalban, or Dom DeLuise. "That's all part of the scene," Bassett said. "I can't guarantee that we can win, but I can guarantee when a fan goes to our games he's going to have a good time." The Bandits' official mascot, the Bandit, was a man in a red-and-black bandana who galloped across the field atop a black horse, Smokey. The animal often excreted large turds in the tunnel where the players lined up to be introduced before a game.

As a favor to Reynolds, Jerry Reed, the famed entertainer, wrote the team's theme song, "Bandit Ball," which blared from the speakers. The colors — silver, red, and black — rivaled the Los Angeles Lakers' for flash, and in 1983 AJD, a company that manufactured hats with sports team logos, sold more Bandit gear than all professional teams save the Dallas Cowboys and Washington Redskins. When Jim Brown, the NFL's all-time leading rusher, spoke with *Sports Illustrated* about making a football comeback at age 48, the first team to call was Tampa Bay. "Anywhere he plays there will be 74,000 people in the stands," Bassett said. "I want it to be with the Tampa Bay Bandits." (Brown wisely remained retired.)

The team was up for anything. There was a $1 million dangling-carrot offer to Buccaneers' owner Hugh Culverhouse for a game between the two teams (it was declined). University of South Florida students were presented with a 10-games-for-$35 ticket package. Bassett came up with the idea of catering the Tampa Stadium press box with food representing the visiting team's home state. When the Breakers came, for example, the Bandits served New England clam chowder. (Recalled Dick Schneider of the *Fort Myers News-Press:* "The scent hung around the press box for weeks. It reached the point where you'd step from the elevator, take a deep breath and order a couple pounds of grouper to go.") There were also giveaways — tons of giveaways. Even if a game were as lopsided as Blitz 42, Bandits 3, Bassett knew how to hold a crowd's attention. So the Bandits hosted a halftime $10,000 diamond handout and a halftime Dolly Parton lookalike contest. There was a Miss Tampa Bay Bandit bikini contest. (Wrote Nick Moschella of the *Fort Myers News-Press:* "It was a skin and sex spectacular. Ninety-eight curvaceous beauties, wearing nothing more than Pepsodent smiles, teeny-weeny bikinis and plunging, one-piece bathing suits parading around in high heels.") There was a promotion where the team offered to pay up to $50,000 of a fan's mortgage. Once, Dodge sponsored a giveaway where six cars were driven onto the field, the stadium lights were dimmed, and attendees' names were announced as the winners. "In full view of everyone, someone opened a car door, sat down, drove it out of the stadium, and never returned," said Zenon Andrusyshyn, the Bandits punter. "You can't make that up."

The most memorable promotional ploy came when the Bandits promised $1 million to a random fan during halftime of a game against the Denver Gold. The winner was a local engineering technician named George Townsend, who — before a crowd of 54,267 — held aloft an enormous cardboard check for $1 million. Townsend soon learned the payout would come in annual increments of $50,000 . . . *beginning in 2005.*

One day, Mackovets received a call from Bob Best, the PR head of the Buccaneers. "God," he said, "you guys are just kicking our asses in the community."

By the season's midway point, the commissioner was able to set the naysayers' concerns aside and see legitimately good signs. Sure, the Pacific Division was odious, with all four teams possessing 4-5 marks. And, sure, the 1-8 Federals were an embarrassment, and Walker's 3-6 Generals little better. But the 8-1 Philadelphia Stars were playing football at a near-NFL level, the Blitz and Bandits were tied atop the Central Division with identical 6-3 marks, and the Breakers, thought by many in the preseason to be the league's worst collection of players, were 5-4. There was also the emergence of young stars, ranging from Philadelphia linebacker Sam Mills and halfback Kelvin Bryant to wide receivers Trumaine Johnson of Chicago and Eric Truvillion of the Bandits. "It was a great league very early on," said Truvillion, whose 15 receiving touchdowns would lead the USFL. "Were there problems? Sure. And they were expected. That's growing pains. But I really thought we had something great in the works."

The USFL was exceeding its rival league's expectations, and the increasingly concerned NFL knew it. The new rules — in particular the two-point conversion — were well received, and the bright uniforms and fresh nicknames felt invigorating and lively. If the 1980s was the era of blissful, color-ful, dynamic excess, the USFL was the football league of blissful, colorful, dynamic excess. Unlike the NFL, the USFL refused to penalize for excessive celebrations. If a player wanted to moonwalk in the end zone, he would be allowed — *no, encouraged* — to do so. Balls were spiked over the goalposts. Pretend grenades were tossed into a circle of pantomiming linemen. Funk-adelic handshakes, head bobs, butt shakes — all embraced by a league in love with televised highlights. "At the time the NFL was the no-fun league," said Charley Steiner, the Generals' broadcaster. "The USFL saw that and flipped it on its head. I'd get calls on occasion from the league office — 'Why don't you come on over?' And we would sit around in Chet's office. And one day they'd say, 'What do you think about two-point conversion?' and 'What do you think about a replay and a red flag?' It was always the same — 'Sure, why the hell not?' It was so cool. We were just shooting the shit. 'What do you think about wide receivers not wearing numbers in the 80s, but single dig-its?' — and that's the way these things evolved. Everything was on the table."

And yet . . .

The league's owners couldn't help themselves. They simply couldn't. Wealth and ego go hand in hand, and neither rests easily. Although every-one was warned that professional sports teams lose money before they make money, and that crawling comes before running, which comes before sprint-ing, some in the USFL desperately wanted to sprint. Patience? To hell with patience. That's why, by a near-unanimous vote, the 12 teams made the deci-sion to *expand* for the 1984 season. From afar, it made no sense. Expansion was one of those moves that happen after, oh, a decade. You establish your name, establish your product, start making money — *then expand.* But now? Here? Hell, organizations like Washington and Chicago were struggling mightily with attendance figures. The Breakers were toying with the idea of a relocation to a bigger stadium in a different city. The Express owners wanted out — and Alan Harmon, who ran day-to-day operations, was humiliating himself with one boneheaded decision after another (he remains the only man in professional football history to fire a general manager — in this case Curly Morrison — after six games). Only one team, the Denver Gold, would end the season with a small profit, and that's primarily because Ron Blanding, the owner, was, in one player's words, "a cheap-ass fucker." Plus, a league already thought (rightly) to host subpar talent would be diluting its own pool. Expansion? Horrible idea.

There was, however, a factor in the pro-expansion movement: greed. The price for a new team, as determined by the league, would be $6 million (payable over three years). That meant money in every owner's pocket. That meant new TV markets, new merchandising opportunities. That meant exposure. So, with the smell of freshly printed dollars wafting through the air, the USFL's owners went from deciding to add two teams to four teams to — *hey, why not?* — *six* teams. "It was insanity," said Randy Vataha, the Breakers' co-owner. "We'd played less than one season. One. That's all. It was all about the money. But when you do something just for the money, it usually comes back to bite you."

Over the course of the second half of the season, the league pinpointed appealing markets. Initial efforts in Minneapolis and Seattle fell short. Pittsburgh, home of the legendary Steelers, was viewed as a near-must, and the stars aligned rightly when Eddie DeBartolo Sr., America's preeminent developer and builder of shopping malls, agreed to bring the city a team. To the USFL's absolute delight, DeBartolo's son was the owner of the San Francisco 49ers — a fact that drove Pete Rozelle, the NFL's commissioner, to the brink of furor.

On April 28, the USFL announced its first expansion team — the Pittsburgh Maulers. Five more would follow, and before long the USFL could state with certainty that it would be playing its 1984 season in the exciting markets of Houston, Memphis, Jacksonville, Tulsa, and San Antonio. "Some good cities there," said Vataha. "And some that had you scratching your head."

Around the same time the USFL was growing, Alfred Taubman, owner of the Michigan Panthers, was making a series of decisions that proved both marvelous and ruinous. At the halfway point of the season, the Panthers were 5-4 and in third place in the Central Division. As teams like Denver and Tampa Bay drew large crowds, the Panthers struggled from the get-go. They sold less than 10,000 season tickets, and the 80,638-seat Pontiac Silverdome often felt like a shopping mall five minutes before closing. Just 13,184 came out for the 34–24 Week 8 win over the Express, which was only slightly bet-ter than the 11,634 who attended a 17–12 triumph over the Blitz seven days earlier. In quarterback Bobby Hebert and wide receiver Anthony Carter, the Panthers had two of the USFL's most dazzling young players. In a helmet de-sign that featured a "royal plum" panther on a "champagne silver" base, the Panthers sported the USFL's sweetest duds.

The team chased attention like a publicity-starved has-been actress — a three-year, $2.5 million offer to Redskin fullback John Riggins; a tryout camp with 740 invitees; meet and greets with players, coaches, cheerleaders. Yet few seemed to care. "People were giving [Taubman] a hard time," said Ehrhart. "I was at one game in particular, and his team was getting crushed, and it was hard to watch. There were different ways an owner responds to those types of situations. He did something unique. I'll give him that."

What Taubman, a man worth $3.1 billion, decided was that the USFL's gradual growth plan was garbage. Simmons warned the league's owners about engaging in a can't-win spending war, but Taubman was no ordinary owner. His ego was enormous and his wallet even larger. "He was in this to triumph — period," said Ray Bentley, the Panthers' standout linebacker. "Everything about the organization was top notch, from hotels to salaries to facilities. Mr. Taubman wasn't there to come in second."

That's why, to the dismay of the other franchises, the Panthers went out and paid enormous money for Ray Pinney, Tyrone McGriff, and Thom Dornbrook, three ex–Pittsburgh Steeler offensive linemen with a combined 11 years of NFL experience and three Super Bowl rings. "The organization used to ask me all the time about how the Steelers did things," said John Banaszak, a Panthers defensive lineman who played in Pittsburgh for seven seasons. "They saw the Steelers as the model." From a purely football stand-point, the transactions made perfect sense. Hebert was sacked 15 times over the first five games and spent much of the early season rushing throws and absorbing hits. Ken Lacy, the rookie halfback out of Tulsa, saw nothing but the helmets and pads of opposing defensive linemen. Then the Steeler rein-forcements arrived. "Through college and now, I've been playing running back for about 10 years and I've never had a line as good as this one," Lacy said at the time. "I think the key to our season was signing those Pittsburgh guys."

He was correct. The Panthers wound up winning 11 of their last 13 games, and by the final weeks crowds in excess of 30,000 were entering the Silverdome. But with the big-money additions, the USFL foundation permanently shifted. It was one thing to sign superstars for large dollars. They drew both fans and media attention. But offensive linemen? Bassett, the Bandits' owner and an experienced sports executive, was particularly galled by Taubman's approach. "John knew the perils of bidding wars," said Ehrhart. "It was short-term thinking versus long-term thinking. John wanted the league to last. Yes, winning was important to him. But survival was the top priority.

"We needed the USFL to survive."

6

TITLE DREAMS

I'm handing football quotes to writers, and they're saying, "Football quotes? There's a police riot on the field!"

— Beth Albert, USFL coordinator of media relations

"TAKE THAT — nigger."

There are insults one can use in the midst of a professional football contest, and there are insults one cannot use. Making fun of someone's mother? Acceptable. Mocking a festering zit? Also acceptable. Calling someone a pussy, a whore, a retard, a bitch, a motherfucker, a cackling douchebag, a no-good chunk of fermented anal discharge? All within reason.

The N-word? No.

On the afternoon of Sunday, June 5, the Michigan Panthers traveled to Philadelphia's Veterans Stadium to take on the Stars in a Week 14 clash of the league's two hottest teams. At 11-2, Philadelphia was about to clinch the Atlantic Division title, and with 1,276 rushing yards Kelvin Bryant, its superlative halfback, was a frontrunner for the USFL's first Most Valuable Player award. "Kelvin would come into the huddle, look the linemen in the face and say, 'They know we're running — I don't care,'" recalled Irv Eatman, the offensive lineman. "He wanted the ball." The Panthers, meanwhile, were riding their revamped offensive line, as well as the Bobby Hebert–Anthony Carter offensive express, to eight wins over their past nine games. It was *the* marquee matchup of the year.

Though only 19,727 sat through overcast skies at the Vet, Panthers-Stars met expectations. Michigan led 14–9 at halftime on the strength of Hebert's 5-yard touchdown toss to tight end Mike Cobb, but the Stars bounced back on three Chuck Fusina scoring passes. Philadelphia wound up winning, 29–20.

Unlike the NFL, the USFL rarely asked its officials to crack down on trash-talking, or eject players for excessive violence, or keep an eye out for trouble. If anything, the behavior was encouraged as colorful must-see entertainment. Linemen set out to twist opposing knees, bite opposing fingers, fire off an extra elbow to the ribs. "I once had a guy spit in my face," said Will Cokeley, a Panthers linebacker. "Just to get an edge." Though the league had its superstars, the vast majority were men making in the $25,000 range; men who knew they were one pink slip away from driving a rig or loading boxes into the rear of a Pathmark; men who were *starving* to survive. "If you were a guy making NFL money, people would really go after you in the USFL," said Pete Kugler, a defensive lineman who jumped from the San Francisco 49ers to the Philadelphia Stars. "There were a lot — and I mean a lot — of cheap shots and different things that the refs refused to call."

Indeed. During the Panthers-Stars game, Bryant took a handoff from Fusina, cut through the Michigan defense, then was leveled by David Greenwood, the hard-hitting rookie safety. Inside the Panthers' locker room, the University of Wisconsin product was viewed both warily and comically. "Like a crazy person," said Ray Bentley, the Panthers linebacker. "But a somewhat nice crazy." Born and raised in tiny Park Falls, Wisconsin, as the eighth of 11 children, Greenwood spent his free boyhood time either playing sports or hunting and fishing. He described his father as "hard-nosed and mean," and spoke glowingly of a grandfather who was "the toughest guy in Park Falls. He used to show off at the tavern by lifting up those cement blocks they tied the horses to."

Greenwood was backwoods. He chewed tobacco and blared Merle Haggard and said things that probably should not be said. "He was rough around

the edges," said Kyle Borland, a Panthers linebacker and Greenwood's roommate. "He was a nice guy off the field, but he could be as mean and ornery
as they come."

"Nasty," said Derek Holloway, a Michigan wide receiver. "Greenwood was
mean and nasty."

During the season, Greenwood (who doubled as the team's punter) engaged in multiple scuffles with a particular African American teammate, and
more than a few blacks believed them to be racially driven. "He was kind of
a prick," said Cokeley. That's why, upon rising from Bryant's corpse, few Panthers were overly shocked when Greenwood glared downward at the African
American runner and said, with a drawl, "Take that — *nigger.*"

What?

Philadelphia was the best and most cohesive team in the USFL. Its players — black and white — were incensed, and several charged toward Greenwood, demanding a fight. "It got very personal very quickly," said Bentley.
"Guys on our team were upset, but we knew what to expect of David. The
Stars — those guys were really pissed off. They wanted us bad."

Greenwood's primary body guard on the Panthers was, of all people, John
Corker, a six-foot-five, 240-pound African American linebacker who was
the safety's polar opposite in all ways but one: he, too, was six cents short of a
dime, and widely regarded as Michigan's craziest player (no small feat, considering kicker Novo Bojovic wore a white glove on his left hand and stashed
a clove of garlic in his right shoe for good luck). Way back in 1980 Corker
had been the Houston Oilers' fifth-round draft pick out of Oklahoma State
(a knee injury caused him to drop from first-round projections), and while
his physical skills wowed the coaching staff, his inability to stay drug-free led
to his NFL demise. "I played on the Steelers with some of the greatest linebackers of all time," said John Banaszak, a Panthers defensive lineman. "Jack
Ham, Jack Lambert. And Corker could have been that good. But he burned
the candle at both ends like no player I'd seen before." When the USFL came
along, Corker found a life preserver. The Panthers signed him because head
coach Jim Stanley had been with him in college, and Corker turned into
the league's version of Lawrence Taylor. Wrote Joe Santoro of the *Fort Myers
News-Press*: "[Corker] is a lean, mean sacking machine; the kind of guy who

could shake Rambo down for lunch money." He led the USFL with 28 sacks for 199 yards lost (both professional football records), but also led in manic mayhem.

Early on during training camp, Corker — nicknamed Sack Man — gathered the team in a circle and guided the Panthers in prayer. "He started praying like a Baptist black preacher," said Dave Tipton, a defensive tackle, "and I thought, *Wow, Corker must walk with the Lord.*" Not quite. Blessed with the world's largest penis, Corker never shied away from showing it off to fellow Panthers. "The biggest johnson in the USFL," said Matt Braswell, the team's center. "We had women reporters come into the locker room, and Corker would position himself so he was in full view of any females. He had this vat of Nivea skin cream, and he would just make sure to completely rub it and moisturize it." Corker operated on a clock that required only two to three hours of sleep per night, and was powered by the dual fuels of alcohol and cocaine. He kept a gun in his car's glove compartment, missed as many meetings as he attended, and proudly pasted his pay stubs to his locker, so that teammates could marvel at the money he was being docked. Once, Hebert drove with Corker from Pontiac to Detroit for a promotional appearance. It was snowing outside, the roads were slippery — "and Corker was driving, smoking one joint after another," said Hebert. "We both walked in reeking of pot." In a USFL urban legend that actually checks out, Corker was once found naked on the ice at Joe Louis Arena in the early-morning hours. He had passed out, and spent so much time on the cold surface that some of his skin had to be ripped off. "That," said Bentley, "surprised none of us."

Yet for all his unpredictability, Corker stood by Greenwood's side, and would no sooner let the Stars attack his teammate as he would go to bed at nine o'clock. "John was a tremendous teammate," said Whit Taylor, the Panthers backup quarterback. "If you needed to count on one guy to have your back, he was it. No matter what."

The Stars refused to forget what transpired. They wanted to see Michigan again. It wasn't your typical clichéd revenge blather that often accompanies professional sports. No, this was very raw and very real.

The 18th and final week of the USFL's inaugural regular season wrapped up on the weekend of July 2 and 3. The Stars won the Atlantic Division with a league-best 15-3 mark, while Michigan and Chicago both finished 12-6 atop the Central (a tiebreaker gave the title to the Panthers). In the Pacific, the 9-9 Invaders captured the most mediocre of the USFL's three divisions — made somewhat acceptable by the team's sound attendance figures and surprisingly rabid fan base. The divisional playoffs would feature Oakland at Michigan and Chicago at Philadelphia, followed by the first-ever USFL championship game on July 17 in Denver.

As most of the owners and league officials expected, there was a whole lot of positive and a whole lot of negative to digest. Simmons put on a triumphant face, and talked about the 6.0 television ratings for ABC games — which was 20 percent higher than the 5.0 predicted by the network. He talked about the Denver Gold averaging 41,736 fans per game; the Bandits (39,896), Generals (35,004), and Invaders (31,211) all averaging more than 30,000; and three other teams exceeding 20,000 per game. The USFL, he repeated ad nauseam, had a far better first season than the American Football League, which averaged a scant 16,538 fans per game (the USFL clocked in at 24,824).

After a rough start, Herschel Walker found his footing and paced the USFL with 1,812 rushing yards. (His Generals, however, won but six games. Wrote *Time* magazine: "Walker set a record of sorts by telling the same speech to reporters week after week: 'I've still got a lot of learning to do.'") The league's two best quarterbacks wound up being Michigan's Hebert and, shockingly, Oakland's Fred Besana, the ex–Twin City Cougar who led the USFL with 3,980 passing yards and placed second with 21 touchdown passes. Philadelphia's Bryant won the MVP trophy by rushing for 1,442 yards and 16 touchdowns while catching 53 passes for 410 yards. "Herschel had all the hype, but Kelvin was the better player," said Irv Eatman, the Stars offensive lineman. "It wasn't even close. He could catch, run, run over people, run by people, run around people, block. He was the best offensive player in that league." As for the season's most endearing moment, nothing could top Greg Fairchild, a Panthers offensive lineman, motioning toward a gaggle of heckling fans during a May 1 game at Nickerson Field in Boston. "I know you guys are booing us, and that's cool," he shouted. "But what are y'all eating?"

Someone held aloft a bratwurst and a hotdog.

"Would you mind getting me one of those?" Fairchild asked. "I'm good for the money."

Moments later Fairchild could be seen lounging on the bench, bratwurst and Budweiser in hand. "Then he tipped his glass to the fan," said Matt Braswell, the Panthers center. "Like they were at a bar."

Simmons raved of all the good things, and could even laugh at brats and beers. But, deep down, he knew there were serious problems. Both the Los Angeles and New Jersey ownership groups wanted to sell, and Boston planned on moving (they drew a league-low 4,173 fans to Nickerson for a Week 11 snoozer against the Gold). The Federals were a joke and the Blitz — for all of George Allen's supposed genius — made barely a ripple in the Chicago market. The franchise that prepared more than any other averaged but 18,133 fans to a largely deserted Soldier Field, and the owner (Dr. Ted Diethrich) also openly pondered a shift to a different location. What the USFL was learning (the tough way) was that cities with die-hard NFL fan bases were slow to embrace a new team. Who needed the Federals and Blitz when the world already had the Redskins and Bears?

Midway through the season Simmons had hired Frank N. Magid Associates, the company that conducted David Dixon's original persuasive "spring football" survey, to submit a follow-up on why the league wasn't meeting or exceeding all of its set expectations. What came back was nothing short of devastating. Clearly, Dixon had fudged some of the initial numbers, because the new survey offered little of the first study's widespread football frenzy. This time 1,127 calls were placed to find 600 fans who were "interested" in football — a 22 percent drop-off rate. A mere 5 percent of respondents had attended a USFL game, and 39 percent said they "planned" on attending. When Dixon first approached owners, he listed that figure at 67 percent. Worst of all, the survey showed that 75 percent of fans viewed the USFL unfavorably when compared with the NFL. Though this was no great surprise, it felt like a dagger to the commissioner's heart. He himself had now watched 18 full weeks of USFL games, and admitted in private that the action was often lacking. Too many contests felt Division II college–like. Some of the coaches were terrific, but others (specifically the two Canadian League imports, Los Angeles's Hugh Campbell and Washington's Ray Jauch) failed to pick up the American flow to the sport. If Simmons never again had to watch the Federals bungle a snap, he'd be a happy man.

The NFL, for its part, was optimistic. Rozelle's initial nervousness over the competition quickly faded. He fully expected the USFL to last for another one or two seasons, then die a quick death. "You will know when they [USFL] are on their last legs because they will bring a lawsuit against us," Rozelle said at the time. "This is how they will try to keep the owners from jumping ship. They will hold out their suit as a way of getting their money back if they stay around."

NFL scouts attended spring games, and their reports were, by and large, painful. Sure, every team had a small handful of players who could hang in the big time. But the vast majority of guys were grunts, better suited to frying eggs. Inside the NFL's New York City headquarters, officials mocked the USFL by referring to it as "the Useless."

The executives from the older league did nothing to help Junior, and everything to hurt it. In anticipation of the title game in Denver, for example, the USFL asked whether Broncos owner Edgar Kaiser Jr. would allow usage of his private luxury box. He steadfastly refused—and the USFL, which leased the stadium from the city, was charged $1,750 by Kaiser for the cost of the seats in the box. Wrote Tom Callahan in *Time* magazine: "The National Football League looked down upon the woes of the new league with lordly science, but its attitude was summed up well enough by Jim Finks, general manager of the Chicago Bears, who referred casually to his Second City rivals as 'the Chicago Blintz.'"

"Were we scared of the USFL?" Gil Brandt, the Dallas Cowboys vice president of player personnel, said years later. "Ha, ha, ha, ha. Hardly. The USFL was sort of a joke to us. A gag. We knew it wouldn't last."

Simmons's concerns were momentarily alleviated by the divisional playoffs—but only sort of. In Pontiac, a USFL-record 60,237 fans packed the Silverdome to see the Panthers demolish the Invaders, 37–21. It was amazing and dynamic and fun and celebratory (fans stormed the field to tear down the goalposts with 25 seconds remaining), and the Panthers played at a high-octane pitch that thrilled spectators. Yet the attendance figure came with an asterisk. Terrified by the thought of a vacant stadium, Taubman slashed ticket and parking prices. The other USFL owners were disenchanted. The wealthiest of wealthy developers could afford these types of discounts. The majority could not. "So instead of taking 35,000 at almost $12 and making money," a USFL employee complained, "he gets 60,000 spectators and loses money. That's good business?" The payoff wasn't financial, but ego. Taubman marched to the center of the Silverdome before kickoff, waved to the crowd, and was saluted with a standing ovation.

In Philadelphia, the Stars and Blitz played what, three decades later, still goes down as one of the great playoff games in football history. After committing seven turnovers through three quarters and trailing 38–17 well into the fourth, the Stars charged back. "It felt like we had nothing left in the tank," said Ken Dunek, Philadelphia's tight end. "And then we exploded." Quarterback Chuck Fusina, a man of little physical or professional football note, somehow fired three touchdown passes within a nine-minute span while hitting on 10-straight completions. With 50 seconds remaining in regulation, the Stars lined up for a second and 10 at the Blitz 11. Chicago rushed five defenders, and Fusina took a quick drop and found Tom Donovan, the slot man, crossing the middle 5 yards away. A former Penn State star who had been cut by five NFL franchises in a two-year span, Donovan grabbed the ball, broke a tackle, headed toward the goal line, stepped out of the foothold of Blitz safety Ted Walton, and hopped into the end zone. He held the football aloft in his left hand, skipped nine steps across the turf, and then, without so much as a pause, flipped head over heels into a full-blown spike. With the extra point, the game, once out of reach by all logical football measures, was tied at 38.

The crowd went wild.

Well, sort of wild.

Well, kind of wild.

Well, not *exactly* wild.

Veterans Stadium held 72,204 people, meaning 56,518 empty seats witnessed the magic. A humiliating 15,686 paid to attend, but with temperatures near 90 and a cheesecake-thick humidity, probably 10,000 remained. When the Stars won in overtime on Bryant's 1-yard touchdown plunge, there were as many football players and staffers as spectators. Which meant two things:

1. Philadelphians missed one hell of a football game. (Ed Sabol, the founder of NFL Films, called it one of the three most exciting games he had ever seen.)
2. The USFL had some serious issues.

Back in the 1980s, the Super Bowl was something of a joke.

It's an easy little detail to forget decades later, but of the seven Super Bowls played between 1977 and 1983, five were decided by 10 points or more. The games were often dull and sloppy, featuring one team far superior to the other. Every time a Super Bowl turned into a snoozer, Pete Rozelle knew the critics would be lurking, anxiously waiting to take a bat to the multibillion-dollar league.

Chet Simmons, sports television impresario before all, considered this to be one of the biggest chinks in the NFL's armor. So, in the advance to the first-ever USFL championship game (nicknamed "the Game with No Name" by members of the media), to be held on July 17 at Denver's Mile High Stadium, the commissioner committed the league to putting on the greatest show imaginable. No, he couldn't influence the on-field quality of Michigan-Philadelphia. But he could make the weeklong lead-up to the event exciting, interesting, engaging. The two teams arrived in Denver on Monday, July 11, and Simmons was determined to have the experience feel exactly like the Super Bowl. That's why he arranged for beat writers to travel on team planes to Denver (alas, as they prepared to board a handful of Panther scribes were bumped because of overcrowding and forced to find their own transportation). That's why he headquartered the working media at the beautiful Regency Hotel at a drastically reduced rate. It's why the league rented a fleet of vehicles for working reporters, and featured an around-the-clock press center fully loaded with food, drinks, assistants. It's why there was a swank prechampionship-game cocktail party on the 72nd floor of a new downtown high-rise. (The building, still under construction, led Simmons to note: "Just like our league. Not quite finished, but a marvelous, marvelous view.") This was followed, one night later, by an open-to-the-public-for-$100 black-tie gala with some of the league's superstars. More than 700 people attended. For many, the week would serve as an introduction to the USFL as a major-league sports and entertainment entity. Everything had to be right.

And then . . .

The calls to reporters came late Monday night, from one sports editor after another. The news had just been leaked that Stanford quarterback John Elway, the No. 1 overall pick in the recent NFL Draft, would be reporting to his new team, the Denver Broncos, for his first day of training camp in Greeley, Colorado, on Tuesday morning. The orders were clear: to hell with the USFL — *get to Elway.* "So all of those cars we rented the media were used to go to Greeley," said Doug Kelly, the USFL coordinator of information. "Meaning we were paying money to help people cover the NFL."

"Every single one of us got in our cars and drove to see Elway for a day and a half," said Harvey Araton of New York's *Daily News.* "During that small time period the USFL got zero attention. It didn't exist." Elway stole any and all coverage, so much so that the Associated Press began the article, headlined ELWAY MAKES BIGGEST SPLASH AS CAMPS OPEN, with, "The first huddle at the Denver Broncos training camp was a massive one and only one player was in it. The rest were media types clawing over each other."

Were that not dispiriting enough, on the same day the AP story led off sports pages across the nation, a smaller piece USFL TITLE CONTEST MAY RESEMBLE REGULAR GAME, also ran. The theme: nobody much cared about the championship clash.

It was true. The vast majority of journalists assigned to the event didn't want to be there ("It was a case where the paper needed somebody — and I was somebody," said Araton), and knew nothing about the Stars and Panthers. The City of Denver hardly helped, refusing the USFL's request to place sod on worn parts of Mile High's field.

Yet as the week progressed, intrigue built. First and foremost, there were rumors of the Greenwood-Bryant conflict, and expectations of on-field violence. (The USFL didn't mind. All hype was good hype.) But there was also color. Tons and tons of color. When reporters covered a Super Bowl, they were shackled by the rigid NFL media mechanisms, which meant almost no potential for one-on-one time with participants. The USFL, on the other hand, was all-access, all the time. The league's PR staff desperately sought attention, and would give journalists everything they requested.

In the lead-up to the title game, much of the media focus was devoted to the two starting quarterbacks—contrasting narratives that offered riveting insight into alternate paths toward professional football. Chuck Fusina, the Stars' signal caller, epitomized the football lifer gifted one last shot by a new enterprise. Back in 1978, he wrapped his career at Penn State (one that saw him go 29-3 as a three-year starter) by finishing second to Oklahoma's Billy Sims in Heisman Trophy voting. Although his successes were undeniable, his mediocre arm strength and genteel nature turned off NFL scouts. By the time Tampa Bay selected Fusina with the 133rd overall pick in the fifth round, the label was clear. He was a lifelong caddy, and little more. "[Fusina is] a go-fer," wrote Alan Goldstein in the *Pittsburgh Post-Gazette*, "who wears the headset and carries the clipboard." In three years as Doug Williams's backup with the Buccaneers, Fusina attempted five passes, completing three. In the lead-up to the 1982 season he was traded to San Francisco. "But they had Joe Montana," Fusina said. "I played in the last preseason game, but they told me they could only keep two quarterbacks, and I was third."

San Francisco released Fusina, but urged him to hang tight until the beginning of the season. "I started wondering if I could play anymore, or if I really wanted to," Fusina said. "I was seriously thinking of going back to school and getting my master's degree in business."

Then, one day, Carl Peterson called. The Stars' general manager had been urged by Joe Paterno, Fusina's coach at Penn State, to give him a shot. "The guy doesn't have the greatest arm strength," Paterno said. "He doesn't throw a tight spiral, he's not real fast, he looks bad in a uniform. But he finds a way to win, and he keeps guys like you and me employed." Peterson heard enough. He reached out to Fusina and spoke about this new league, with cool duds and inventive ideas and a team—the Stars—located in his home state. "You'll get an opportunity to play," the general manager said.

"I'm in," Fusina said. "That's all I need to hear."

He captured the starting job, and played with the pedestrian gusto of an accountant. Fusina threw for 15 touchdowns and 10 interceptions, and the most exciting thing about him was his ode-to-John-Oates mustache. Otherwise, his game was slants and timing patterns and screens. "Chuck couldn't throw the ball 80 yards down the field," said Scott Fitzkee, his No. 1 wide receiver at Penn State and with the Stars. "But he had the smarts, he had the feel, he read defenses very well. Maybe he didn't have NFL intangibles. But he was the perfect USFL quarterback."

The Stars were the USFL's most intriguing franchise. The offensive line featured the six-foot-seven, 300-pound Irv Eatman. Bryant, the magnificent halfback, had a diamond stud in his left ear and a game just as flashy. Sam Mills was the USFL's answer to Dick Butkus; he paired him with fellow linebacker Vince DeMarinis to give Philadelphia two starters from tiny Montclair State. "We played in front of 150 people a game in college," said DeMarinis. "On a good day." Even the Stars' sales office would prove noteworthy — sitting at side-by-side desks, charged with selling tickets, were former Eagles wide receiver Vince Papale and future Villanova men's basketball coach Jay Wright. "It felt like starting something great, where we were all in it together," said Wright. "I played on the Stars basketball team, I hung out with Vince. I even met my wife [Patty, a Stars cheerleader]. Best time ever." There was also a pair of brothers on the offensive line, Brad and Bart Oates, who lived together in Cherry Hill, New Jersey, and commuted to practice in a junky Chevrolet they bought for $300. "One day in South Philly the tires caved in," said Brad, a starting tackle. "We got out, looked at the car, grabbed our stuff, and just left it there abandoned in the middle of the road. That was quite a moment."

In Denver the attention went toward Fusina, as it did his 180-degree opposite — Bobby Hebert (a.k.a. the Cajun Cannon) of the Panthers. If the Philadelphia quarterback's arm was a tricycle, Hebert's was a Bugatti Veyron Super Sport. If the Philadelphia quarterback's story was rags to riches, Hebert's was slums to even greater riches. He was born and raised 32 miles south of New Orleans in the town of Cut Off, Louisiana (population: 3,500), and after leading South Lafourche High to a state championship, he attended Northwestern State–Louisiana, a Division II school in Natchitoches. Because 1983 was the year of the quarterback (five were selected in the first round of the NFL Draft), Hebert fell between the scouting cracks. Teams came to Natchitoches to watch him throw, but the competition was suspect and the other available quarterbacks (beginning with Stanford's Elway and Pittsburgh's Dan Marino) dazzling. "I went to the NFL Combine in Florida," Hebert said. "Gil Brandt was there, and he told me if I did well I might be drafted in the second or third round."

"Well, is there anything the NFL can guarantee me?" Hebert asked.

Brandt laughed. "That's not how it works," he replied.

When the Panthers selected Hebert with their third-round pick, then offered an $80,000 signing bonus and $70,000 contract, he jumped. There were 13 quarterbacks at the team's Daytona, Florida–based training camp, almost all of whom had played at bigger college programs. Yet Hebert was, literally, hungry. He and his wife, Teresa, had an infant daughter, Ryan, and lived on food stamps. "I was broke," he said. "My father's a civil engineer and my mother's a schoolteacher. My daddy would have supported me but I believe when you get married you should break away from your family."

Jim Stanley, the Panthers coach, named Hebert the starter based on his otherworldly arm strength and linebacker size (six foot four, 210 pounds), but his rise came accompanied by a unique problem. Like many Cajuns, Hebert's family was largely French-speaking. His grandmother Birdie was incapable of communicating via English and his father, Bobby Senior, did not learn the language until reaching elementary school. "My dad graduated from LSU, is a wizard in trigonometry, calculus," Hebert said. "But you hear him speak and wonder if he even went to high school." As was the case with his father, words from Bobby Junior's mouth appeared to be spoken in some far-removed Klingonian dialect. His roommate with the Panthers was Mike Hagen, a fullback out of the University of Montana. At the end of his first day in camp, Hagen called his wife, Ann, and said, "It's weird. The guy plays football but he's not from America." In the huddle, wide receivers and halfbacks stood irregularly close to their quarterback, and often broke without a clue of what to do next. In the case of an opposing defense's safety blitz, for example, Hebert was supposed to yell, "Snake! Snake!" Only he was unable to enunciate the s before the n, and "Snake! Snake!" became "Nake! Nake!"

"What the hell is a nake?" said Derek Holloway, a Panthers wide receiver. "We had to change the word because he just couldn't say it."

"I was from Georgia, so that made me the closest guy, geographically, when it came to understanding Bobby," said Matt Braswell, the center. "He wasn't Cajun, he was straight coon-ass. If the game was getting closer, he'd get more excited. And if he got more excited he became harder and harder to understand. We'd be in the huddle, Bobby would kneel down and say what he had to say. Nine guys would look at him and as soon as he was done talking they'd turn to me to translate."

Before long Hebert was the USFL's best quarterback. He ranked first in touchdown passes (27), third in passing yardage (3,568), and third in completion percentage (57 percent). Even if reporters only grasped every third word, he still made for top-shelf copy. His outfit at press conferences always included a Mickey Mouse T-shirt. He told teammates that, in his neck of the woods, shrimp boots were called Cajun Reeboks. "He was the friendliest guy around," said Paul Keels, the Panthers' play-by-play announcer. "If you didn't like Bobby Hebert, you didn't like people."

On tape, Mora didn't like Hebert. Or his fleet wide receivers, Holloway and Anthony Carter. Or halfback Ken Lacy, who ran for 1,180 yards and six touchdowns. Or Greenwood and Corker, whose hard hits too often felt dirty and intentionally injurious. Michigan was loaded, and Curt Sylvester, who covered the Lions and Panthers for the *Detroit Free Press*, was convinced the city's USFL team would trounce the NFL team. That's why Mora, the Stars' leader, was terrified. That's also why, in the days approaching kickoff, he worked his team excessively hard. It was Young Coach in a Panic 101, and the small handful of the Stars who had once played for the Philadelphia Eagles recognized the look. Two years earlier, in the lead-up to Super Bowl XV in New Orleans, Eagles coach Dick Vermeil put his men through, in the words of *Philadelphia Daily News* columnist Ray Didinger, "long meetings, longer practices and an unrelenting passion that left his players numb." By the time kickoff arrived, the Eagles were mentally and physically drained. They lost to the inferior Oakland Raiders, 27–10. "I did the same thing," Mora said years later. "I worked us too hard. I wanted to establish a mind-set of toughness, of keeping what we had going. Instead, I tired them out. That's on me."

To the NFL's surprise, 59,906 fans entered Mile High Stadium on a perfectly mild and gorgeous Sunday evening. The USFL championship game was, at long last, a thing, and even though it wasn't a sellout ("The first Super Bowl didn't sell out, either," a league official told members of the press), and even though the local fire marshal rejected Simmons's request for fireworks, the feeling was . . . "big and important," said Fusina. "It felt like we were there for a special moment in time."

After a handful of skydivers landed in the middle of the stadium and cheerleaders from the Gold, Panthers, and Stars performed a choreographed routine and trumpeter Al Hirt played the national anthem, the captains met at midfield for the toss of (inexplicably) a coin from the 1984 Los Angeles Summer Olympics. Bill Parkinson, the referee, flipped the money into the air and the Stars, in red jerseys, chose to receive the kickoff. As Novo Bojovic placed the football on a red tee and stepped back, the attendees let loose with a thunderous sound. "The stands were made of wood," said Ray Bentley. "And the fans were stamping hard on that. It got really loud." Fusina was right — this did feel big and important, surely not altogether unlike the first-ever Super Bowl 16 years earlier.

The greatest fear among Simmons and his loyalists was a subpar game. The USFL would find a way to explain disappointing ratings or under-whelming attendance. But an up-and-down season *could not* wrap with a 43–7 yawn fest. It just couldn't. "The league wanted a show," said Hebert. "That was clear."

It took a while. The Panthers jumped out to a seemingly insurmountable 17–3 third-quarter lead on the strength of Hebert's two scoring passes to Holloway. Unbeknownst to the media, during the week Hebert had received a telephone call at the hotel from a man who said, come Sunday, the quarter-back would "die by high-powered rifle." The Panthers were concerned, but Hebert was not. "There was nothing I could do about it," he said. "I wasn't gonna play in a bullet-proof vest."

The Stars, on the other hand, were preoccupied with the high-flying, body-crunching Greenwood, who seemed to go out of his way to hit late and low. Mora was a strict disciplinarian, and warned his men about going after Greenwood, or talking trash to Greenwood, or even paying Greenwood the slightest bit of mind. Were the Stars still thinking about Greenwood's racist slur toward Bryant? Yes. "How do you forget that? It had to bother them," said Bentley.

Then, something snapped. The Stars drove down the field midway through the third period, endured a missed 34-yard field goal from David Trout, but moments later had another drive, capped by a successful 28-yard boot. It was the first sign of positivity for Philadelphia, and the momentum continued when, 8:49 into the fourth quarter, Willie Collier, a wide receiver out of Pittsburgh, soared through the air to snag a 21-yard touchdown throw from Fusina. With the successful two-point conversion, the score was 17–14 and the feel of the game shifted. The sun had set. The lights were bright. The stadium grew louder. People had waited for the Stars to play like champions, and finally they were. In the ABC booth, Keith Jackson and Lynn Swann shifted in tone and approach. A blowout was no longer a blowout. Philadelphia had arrived, and they were about to take over.

With 3:11 remaining in the game, the Panthers found themselves with the ball at the Stars' 48-yard line. It was second down and 10, and one play earlier Lacy had taken a handoff from Hebert, only to be slammed to the ground by Mike Lush, Philadelphia's bone-shaking safety. Lacy spent the next few minutes writhing on the turf in pain. On the Stars' sideline, Fitzkee told ABC's Tim Brandt that his team was about to win. "We have the confidence," he said. "If our defense gets the ball, I think we'll go down the field and score."

Now, after the injury delay and a Stars' timeout, Hebert stepped to the line. Holloway, quick and undersized, stood wide left. Carter, speedy and lean, went far right. Halfback John Williams and fullback Jim Hargrove positioned themselves behind the quarterback. The play call was split right A44 pass corner 2 — a quick sideline throw to Carter. With the snap, Lush came charging across the line as Hebert looped to his right, cocked back his arm, and, after taking two quick stop steps, fired a laser to Carter 30 yards away. The fleet wide receiver, weighing but 168 pounds and wearing uniform No. 1, caught the ball, spun inside, stutter-stepped, and burst past strong safety Scott Woerner and cornerback Antonio Gibson, both of whom tumbled to the ground. Carter dashed down the field and, as he crossed into the end zone, raised the pigskin in his right hand atop his helmet and into the darkness. "Nobody could catch him," said Eatman. "He was lightning." Carter leapt into the arms of center Wayne Radloff and tossed the ball into the stands. ("You know, I think I should have kept that," Carter later lamented.) The rest of Michigan's offensive players joined the party, with Bojovic, all 5-foot-10 and 170 pounds of him, clinging to the back of a lineman like a koala to a tree. Jackson's call — "Hebert gets it away . . . Carter's there . . . first down . . . he's loose . . . he's got some help . . . he's going for the corner . . . touchdown! Whoa!" — was perfect in its simplicity, and in the blink of an eye the oft-maligned USFL had its signature moment. Hebert would win the game's MVP award, but Carter (9 catches, 179 yards) was the hero. "I am No. 1 and the team is No. 1," Carter said afterward. "There is nothing better than that." A late Stars' touchdown mattered not, and Michigan, with a 24–22 triumph, was the USFL's first champion. After his team stormed into the locker room, Stanley received a congratulatory call from President Ronald Reagan aboard *Air Force One*. Magic Johnson, the Michigan native, entered to supply hugs and high fives. There was champagne aplenty.

"My grandma bought 27 tickets to the game, so I had everyone there," said Hebert. "She'd never even been on a plane before, and now she's watching me win a championship. The win, the ring, the togetherness — it was all that a guy could imagine."

For Hebert, the Panthers, and the young league, the night could not have gone better.

But . . . *wait*.

With the USFL, something generally *had* to go wrong. It was all but required by law. So, yes, the game drew a solid 10.1 rating. And attendees were entertained. But in the seconds before the final gun sounded, some 1,500–2,000 people charged onto the field and attacked the goalposts. Denver's police force, out en masse, was under strict orders from the city that no damage could be done to the field.

Hence, officers maced the fans.

There were snarling police dogs and bloody heads and chants of "Bullshit! Bullshit! Bullshit!" Large objects were thrown onto the field — "a big ol' ice chest landed right next to me," said Jim Van Somersen, the Gold's public relations director. "There was a man in jeans but no shirt, who got in a cop's face, and the cop let his dog go and the dog went after the guy." One fan was re-moved on a stretcher. Four were cuffed. "If it had been my decision, I would have let them have the goalposts," said Jerry Kennedy, a police captain. "My officers were attacked and forced to use mace to quell the disturbance. In my 20 years at Mile High Stadium, I've never seen fans act like that."

Some of the blame was assigned to idiot rioters, 12 of whom were arrested ("There were a load of drunk people," recalled Gregory Clow, the league's art director, who stood along the sideline). But a great deal was directed toward the league, which hosted a Miller-sponsored free-beer handout in the park-ing lot 90 minutes before kickoff. "The response to that give-away was enor-mous," William N. Wallace wrote in the *New York Times*. It hardly helped that alcohol concessions stretched through the fourth quarter.

In the press box far above the turf, members of the media and league officials watched the free-for-all in horror. Reporters came back from the field with their eyes burning from the spray. Doug Kelly, trying to navigate through the scrum, was pushed down and had his eyeglasses shattered. Sim-mons wanted to hide. "It was the USFL's worst nightmare," said Araton, the *Daily News* reporter. "It went from being this great championship game to a riot."

Like many papers across the nation, the following day's *New York Post* chronicled little of the game, and much of the melee. Beneath the head-line ROCKY HORROR SHOW was the subhead COPS, WILD FANS IN UGLY CLASH AFTER PANTHERS COP USFL TITLE. Steve Serby, the *Post* writer, referred to it as "a bloody 20-minute war."

"Talk about a momentum crusher," said Kevin Noonan, who covered the Stars for the *Wilmington News Journal*. "Amazing game, amazing capper to the season, and the final image is cops teargassing fans.

"The NFL must have been laughing its ass off."

ACT TWO

It was such a wild league. We were playing Denver once, and a guy and I were talking back and forth. I said to him, "The first one to cry is a sissy." We're pushing, and he hits me late at the end of a running play. The next play we're running again, and I come behind him and hit him and knock him over the pile. He's really mad, and he gets up and starts cussing. I'm like, "What, you quit already? You want your mama?" As I was walking back to huddle he took his cleat and stepped on the back on my Achilles. That hurt — oh boy. So I turned my elbow pad around and I threw the elbow to hit him, but I missed and hit him in the chest. After the game he was looking for me and I looking for him. We both laughed. That was the USFL.

— Gordon Banks, wide receiver, Oakland Invaders, 1983–85

7

CRAZINESS

The NFL was boring. The games were boring, the practices were boring. Everything was boring. The USFL wasn't as good as the NFL. But it was 10,000 times more fun.

— Jairo Penaranda, running back, Memphis Showboats

THE WEIRDEST OFFSEASON in the history of organized sports begins . . . *where?*

That's not a question to be taken lightly. Truly, where does one start? Chronologically? Or in order, *oddest* to *slightly less odd* to *not so odd* to *relatively normal but sort of odd?* Should we just throw the 100 different examples of outlandish, nonsensical zaniness into a hat, then blindly reach in and choose?

Hell, what would you do?

Really, *what would you do?*

For the sake of cohesion and simplicity, let's start with the basics: thanks to a (terribly misguided and ultimately murderous) expansion plan, the USFL would jump from 12 to 18 franchises in 1984. The new teams were the Pittsburgh Maulers, the Memphis Showboats, the Jacksonville Bulls, the Houston Gamblers, the San Antonio Gunslingers, and the Oklahoma Outlaws. With expansion, the USFL ballooned from three to four divisions. In the Eastern Conference, the Philadelphia Stars, the New Jersey Generals, the Washington Federals, and the Maulers comprised the Atlantic Division, while the Birmingham Stallions, the Tampa Bay Bandits, the New Orleans Breakers (yes — *New Orleans*; we'll get to that momentarily), the Showboats, and the Bulls called the Southern Division home. The Western Conference's Pacific Division belonged to the Los Angeles Express, the Arizona Wranglers, the Denver Gold, and the Oakland Invaders, while the Central Division boasted the Chicago Blitz and the Michigan Panthers, the Gunslingers, the Outlaws, and the Gamblers.

Deep breath.

The Boston Breakers, sold to a Louisiana real estate developer named Joe Canizaro, moved to New Orleans, where they would play in the 69,658-seat Superdome (as opposed to the 21,000-seat Nickerson Field) and threaten the NFL's Saints, who in 17 years of existence had yet to enjoy a better-than-.500 season. "The city wanted a winner," said Canizaro. "I figured we could out-perform the Saints. Really, at that point *who couldn't* outperform the Saints?" The Boston to New Orleans transfer was greeted with mixed emotions inside the USFL offices, in that the franchise was relocating from America's sixth-largest television market to its 50th. It paled on the bizarre-o-meter, however, compared with what happened on September 30, 1983, when the roster of the Chicago Blitz and the roster of the Arizona Wranglers were traded for each other.

Yes, traded for each other.

It was, unofficially, the largest singular professional sports transaction of all time. Wrote Jim Litke of the Associated Press:

> It's a confusing trade. The Blitz' name will stay in Chicago, but the Blitz' players will become the Wranglers in Phoenix. The Wranglers' name will stay in Arizona, but the Wranglers' players will become the Blitz in Chicago . . . At a news conference in Phoenix, Dr. Ted Diethrich, former Chicago owner and now owner of the Wranglers, said, "I believe this is the first time in football history that a franchise with all of its players have been transferred. We are the new Arizona Wranglers, but we hope to carry on the new winning tradition that we had in Chicago.

Litke's piece failed to do justice to a deal Tim Tyers, the Wranglers' *Phoenix Gazette* beat writer, called "the definition of lunacy." The new Blitz owner was a Milwaukee-based cardiologist named James F. Hoffman Jr., who paid $7.2 million for a franchise with money he did not have (his listed net worth, $18 million, was slight for the owner of a sports franchise). George Allen, the coach of the Blitz, and his son Bruce, the general manager, relocated to Arizona, and Chicago hired Marv Levy, former head coach of the Kansas City Chiefs, to run its team. "I've rebuilt four franchises," George Allen cracked, "but I've never gone through this type of thing." In all the hype and hoopla, Hoffman forgot to inform Levy that he was taking on the sludge Blitz, not the stellar Blitz. The club he was inheriting had not finished 12-6 in 1983, but 4-14. "I had no idea," said Levy. "I thought I was getting a lot of talent. I actually got almost no talent." Diethrich initially wanted to make the Arizona Wranglers the Arizona Blitz, but Hoffman insisted the Chicago Blitz remain the Chicago Blitz, so the Arizona Wranglers became the "New Arizona Wranglers."

When asked by members of the media why, exactly, the Blitz and Wranglers swapped rosters, league officials shrugged, stammered, stuttered. *It was, you see, a complicated decision based upon the financial implication of . . . um . . . eh . . . ah . . .*

"It's pretty simple," said Bruce Allen. "Dr. Diethrich lived in Arizona, and he didn't want to fly to Chicago for games."

Put differently: because the multimillionaire owner of the Chicago Blitz preferred not to take nine three-hour weekend flights via private jet, two teams (and 100 total players) were dealt for each other. Lives were uprooted. Houses were sold and bought. Children changed schools. The USFL endured a week as the nation's laughingstock. "It was very bad for the league," said Bruce Allen. "First, it looked amateurish — which it was. Second, it left a bad taste in the mouths of a lot of fans. We had to rebrand two franchises."

"Honestly, it was a mistake," Diethrich said years later. "I thought it would be a good idea, but we left a great market, Chicago, for an OK one. That's my fault. Bad call."

Worst of all, the USFL somehow failed to notify many of the players about the transaction. "They never told us anything," said Mark Buben, a defensive lineman who went from Chicago to Arizona. "You'd think maybe a phone call, right?" Dan Manucci, the Wranglers' backup quarterback in

1983, caught wind of the trade and called Bill Polian, the new Blitz director of player personnel, two weeks before the opening of training camp. "Mr. Polian," he said, "when should I report?"

"Report where?" Polian replied.

"Um, to Blitz camp," said Manucci.

"Oh," Polian said. "We traded a lot of the guys we got from Arizona to Memphis. You're with the Showboats now." Indeed, in the *second*-largest deal in USFL history, the Blitz sent 21 players to the Showboats for a late-round draft pick. Five days later Memphis released 20 of the 21, keeping only Clint Davenport, an offensive lineman.

A stunned Manucci asked his father to reach out to Pepper Rodgers, the Showboats' newly named head coach. "Ah, hey, Mr. Manucci, thanks for calling," Rodgers said. "But we cut Dan a week ago."

While Wranglers-Blitz was unsettling, the USFL's expansion efforts paid early dividends. First, there was the much-needed infusion of money ($6 million a pop) from the six teams. But more important, most of the owners and general managers seemed genuinely interested in building winning franchises and improving the quality of the league. The original USFL blueprint called for slow growth and wisdom before impulse. The new USFL blueprint was far more confusing. Having tiptoed through much of the debut season, often getting by with subpar talent at key positions, many USFL decision makers decided it was time to follow the Michigan Panthers and spread around some dough.

"Not smart, not wise," said Steve Ehrhart, the league's executive director who left the USFL office to run the Showboats. "But there was pressure to compete."

Befitting the franchise moniker, the Houston Gamblers — more than any of the other expansion teams — aspired to be immediate players. The club was owned by four men (Alvin N. Lubetkin, Bernard H. Lerner, Dr. Jerry Argovitz, and Fred M. Gerson), one whose name raised eyebrows. Argovitz happened to be the famed dentist turned players' agent who helped woo Herschel Walker to the USFL and who also *still* represented a handful of top-level NFL talent. The conflict of interest deeply troubled Simmons, even as

Argovitz promised to cease his agent work within a year's time. "Representing players wasn't my true love, and NFL teams weren't as open to dealing with me as they had been," Argovitz said. "I had been a hard negotiator, and they didn't like that. So they punished me for it."

Argovitz was one of the first to broach Houston as a potential USFL city, and David Dixon, the creator of the league, saw in him a kindred spirit. Both were salesmen. Both were dreamers. Both were southerners with money. As part of his contractual deal with the league he founded, Dixon was guaranteed a franchise in the unoccupied city of his choice. Yet as salaries escalated and ego seemed to corrupt patience, he lost interest, and chose to sell that right to Argovitz. "I don't think it was ever David's dream to own a football team," Argovitz said. "He was an entrepreneur who wanted to create, not own." The Houston group paid Dixon $4 million for the franchise rights, a figure that infuriated the other owners, who were charging $2 million per expansion club and saw the move as a devaluing of product. "David was smarter than most of the USFL guys," Argovitz said. "A lot of those people were businessmen without sports insights. David understood the hunger for sports, and he made money off of that. Good for him."

From the beginning, the Houston operation prided itself on a sort of Billy the Kid, stagecoach-robbing approach to football. The uniforms were black, red, and white, with a Texas-shaped *G* on the helmet. The name "Gamblers" was selected by ownership, and it fit. ("People were coming to Houston during an oil boom, gambling with their lives," said Argovitz. "Plus, I loved the Kenny Rogers song.") The organization was elated when Roone Arledge, the head of ABC Sports, sent a telegram complaining of the tie between gambling and football and threatened to refuse to televise the team's games. Argovitz fired back an angry note asking how a network that employed handicappers to offer betting odds could make such a claim. The Gamblers paid for a full-page advertisement in the *Houston Chronicle* that reprinted the Arledge letter and asked, SHOULD WE CHANGE OUR NAME? "Eighty percent said we had to keep it," Argovitz said. "So we did."

Whether he would be a good owner or bad owner was up for debate. One thing Argovitz grasped, though, was the importance of star power. Houston had hired its head coach, veteran Jack Pardee, as well as much of its staff in the late spring of 1983, and shortly thereafter the team made a move nearly as bold as the Generals' play for Herschel Walker.

The first round of the 1983 NFL Draft had been held on April 26 at the New York Sheraton Hotel, and five of the six selected quarterbacks were content. John Elway, plucked first overall by the Baltimore Colts, would be traded to Denver — a perfect landing spot. Penn State's Todd Blackledge went seventh to the Chiefs, Illinois's Tony Eason 15th to the Patriots, Ken O'Brien of Cal–Davis 24th to the Jets, and Pittsburgh's Dan Marino 27th to the Dolphins. The only truly unhappy signal caller was the University of Miami's Jim Kelly, who — in the lead-up to the draft — listed the NFL teams he did not want to play for. Topping the chart was the Buffalo Bills, who (a) finished 4-5 in the strike-shortened 1982 season, (b) existed in the freezing cold, and (c) had never reached the Super Bowl. Despite being aware of Kelly's feelings, and despite Kelly missing all but three games of his senior season with a separated right shoulder, the Bills grabbed the quarterback with the 14th overall selection. Kelly felt trapped. Yes, the USFL's Chicago Blitz took him with the 163rd overall pick, but that, too, was a cold city in an unfamiliar land. It did not appeal to him.

On May 12, Kelly and Greg Lustig, his agent, traveled to Buffalo to negotiate with the Bills. The two sides agreed in principle to a four-year, $2.1 million deal, and Blitz GM Bruce Allen learned that an offer was on the table. "He's a guy that I thought could really make this league special with his swagger and his talents," Allen said. In other words, he believed Kelly could be to the USFL what Joe Namath had been to the AFL. On the day Kelly was set to sign, Allen dialed the number for the Bills' offices, and a receptionist answered.

"Can I speak to Greg Lustig?" he asked.

"He's in a meeting," she replied. "How can I help you?"

"Please put Greg on the phone," Allen said. "We're having a family emergency."

He was placed on hold, and in a couple of seconds Lustig picked up. "Greg," he said, "this is Bruce Allen. I know you're there with Jim Kelly to sign a contract with the Bills. If you sign with Buffalo, you're making the biggest mistake of your life, Jim's life, and his family's life. He doesn't want to play in that weather, and the USFL is prepared to make you a deal you can't refuse for Jim. When you hang up this phone, don't sign anything, walk out of that room right now."

"Are you serious?" Lustig whispered.

"I'm dead serious," he replied.

Standing alone in a hallway, Lustig was stunned. He filled Kelly in. "Had that secretary not walked into the room," Kelly said, "I probably would've been a Buffalo Bill." The pair talked their way out of the complex and called Allen. The Blitz GM spoke on behalf of the league when he asked, "Who do you prefer to play for?" Over the next couple of days Kelly proceeded to meet with the Bandits, Blitz, and Gamblers. In Houston, the fast-talking Argovitz told Kelly he wanted him to be his quarterback, but he first needed to see whether his arm was fully recovered. The two retreated to Memorial Park, where Argovitz removed a football from his trunk, handed it to Kelly, and began running pass patterns. "I must be 40 . . . 50 yards down the field, and he hit me with a zinger and broke the pinkie on my left hand," Argovitz said. "I popped it back. I said, 'Jim, I think your arm is fine.'"

To the joy of Houston and the dismay of Buffalo, on June 10, 1983, headlines across America trumpeted the signing. The Gamblers traded four future draft picks to the Blitz for Kelly and running back Mark Rush, his University of Miami teammate (and roommate), then signed the quarterback to a contract that paid $1 million up front and $500,000 per season for five years (and included a clause that guaranteed Kelly would always be one of the USFL's three highest-paid quarterbacks). Argovitz was gleeful — as an agent he had resented Buffalo for years, dating back to the way the team mistreated his client, halfback Joe Cribbs. The Bills, on the other hand, were humiliated. "We considered three different offers that they threw at us and they were happy with the offer we made to them," said Pat McGroder, the Bills' executive vice president. "We are very, very disappointed in not signing Jim Kelly."

And, with that, the NFL was no longer laughing. The Useless wasn't useless. The Herschel Walker signing had been a setback to the old league, but it involved a player who was not even eligible for the NFL Draft. Kelly was different — he had a choice where to go. And he picked the Houston Gamblers over the Buffalo Bills. "It wasn't as hard a decision as you might think," Kelly said. "I *did not* want to go to Buffalo."

Exactly two months later, another expansion team made another move that simultaneously petrified and infuriated the NFL. In Tampa Bay, the Buccaneers were known as the league's sorriest franchise. They were cheap, they were sloppy, and their owner, Hugh Culverhouse, was a notorious racist who treated African Americans with minimal respect. Wrote Doug Williams, the team's quarterback, in his autobiography: "If someone wanted a perfect redneck asshole, Culverhouse would be a top candidate."

One of the few bright spots for the organization was Williams, a former first-round draft pick from Grambling who, from 1978–82, started for Tampa Bay. Williams was a natural leader with a nuclear arm and unrivaled work ethic, but he ended 1982 as the league's 43rd-highest-paid quarterback, and Culverhouse refused to surrender fair market value. Entering 1983, Williams — a free agent — asked for a five-year contract at $600,000 per season. The Buccaneers would go no higher than $400,000. Williams wrote, "I wasn't going to be their slave anymore . . . Culverhouse thought that I was a good ol' black boy who never had nothin' and would go along with whatever Mr. Culverhouse wanted. He paid me like a slave, but I wasn't that much of a slave to sign that deal."

Pre-USFL, Williams would have been stuck without options. Sure, a couple of players went to the Canadian Football League, but the money was rarely as good. And sure, you could demand a trade — but without unrestricted free agency, and with the owners aligned in cahoots against the players, the request usually fell on deaf ears. At one point the Raiders had wanted to deal for Williams, but Culverhouse refused to so much as listen, because he hated Al Davis, Oakland's owner. "I had no power," said Williams. "And no say."

In 1984 the Oklahoma Outlaws came calling.

They were, along with the San Antonio Gunslingers, the least probable of the 18 USFL outfits. William Tatham Sr., the franchise's owner, lived in Fresno, California, where he served as the president and chairman of the board of Consolidated Industries. Like Tampa Bay's John Bassett, Tatham had been the owner of a World Football League club (the Portland Thunder) and believed sports to be a worthwhile gamble. Tatham teamed up with his son, William Junior, to try and start a USFL organization in San Diego, and at one point in June 1983 even came to terms with Dan Fouts, the Chargers' four-time All-Pro quarterback, on a contract. "It was a done deal," said Tatham Jr. "We had a big press conference planned, the numbers were agreed upon. But this is what type of man my dad is. We were talking with Dan, and he was talking about how close the Chargers had come to the Super Bowl, and how great his teammates were. And Dad was feeling genuinely sad for him. And he told Fouts, 'Take our offer to the Chargers. If they match it, stay in the NFL.'" The next day, Fouts sat down with Gene Klein, owner of the Chargers, told him of the USFL numbers and asked whether the team would match. They did — with a six-year deal that paid $1.2 million annually. "It's the highest paying [contract] in the history of the NFL," Klein told reporters. "Dan is deserving of it."

Unfortunately for the Tathams, the San Diego Stadium Authority rejected a bid to use Jack Murphy Stadium, then the city's city council (with tremendous pressure from the Chargers) denied an appeal. So the USFL, anxious to develop the Southwest, suggested Tulsa — a location the Tathams had never visited or considered visiting. It was a strange choice, what with its status as America's 60th-largest television market, as well as a facility, Skelly Stadium on the University of Tulsa campus, that held only 40,000 and needed significant upgrading. But the school's football team, the Hurricanes, drew 35,000–40,000 fans every home game, which was encouraging. "So everything's done, papers are signed, we're really excited — and come to find out the school has been inflating figures," said Tatham Jr. "They were drawing around 17,000. We were screwed."

Enter: Williams.

Despite all the nonsense with the Buccaneers, the quarterback presumed he'd suck up his pride and return to Tampa Bay. But a tragedy changed everything. On April 17, 1982, Williams married Janice Goss, a Gainesville, Georgia, native he began dating at Grambling. The following year, while making an Easter visit to Williams's hometown of Zachary, Louisiana, Janice was crippled with excruciating headaches. A CAT scan detected a grapefruit-sized tumor in her brain. Williams wrote of his wife's final night in the hospital: "It was storming outside, and she started screaming again. The headache had come back, and she couldn't keep fluids down. She was in tremendous pain. I tried to console her, but it didn't help, so we had to call the doctor. They ordered us out of the room. While they were trying to help her, fluid built up in her chest. Finally, her lungs collapsed and she died."

Williams was a widower at age 27, charged with raising a three-month-old daughter, Ashley. He was heartbroken and alone and angry — and no longer felt compelled to put up with Hugh Culverhouse's garbage.

Kelly had recently signed with the Gamblers, and Argovitz knew Tatham needed a marquee star. He mentioned that Williams was a free agent, so Tatham Jr. called Jimmy Walsh, the quarterback's representative. They agreed to financials via phone, and the three met at the Oral Roberts Hotel in Tulsa under assumed names. "That night I handed Doug a check for $1 million, which was his signing bonus," said Tatham Jr. "He could not have been happier."

"The Tathams approached me, and from the beginning they never treated me like a piece of cattle in a stockyard," Williams said. "They brought me in, they introduced me to their family members, they treated me like a human, and they wanted me to be a big part of the organization. I didn't know much about the USFL, but I finally knew what it felt like to be appreciated."

On August 9, 1983, Williams signed a five-year, $3 million contract that made him one of the highest-paid players in professional football. At his press conference in Tulsa, a beaming Williams cited the "great feeling" of being an Outlaw, and wished the Buccaneers nothing but failure.

"I was euphoric to leave," said Williams. "It was a new start for me. A fresh start."

Kelly and Williams were the first two big splashes of the expansion movement, but the earth shaker was still to come. Approximately two weeks before the USFL's January 4, 1984 collegiate draft, George Heddleston, the general manager of the Pittsburgh Maulers, was summoned to New York for a meeting with Chet Simmons and Peter Hadhazy, the league's director of operations. The three sat down and the commissioner did not mince words.

"As you know, we got Herschel Walker last season and it really helped our television ratings," he said. "And ours is a television game, because spring isn't big crowd turf."

The Mauler GM wasn't sure where Simmons was heading. "What are your personnel people telling you about Mike Rozier?" Simmons asked.

Heddleston paused. A couple of weeks earlier Rozier had won the Heisman Trophy after rushing for 2,148 yards and 29 touchdowns at the University of Nebraska. Though only five foot ten, Rozier's body was a 210-pound granite carving. That said, Heddleston's personnel people told him nothing about Mike Rozier. The running back had never before entered his mind. With all that came with building a franchise, long-shot acquisitions weren't exactly a focus.

"Um, do you want me to sign Mike Rozier for the Pittsburgh franchise?" Heddleston asked.

"Well," Simmons said, "we obviously can't tell anyone this, but if you can, Pittsburgh will get the first pick in the draft."

"Oh, we can sign Mike Rozier!" he said. "Definitely!"

He had no idea.

A couple of days later, the *Pittsburgh Press* sports section led with the boldfaced headline MAULERS GET FIRST PICK IN DRAFT. According to Ron Cook's piece, the USFL held a lottery among the six expansion franchises, and Pittsburgh lucked into the top slot. It was blockbuster news, and Heddleston, ever the good soldier, went along with the narrative. "What was I supposed to do?" he said years later. "Admit it was all prearranged?"

That's why, in the waning days of 1983, Heddleston found himself in Miami, Florida, where Nebraska and Miami were preparing to meet in the January 2, 1984, Orange Bowl. Although it was technically against NCAA bylaws for players to be represented by agents, Rozier was aligned with Mike Trope, and had been throughout his senior year. Trope was known in professional football circles for his annoyingly aggressive and obnoxious negotiating tactics, and he told Heddleston to come to Miami if he genuinely wanted a shot at his client. "Trope was the agent we all loved to hate," said Heddleston. "He got his clients big bucks, but he was very difficult. The first time I met him he said to me, 'Mike will do what I tell him to do. If you pay him the money, he'll play in the USFL.'" Heddleston never met Rozier in the days before the Orange Bowl, but he and Trope went back and forth with financials. Finally, the general manager called Eddie DeBartolo Sr., the Maulers owner.

"I think we can get Rozier," Heddleston said.

"How much will it cost?" DeBartolo replied.

"Well," Heddleston said, "there haven't been many $1 million athletes."

A lengthy pause. One million dollars was no joke. And Rozier, for all his notoriety, was a bit of a gamble. First, his size was a concern. Second, he was known to be of limited intellect. Third, there were rumors of a cocaine problem. (Said Heddleston: "I said to [team president] Paul Martha that I was worried Mike used cocaine. And he said, 'Come on George, everyone's doing it. And it might make his ankle feel better.'") Fourth, his attitude and outlook were a bit iffy. Herschel Walker approached football with love and artistry. Rozier, on the other hand, was a product of the mean streets of Camden, New Jersey, annually ranked as one of America's 10-most-depressed cities. He played football for survival; for release; for money. There was nothing wrong with such motivations. But they hardly inspired confidence. "I told the Maulers not to draft Mike," said Rick Buffington, the team's director of scouting. "I told them it was a bad idea; that he wasn't the player they thought he was, and there were some real holes."

Regardless, DeBartolo promised Heddleston that he would pay the requisite dough for Rozier—but not a penny more.

"He wrote a $1 million check to give to Mike up front," Heddleston said. "It was hard for me to keep a straight face. If I couldn't get him signed for $1 million, I had no business doing that job."

On the night of the Orange Bowl, Heddleston checked into a room on the 17th floor of the lush Fontainebleau Miami Beach, the University of Nebraska's team hotel. He was told that, as soon as the game ended, Trope's assistant, a man named Ivory Black, would deliver Mike Rozier. "Well," Heddleston said, "nothing ever goes as planned." The Cornhuskers entered the matchup 12-0 and ranked No. 1 in the nation. In the fourth quarter, Rozier was knocked out with an ankle sprain, and Miami went on to win a 31–30 nail-biter. "I'm in my room, and it's freezing because the heater isn't working. I mean, I'm literally shivering for hours," said Heddleston. "But I won't leave, because he's coming any minute and I've got this $1 million check in my pocket. I'm waiting and waiting and waiting, all night, and Mike never comes. They don't deliver him, because apparently he's holed up with some broad in another room."

At one point, Heddleston took the elevator to the lobby to complain about the broken heater. He was beginning to panic — the USFL Draft was in two days, and he knew not whether Mike Rozier even wanted to join the league. "I get back to the elevator and Dudley Moore and Susan Anton are there, arm in arm," he said of the famed actors. "They had had a few and they say, 'Join us! Have a drink!' So we have a drink together at the hotel bar — me, this little British man, and this super tall blonde. It was the weirdest night of my life. I'm really upset about this whole mess, but I'm with these two stars. At one point they asked why I was there and I said, 'Any chance you guys have ever heard of Mike Rozier?'"

They had not.

The following morning, at approximately 10 o'clock, Heddleston was sitting with Martha, who had arrived earlier from Pittsburgh. They heard a knock at the door. It was Ivory Black, accompanied by a red-eyed Rozier.

"He signed the contract," said Heddleston of a three-year, $3.1 million deal, "and just like that, the USFL had the back-to-back Heisman Trophy winners."

Actually, not quite *just like that*. Trope, Black, and Rozier asked that the agreement not be announced, so the running back could maintain eligibility for the Japan Bowl, a college all-star game scheduled for January 16 in Yokohama. Word inevitably leaked, however, and on January 9 the Maulers finally addressed rumors in a formal press conference at the Pittsburgh Civic Arena. It was the sort of WELCOME TO TOWN shindig franchises build their reputations on. Picture Reggie Jackson holding up a New York Yankees jersey, or Tony Dorsett being introduced as a Dallas Cowboy. Cameras flashing, pens scribbling, smiles and optimism and dreams of heady days.

Yet Mike Rozier did not attend. Instead, members of the press could speak with Glenn Carano, the team's new quarterback (and former Cowboys third-stringer), who said, "This is wonderful! This is terrific!" before fading off into the oblivion that was his general career. Yes, the franchise crowed, Mike

Rozier was now a member of the Pittsburgh Maulers. No, they had done nothing wrong. Yes, Rozier was excited. *Um, anyone want some pigs in a blanket? They're on a table in the back of the room . . .*

Wrote Bruce Keidan of the *Pittsburgh Post-Gazette* in a column dripping with sarcastic damnation: "The marriage had been consummated a week ago today, as it turns out. But the player, we are told, was too shy to allow the announcement to be made public until he was safely outside American territorial waters."

The story that never ended refused to end. Rozier sat out the bowl game. Six days after the announcement, the *Boston Globe* reported that the new Mauler fired Trope in a meeting at the Rozier home in Camden.

"The [Rozier] family is upset," said Art Wilkinson, his new attorney. "Mike Rozier never read either of his personal services contract with Trope, or his football contract with the Maulers. All he did was sign the back of both." Trope, it was revealed, made $300,000 off of the USFL deal. He was fired by Rozier, but the damage had been done.

The contract was binding.

Like it or not, Mike Rozier was a Pittsburgh Mauler.

Jim Kelly, Doug Williams, and Mike Rozier were huge additions for the USFL. But they were merely three of many.

Although a good number of voices within the league office wanted to take baby steps, a handful of organizations were swept up by big-money, big-name football fever. In Birmingham, for example, the Stallions ridded themselves of Reggie Collier, the young but erratic quarterback out of Southern Mississippi, and replaced him with Cliff Stoudt, a longtime Pittsburgh Steelers backup. Stoudt told the Associated Press that it was "a great career move for me" — and few could argue. His three-year, $1.2 million contract was guaranteed, and four times more money than he made in the NFL. That addition was official on January 12, 1984, and soon enough Stoudt would be handing the ball off to Joe Cribbs, who had spent the past four seasons as a star halfback with the Buffalo Bills.

A product of Auburn University, Cribbs was represented by Argovitz, and their dealings with the Bills were routinely awful. As a rookie in 1980, Cribbs had a clause in his contract that called for a $10,000 bonus should he exceed 1,200 rushing yards. In the season's final regular-season game at San Francisco, he rushed for 128 yards on 18 carries — then was pulled early on order of management, which knew he needed but 15 more yards to earn the loot. The next day Argovitz called Stew Barber, Buffalo's vice president, to politely ask if the organization might reward his client for a season well played. "I was told in no uncertain terms that the team simply wouldn't," Argovitz later wrote.

The following year, Cribbs again paced the Bills in rushing and touchdowns, but suffered a knee injury in a playoff loss at Cincinnati. He was placed on crutches and ordered to skip the Pro Bowl. Most teams flew their honorees to Hawaii, injured or healthy. The Bills refused. "He earned the Pro Bowl," Argovitz told Barber. "At least send Joe and his wife to Hawaii as your way of saying thanks."

"We have policies against such things," Barber said. "No deal."

From that point on, Cribbs viewed the Buffalo Bills not as *his* team, but as a hostile business partner. So when the USFL came along in 1983, Argovitz made a devilishly delicious suggestion to his client: wrap up the season with Buffalo while also signing a futures contract with the new league. In particular, Argovitz wanted Cribbs for his Houston Gamblers, but the running back belonged, territorially, to the Stallions. "I told Jerry that he lost his mind if he thought we were going to let Joe Cribbs go to another USFL team," said Jerry Sklar, the Stallions' president. "Birmingham would have run me out of town." Sklar met with Marvin Warner, the team owner, and said, "We have the opportunity to sign Joe Cribbs."

"Big deal," Warner replied.

"I don't think you understand," Sklar countered. "This will be expensive, but it'll be the greatest opportunity the Stallions will ever have."

Birmingham offered a five-year, $2.45 million contract with a $650,000 bonus, which Cribbs signed after little debate. The Bills sued unsuccessfully (Cribbs had a right of first refusal clause in his deal, but it was agreed upon before the USFL's existence), and suddenly the Stallions featured an NFL-caliber backfield. "He was a beast," said Ray Bentley, the Panthers linebacker. "When Cribbs ran he was trying to hurt you and your family."

From afar, the NFL attempted to maintain its stoicism. Owners and general managers were under strict orders to publicly ignore the junior league; to pretend as if it were little more than a pesky gnat on a giant's shoulder. But that was becoming increasingly difficult. Two decades earlier, when he was out of football, Paul Brown was one of the earliest backers of David Dixon's USFL vision. Now, however, he was the Cincinnati Bengals' general manager, and his support turned to hostility. The anger first percolated on June 27, 1983, when Cris Collinsworth, the team's star wide receiver, agreed to a futures contract with the Tampa Bay Bandits after his NFL deal expired in two years. Collinsworth made the announcement at the halftime of a Gold-Bandits game. He was carted onto the Tampa Stadium field inside a horse-drawn carriage, waved to the crowd, and chose the name of a lucky fan who was to have his mortgage burned. To Brown, the whole display reeked of treason. Were that not bad enough, the Jacksonville Bulls hired Lindy Infante, Cincinnati's much-ballyhooed offensive coordinator, to be their head coach. For Brown, there was a right way and a wrong way—and this was terribly wrong. Infante planned on remaining with the Bengals until he needed to report to Jacksonville, and Brown called it akin to "a fox in the hen house."

"Our inclination is that we have no choice," Brown said. "We can't have a coach working with us who has made this announcement." He ordered Infante to pack up his things and leave immediately. The coach, blessed with a sharp sense of humor, returned to the Bengals' facility a few days later, this time disguised in a beard, mustache, baseball cap, and sunglasses. He stood outside a fence and begged players for autographs. Collinsworth and quarterback Ken Anderson both signed before realizing it was, in fact, Infante. The trio retreated to a nearby bar for some beers.

The hits kept coming. Of the 23 players selected in the first round of the USFL Draft (the six expansion teams were given an extra pick), 10 inked contracts with the league. One club that made a particularly impactful inroad was the Memphis Showboats, who signed Reggie White, the University of Tennessee defensive end who had been both an All-American and the Southeastern Conference's Player of the Year, to a five-year, $4 million deal, then added Alabama quarterback Walter Lewis for $1 million over three seasons (plus a $400,000 bonus).

Although he was deemed a midround NFL prospect, the Lewis addition via the territorial draft exemplified much of the USFL's threat to the football universe. Following a season in which Lewis led Alabama to an 8-4 record while throwing for 1,991 yards and 14 touchdowns, NFL scouts, general managers, and coaches agreed he could make a marvelous defensive back or, perhaps, wide receiver. It was the way the league operated at the time — with rare exception, black college quarterbacks were labeled "athletes" (code: not smart enough to run a team) and switched to other spots on the field. From its inception, the USFL refused to subscribe. In its first season, four African American quarterbacks populated rosters, and now both Williams and Lewis were joining the show. "The NFL just wasn't willing to take a risk on an athletic black quarterback," said Lewis. "They saw me as a fifth-rounder. Well, thanks to the United States Football League I no longer had to settle."

The Showboats were owned by William Dunavant, a Memphis cotton maven, but many of the decisions were steered by Logan Young Jr., a minority investor and longtime Alabama booster. He insisted the Showboats *needed* Walter Lewis, then had the team significantly overpay for his services. "If we're being honest, the demand for Walter wasn't high," said Leigh Steinberg, his agent. "But who was I to argue?"

On February 5, 1984, the Showboats held a press conference at Memphis's historic Peabody Hotel. A high school marching band, replete with dancing pom-pom girls, guided the quarterback into the lobby, where he was introduced and presented with a duck-shaped key to the city. After formally signing the contract and promising to donate money for every touchdown pass (there would not be many), the $1 million quarterback no one else wanted smiled and waved to an adoring crowd.

"It was surreal," Lewis said. "I didn't know what I was in for."

He was far from alone.

8

THE SAVIORS

Chet Simmons called me into his office one day, handed me an index card with a name written on it, and said, "Do me a favor — go to the New York Public Library and find out everything you can about this guy." I looked at the name, but had never heard of him. I spent the entire day xeroxing articles about Donald J. Trump. When I got back Chet asked if I was familiar with him. I shook my head. "Yeah," Chet said, "he's well established in New York, but outside the city nobody knows who he is."

— Doug Kelly, USFL director of communications

THE PRESS CONFERENCES were held exactly three months and 2,805 miles apart, and both were dizzying, luxurious, and, for the USFL, potentially life-saving.

On the afternoon of September 22, 1983, inside the atrium of the 58-story Midtown Manhattan skyscraper he humbly named Trump Tower, Donald J. Trump, a relatively obscure New York real estate developer, introduced himself as the new owner of the New Jersey Generals.

On the afternoon of December 22, 1983, inside a ballroom at the Beverly Wilshire Hotel in Beverly Hills, J. William Oldenburg, a relatively obscure chairman of the board of a billion-dollar mortgage banking company, introduced himself as the new owner of the Los Angeles Express.

Happy days were here again.

Chet Simmons, the USFL's second-year commissioner, couldn't believe his good fortune. While the league's first season featured many successes, the New Jersey and Los Angeles franchises — cornerstones that *needed* to soar — largely flopped. At the Meadowlands, the Generals drew pretty well on the strength of Herschel Walker's presence, but went a forgettable 6-12. And at the Los Angeles Coliseum, the Express finished 8-10 before a proud crowd of invisibles. Neither team's ownership was particularly enamored by the idea of continuing with the grand USFL experiment, especially considering the two franchises lost a combined $6 million.

That's why, when Trump offered J. Walter Duncan nearly $10 million to purchase the Generals, the 67-year-old Oklahoma oilman (who would have gladly taken $8.5 million) didn't have to think twice. "I found that, for me, owning a professional business in the New York–New Jersey area was just too much to handle personally," Duncan said. "I live in Oklahoma City, and I had to attend 18 games about 1,500 miles from home. It was very difficult for me to be at the games, attend league meetings, and look after the organization the way I felt it should have been looked after. The Generals are too important a franchise to the league to have absentee ownership." Translation: *Yip! Yip! Yippee!*

That's why, when Oldenburg offered Bill Daniels and Alan Harmon $7.5 million to purchase the Express, the two cable TV executives didn't have to think twice. They took their $2 million profit and bolted. "It is not practical to be an absentee owner in this day," Daniels said from his Denver home. "This was something we needed to do." Translation: *Yip! Yip! Yippee!*

Though Trump and Oldenburg had never met, commonalities existed. At 37 (Trump) and 45 (Oldenburg), they were two of the USFL's younger owners, as well as the most bombastic and narcissistic. They both fancied blonde, large-breasted women, automobiles that cost more than most houses, and anyone (and everyone) to quiver in their presences. They reveled in being referred to as "Mister," and would happily tell you how much they spent on this, how much they spent on that. Trump's family crest was stitched onto every piece of clothing he wore. Oldenburg's initials were stitched onto the sleeve of every shirt he wore. For Trump and Oldenburg, linen was never merely linen. It was imported 1,600-count Greek Utopian.

"The first time I met Donald Trump was on a flight to a league meeting," said Vince Lombardi Jr., son of the legendary Packers coach and president/general manager of the Michigan Panthers. "I didn't know who he was, just that he was a busy young guy. He started asking me all these football questions, and then he told me I had no idea what I was talking about. I thought, 'Who the hell is *this* guy?' Later on, I noticed one very unique thing about the people who worked for him. They only knew two words — 'Yes, Donald.'"

People paying a visit to Trump's office on the 26th floor of Trump Tower didn't merely sit in a bland reception area, waiting to meet the big man. No, they first had to endure an eight-minute film that chronicled the greatness of Donald J. Trump, New York icon and all-around amazing guy at absolutely everything. The video lavished plaudits upon "this visionary builder." Read a deep-voiced narrator over shots of Trump Tower: *This is Manhattan through a golden eye, and only for the select few . . . Any wish, no matter how opulent or unusual, may come true . . .*

"It was ridiculous," said Barry Stanton, who covered the Generals for the *Journal News,* a Westchester, New York–based paper. "Part of the film was actually a sales pitch for condos at Trump Tower. And it wasn't optional. If you wanted to speak with the Donald, you had to hear how amazing he was." It wasn't lost upon the attention-obsessed Trump that while his real estate deals might (on occasion) land his name on some inside page of the *Post* and *Daily News,* the Generals announcement placed him on the front and back. "He loved being talked about," said Jerry Argovitz. "It was his fuel."

Oldenburg was no different. His firm, Investment Mortgage International, was based on the top floors of San Francisco's Transamerica Pyramid building, and the chairman had his own penthouse suite. When the office first opened in November 1983, Oldenburg gave a welcoming speech as, behind him, mist rose from blocks of dry ice. The walls were painted in gold, a gong sounded with every $1 million transaction, the doors and windows were voice-operated. "You would say, 'Wall, open,' and the damn wall would open," said Steve Ehrhart. "It was awfully impressive. You'd be staring at a wall, something would rotate and it'd suddenly become a full bar." A scrolling digital ticker tape, à la New York City's Times Square, greeted visitors in the lobby. There was a wall of clocks that doubled as a crystal sculpture, as well as a large Jacuzzi. A press release described the facility as "the most spectacular office space ever seen in San Francisco (or perhaps anywhere in the world)."

"It was so opulent," said Leigh Steinberg, the agent. "Glitter everywhere, trying to show his importance."

Though Simmons wasn't one for eccentric millionaires and their annoying excesses, the USFL needed stable ownership. So he could overlook the warning signs, like the NFL having ignored Trump in approximately 2.7 seconds when he made overtures to buy the Baltimore Colts in 1981. Or when Trump boasted — without a sliver of truthfulness — that "I was offered NFL teams. I could have bought NFL teams in the course of the last three or four years." Or when, in the lead-up to the purchase of the Express, Simmons and Steve Ehrhart invited Oldenburg to New York for a sit-down — only to be confronted with a stiff rebuke of "Why? So a bunch of guys can investigate my cock?" Or when, upon officially taking the reins of their franchises, Trump and Oldenburg promised to spend money. Lots and lots and lots and lots of money.

For a commissioner who aspired to keep the USFL on a steady uptick, this is what most terrified him. In Chicago and Washington, for example, the league had organizations that struggled to pay bills. The Federals made, literally, no noteworthy signings between the first and second seasons, while the Blitz (né Wranglers) tried to fool people into thinking that the addition of a lousy ex-Bears quarterback named Vince Evans was somehow a transcendent move. Meanwhile, the Generals and Express wasted little time. Within the blink of an eye, Trump — who wrongly boasted that "top NFL players are willing to sign with us for less than they're getting in the NFL!" — added quarterback Brian Sipe, the 1980 NFL MVP with Cleveland, with a $1.9 million contract, as well as ex–Kansas City Chiefs safety Gary Barbaro (three years, $825,000) and ex–Seattle Seahawks defensive back Kerry Justin (four years, $800,000). A slew of additional NFL refugees joined the Generals, and the collective salaries reached well into the millions. Mark Gastineau, the New York Jets' ferocious and flamboyant defensive end, was summoned for a meeting (said Leigh Steinberg, his agent: "The first thing we had to do was sit through that stupid promotional video about all his achievements"), as was Dallas Cowboys defensive tackle Randy White, who was offered a $400,000 signing bonus and $700,000 annual contract. Neither man switched leagues. Warren Moon, the Canadian Football League megastar quarterback, also met with the Generals, but ultimately signed with the NFL's Houston Oilers.

Trump fired Chuck Fairbanks and planned on replacing him as head coach with Don Shula, the immortal Miami Dolphin who won 226 games over 21 seasons. Shula had a hostile relationship with Joe Robbie, Miami's penny-pinching owner, and his NFL salary was a little more than $400,000. Trump stalked Shula, calling him every Monday night between tapings of the coach's TV highlights show and the start of *Monday Night Football*. "Trump was sell, sell, sell," Shula said. He offered a five-year, $5 million contract — a staggering figure for any coach in any league. Shula said he would only take the job were Trump to throw in a rent-free Trump Tower apartment (value: $1 million). Because he both loved attention and lacked a filter, on October 23 Trump went on the CBS program *NFL Today* to crow that Shula was basically locked in to run the Generals, but that the coach's apartment demand was a bit of a hang-up. The Dolphins beat the Colts that afternoon, and in the postgame press conference the third question to Shula concerned whether he was, indeed, headed to New Jersey. The prideful Dolphin was livid, and announced the next day he was staying in Miami. The Generals released a statement that said, in part, "Trump felt the arrangements to attain a coach like Don Shula were just too complex and time-consuming at this point. A prime example of this was the possibility of an apartment at Trump Tower as part of a contractual agreement."

It was merely one of endless lies stated by Trump in his USFL run.

"Knowing Don as I do, I'm pretty sure he would have taken the job," said Larry Csonka, the Jacksonville Bulls senior vice president and a former Dolphins running back. "But he's a man of principle, and he did not appreciate being thrown out there to the press the way he was."

The Generals briefly pursued Penn State's Joe Paterno, and when he declined the team hired Walt Michaels, who coached the New York Jets to a 1982 AFC championship game loss to Shula's Dolphins. Michaels received neither $5 million nor a Trump Tower pad, but his three-year, $500,000 salary made him among the league's highest-paid coaches. "Look, Donald Trump was a marketer first and foremost, way before he knew anything about football," said Dave Lapham, a New Jersey offensive lineman who was signed away from the Cincinnati Bengals. "He was trying to collect NFL players to show the country something about his status. That's my guess, and I can't complain. It made me and the other guys a lot of money."

Oldenburg wasn't keen to being outdone, and he would not be here. Of the Generals' owner, he told *Sports Illustrated:* "I'm used to winning, to nothing less than becoming the best. Donald Trump can get all the press he wants, but when it comes to business, he can't carry my socks." After buying the Express, Oldenburg's first order of business was naming Don Klosterman, the former Los Angeles Rams executive and one of the sport's legendary figures, as his new general manager. It took two words — *Open. Checkbook.* — to sway Klosterman into joining the young league. Nicknamed "the Duke of Dining Out," Klosterman built the 1979 Rams into a Super Bowl qualifier, and his reputation in the field was gold. "When the Rams fired him [in 1982], he was not happy," said Ehrhart. "He really wanted to stick it up their asses, and running the Express — in the same city — was a pretty great way to do it." Oldenburg's second order of business was to have Klosterman acquire superstars. He saw what Trump was up to on the East Coast, and aspired to destroy him. As Matthew Barr wrote in *Los Angeles Magazine,* "Klosterman's directive from Oldenburg was simple. Get the best, regardless of cost."

"That's exactly what he did," said Tom Ramsey, Los Angeles's quarterback. "No doubt about it."

Much like the Michigan Panthers a season earlier, the Express devoted ludicrous money toward its offensive line, spending $8 million to rope in four of college football's top blockers (Mike Ruether of Texas, Gary Zimmerman of Oregon, Mark Adickes of Baylor, and Derek Kennard of Nevada), then adding Jeff Hart, the Baltimore Colts' veteran tackle, for $500,000. "[Oldenburg] wanted me to design a car to go 180 miles per hour," Klosterman once told *Sport* magazine. "I told him what it would cost. He never backed off." The negotiations with Adickes were particularly telling. During a dinner-table conversation with the All-American tackle and his agent, Perry Deering, Oldenburg turned to Klosterman and said, "I like this kid — give him anything he wants."

"Well," Klosterman asked Adickes, "what do you want?"

Deering removed a pen from his pocket and wrote a figure on a napkin: $700,000.

"My nickname was 'Limo,'" Adickes said years later. "Because they drove me right out on the practice field in one."

Before long, the Express didn't merely have the USFL's most expensive offensive line — it had *professional football's* most expensive offensive line. "Money seemed to be no issue," said Hart. "Oldenburg wanted to win, and he would sign whoever they needed to in order to be great. It was encouraging." When it became clear the Express needed a strong special team, Klosterman, with the owner's blessing, doled out $100,000 for Tony Zendejas, the record-setting kicker out of the University of Nevada. Kevin Nelson, the former UCLA halfback, was presented with a $1.150 million signing bonus — nearly $630,000 more than his older brother, Darrin Nelson, was making annually to play the same position for the Minnesota Vikings. "I've never been in a parking lot that had so many brand-new Mercedes, BMWs, and Porsches [as the Express players' lot]," said Zendejas. "Never."

Oldenburg didn't spend much time at team headquarters, but those who met him recalled a volatile, erratic, simple, and clinically insane man. He was a tiny figure to behold, five foot six in shoes, with thick brown eyebrows and a disconcerting, almost sinister smile. He referred to himself as Mr. Dynamite, and spoke in a booming voice that filled 100 auditoriums. To know Bill Oldenburg was to dislike Bill Oldenburg. "He was an absolute idiot," said Jerry Sklar, general manager of the Birmingham Stallions. "He didn't have the slightest idea what he was doing."

"He came off as a hustler," said Paul Sandrock, the Express's treasurer. "He tried to play with the big boys, but he wasn't one of them."

To the other owners, Oldenburg remained a mystery until, on the night of January 17, 1984, he arrived in New Orleans for a USFL meeting. It was the first time many came face-to-face with Mr. Dynamite, and if there were expectations of a carnival, no one was disappointed. Joe Canizaro, owner of the New Orleans Breakers, hosted a dinner for his peers, and Oldenburg was the only one to enter with an entourage (which included Wayne Newton, the famous Las Vegas performer). As the entrées arrived and the alcohol flowed, Oldenburg went from agreeable to obnoxious to *One Flew over the Cuckoo's*

Nest psychotic. At one point he stood up, ripped open his silk button-down dress shirt, pulled down his pants, and, pointing to Newton, promised his team would open to a pregame concert, a win, and a sellout crowd. He then bellowed, Bobcat Goldthwait–like, that the Express would "beat the shit" out of everyone in the USFL. "I believe that you do not save souls in an empty church!" he screamed. "If you want to boogie-woogie with the king of rock and roll, you better get some dancers!" Oldenburg bragged that, thanks to his power and influence, the USFL had just been given a betting line in Las Vegas. No one in the room had the heart to inform their new colleague that the betting line already existed.

"What can I say?" said Fred Bullard, the Jacksonville Bulls' owner. "He was unstable."

Trump and Oldenburg were big-game hunters surrounded by many peers armed with peashooters. As the Chicago Blitz were signing Evans, and as the San Antonio Gunslingers were adding Rick Neuheisel, UCLA's good-not-great former quarterback, the Generals and Express sought grizzly bears and lions. It rubbed many of the other owners the wrong way. In Philadelphia, Myles Tanenbaum watched Trump — who challenged the New York Giants to an exhibition game with the $1 million grand prize going to charity — from afar with bitter indignation. The Stars had been the class of the East in 1983, and now some New York blowhard was stealing the thunder. He confronted New Jersey's owner in a face-to-face encounter that ended with Trump saying, "I'm in the media capital of the country. When you're in New York, you have to win."

"Donald," Tanenbaum replied, "in Philadelphia you have to win, too."

"Oh," Trump said, "I have to win more."

Tanenbaum, a decent and agreeable man, wanted to punch Donald Trump in the teeth. The anger hardly subsided when the Generals announced they had signed Lawrence Taylor, the New York Giants' all-everything linebacker, to a four-year, $3.25 million contract that would begin in 1988, after his NFL deal expired. The day before the transaction Jimmy Gould, New Jersey's president, invited Stanton, the *Journal News* beat writer, to Trump Tower for a chat about the team's direction. "I get there, and Jimmy brought me a box of truffles, and we're sitting in the atrium," Stanton said. "Gould's telling me this and that, and as we're heading out we see Lawrence Taylor walking into Trump Tower. It was all this big pathetic setup, so I could witness the great Donald Trump bringing in L.T. It was all about ego."

Trump offered Taylor a $1 million signing bonus, and promised to wire the money immediately should he agree to terms. Just 24 and raised by a lower-middle-class family in Williamsburg, Virginia, Taylor didn't have to be asked twice. "He had me call my bank, and sure as shit, 30 minutes later he wired a million dollars into my account," Taylor said. "I was like, 'Thanks, Don.'" I respected that he put his money where his mouth was."

It was typical Donald Trump, who assured Taylor word of the contract would not get out until he was ready — then immediately had his publicist, "John Barron," call the various New York newspapers to supply them with the "alleged" news. Although it went unsaid, many of the reporters receiving the information knew damn well "John Barron" was actually Donald Trump, futilely attempting to disguise his voice and score tabloid headlines. "He created a publicist," said Stanton. "What the hell?"

News of the Taylor signing filled America's sports pages. In Philadelphia it evoked pure outrage. Though he had been a Giant for three years, Taylor played his college ball at the University of North Carolina, which was a Stars' territorial school. Tanenbaum cried foul to Simmons, who shrugged and nodded and said little. It all ultimately mattered not — Trump knew Taylor would likely never be a General. The whole affair was pure PR bluff. Why, two weeks later Taylor — who received a new six-year, $6.55 million offer from the Giants — exercised an out clause in his USFL contract. He returned the $1 million loan to Trump, along with $10,000 interest and an agreement to make a $750,000 settlement over several years. "We got on the phone with [Giants GM] George Young . . . and we sold him back the contract," recalled Gould. "Trump and I effectively took a two-week option and made a $750,000 profit. And it had never been done before. Taylor was happy because he got a new contract from the Giants for an enormous amount of money, and Trump's attitude was, 'Look, I never wanted Lawrence Taylor to not play for the Giants. I never wanted him to be unhappy. I'm Mr. New York.' So he really was shrewd beyond what I had ever seen before. In his mind he was actually keeping the best player in the league in New York and happy."

That happiness wasn't universal.

"You saw that, and on the outside it must have looked like, 'Man, the USFL is kicking some serious ass,'" said Ehrhart. "But the spending was so destructive." In California, Oldenburg watched the Trump-a-Thon with corresponding feelings of hatred and envy. This was supposed to be *his* show and *his* league. Not only were the Generals building a hell of a team, they were gobbling bold headlines. "There was a lot of resentment to what Trump was trying to do," said Argovitz, the Gamblers' owner. "It seemed like he was trying to steamroll the rest of us."

A couple of weeks before the USFL's January 4, 1984, draft, Ehrhart was at his home in Greenwich, Connecticut, when he stumbled upon a small article in the *Greenwich Time*. Steve Young, Brigham Young University's transcendent senior quarterback, was in town for the holidays, staying with his family at their abode on Split Timber Place. "I'd worked as a player lawyer, so I knew the ropes," said Ehrhart. "So I called the Young house and I said, 'Steve, I want to talk to you about the USFL . . .'"

The quarterback strolled over to Ehrhart's home for an informal meeting. "If you could play anywhere, where would it be?" Ehrhart asked. "Would you want a small market, a bigger market? One of the coasts?" Though Young was unwilling to commit to the new league, he told Ehrhart he would prefer to be in Los Angeles, land of palm trees and flip-flops. A few days later, Ehrhart was speaking with Howard Cosell, the legendary announcer, who saw the USFL as something of a pet project. "I'm trying to get Steve Young to consider us," Ehrhart said. "But . . ."

Cosell interrupted. "Steven," he said, "leave it to me."

"Howard started calling Steve and his father regularly," said Ehrhart. "He told them, 'If Steve signs with the USFL, he'll be the Joe Namath of the league . . .'"

The draft's two best prospects were Mike Rozier and Young. Yet while the Nebraska running back went first overall, BYU's quarterback mysteriously lasted until the 10th slot, when he was selected by (*gasp!*) the Los Angeles

Express. "Look, it was prearranged," said Ehrhart. "In a normal world Steve is the top pick, without much debate."

Young returned to Provo, Utah, to complete his degree in international relations at BYU. While there, he was visited by Sam Wyche, head coach of the Cincinnati Bengals, who had Young work out at Cougar Stadium. By virtue of a trade with the Tampa Bay Buccaneers, Cincinnati held the No. 1 pick in the upcoming NFL Draft. "We'll get back to you," Wyche told Young — and the quarterback felt as if he had swallowed a poisonous frog. *Cincinnati?* The team was notoriously cheap, the weather was terrible, and the incumbent quarterback, Ken Anderson, was a four-time Pro Bowler. It wasn't enticing.

Steinberg, Young's agent, suggested he consider the Express. The team's new coach, John Hadl, had been a star quarterback with the Los Angeles Rams and San Diego Chargers, and its assistant coach, Sid Gillman, revolutionized the modern passing game. Klosterman was in the midst of piecing together an NFL-caliber roster. On February 19, 1984, Young drove to a small airstrip outside Provo and boarded Oldenburg's private Gulfstream V. The flight to Los Angeles lasted an hour, and Steinberg greeted him at the airport and took him to meet Klosterman at the Bel-Air Country Club. "Afterward we had dinner in the clubhouse restaurant," recalled Young, "and every few minutes somebody famous dropped by our table to say hello — former California governor Pat Brown, Los Angeles Olympic Committee president Peter Ueberroth, more actors. I kept telling myself, 'This is big-time.'" Young toured the Express's facility at Manhattan Beach, and the next night Klosterman invited him to his home in the Hollywood Hills for dinner. The owner of the neighboring house was none other than Joe Namath, and after dessert Klosterman, Steinberg, and Young visited Broadway Joe for a game of pool on his basement table. Namath told Young the story of a college quarterback out of the University of Alabama who shocked football by choosing the AFL over the NFL. He talked about the unforgettable Super Bowl III triumph over the Baltimore Colts, about carrying an upstart to glory. "Don't be afraid to join the USFL," Namath advised.

The more Steve Young thought about it, the more the Los Angeles Express made perfect sense. He would play in warm weather for two offensive masterminds on a well-stocked team in a city he enjoyed. He could start immediately and, worst-case scenario, later join the NFL as a free agent, thereby picking his team.

The next day, with Young back in Provo, Steinberg returned to the Express offices to meet with Klosterman. One year earlier the Denver Broncos signed John Elway for $5 million over five years, and the Express would have to surpass the figure to land their dream quarterback. "Honestly, I assumed Steve was going to the NFL," said Steinberg. "He grew up with a picture of [former Dallas Cowboys quarterback] Roger Staubach on his wall. He wanted to play in that league. I figured the odds of him truly going to the USFL were about zero." Still, Steinberg listened. He and Klosterman began their session at 9:30 in the morning. Later in the afternoon they retreated to Klosterman's house. "First Don talks about establishing a scholarship fund at BYU for Steve, which is music to my ears," said Steinberg. "Then he tells me he'll guarantee Steve getting into law school at Loyola, because he knew Steve wanted to go. He starts talking about all the judges he knows, the people he knows in law. After that we worked out the contract figures — it was going to be essentially $1 million a year for four years." Every few hours Steinberg took a break to call the Youngs, and every few hours the Youngs would make certain requests. "We kept going and going and going," said Steinberg. "But the tax system was so onerous at that point that Steve would be paying a top rate, like 90 percent. So we decided to try an annuity."

After much back-and-forth, interrupted by phone calls to the Youngs and dives into Klosterman's pool, the two sides agreed to a deal worth a staggering $42 million. "I was used to negotiating with the NFL, where you were lucky if you even got a bonus," Steinberg said. "This was quite a change." The specifics of the contract were creatively unprecedented:

- Young would receive $4 million up front — $2.5 million as a signing bonus, $1.5 million as a tax-free loan.
- Young's annual salaries were but $200,000 in 1984, $280,000 in 1985, $330,000 in 1986, and $400,000 in 1987.
- Young would begin to receive deferred payments, totaling $30 million, in six years, and they would not stop until he reached his 65th birthday. The money was guaranteed, whether Young stayed with the

Express after the completion of the four years or jumped to the NFL. With age, the payments would escalate. He would make $1 million in his 53rd year; $2.4 million in 2027, the contract's last payout period.

- Young would be paid $100,000 per year for four years as part of an endorsement deal with the State Savings and Loan Association of Salt Lake City.
- BYU, Young's alma mater, would receive a $180,000 contribution for a scholarship fund, paid over 20 years.

For good measure, the Express signed Young's best friend, BYU tight end Gordon Hudson, to a two-year, $530,000 deal — even though Hudson would miss the first season recovering from a knee injury. "They wanted Steve badly," said Steinberg. "It was no joke."

Klosterman had been around for decades, and he knew how these things worked. He was far from offended when Steinberg told him he had to speak with Mike Brown, the Bengals' assistant general manager, and present him the opportunity to match the Express offer. The agent excused himself and called Cincinnati from Klosterman's home phone.

Steinberg: "Mike, the USFL is offering Steve a package worth about $40 million, and I wanted to see if your team wants to counter . . ."

Brown: Silence.

Steinberg: "Mike, you there?"

Brown: "I'm here. We'll match — as soon as we discover oil beneath Spinney Field."

Steinberg: "I'll take that as a no."

Brown: "Look, I'll pay you the biggest contract in the history of the NFL for a rookie, but not the money they're offering. That's insanity."

Steinberg again called the Youngs. It was seven o'clock in the morning. "The money is legitimate," he said. "They have a huge Mormon population in Los Angeles, and Steve can live in Manhattan Beach. I never tell a client what to do, but . . ."

Steve Young agreed. He would sign with the Express.

Two days later, Young and Steinberg found themselves entering the San Francisco–based headquarters of Investment Mortgage International, Oldenburg's company. Scrolling above the front doorway on the electronic ticker were the words: WELCOME STEVE YOUNG, QUARTERBACK EXTRAORDINAIRE . . . MR. BYU, MR. UTAH, MR. EVERYTHING . . . WELCOME LEIGH STEINBERG, LAWYER PAR EXCELLENCE. Steinberg and Martin Mandel, the company's chief corporate counsel, excused themselves to talk contract, and Young found himself alone with Mr. Dynamite. Recalled Young: "First he showed me his trophy room, which included pictures of him with everyone from Henry Kissinger to Captain Kangaroo to Wayne Newton. Then he took me to his office. It was more like a luxury suite. There was leather furniture, a fully stocked bar with a marble countertop, and big windows that offered a stunning view of the city."

Although negotiations in professional sports are often portrayed in television and cinema as sexy and exciting, they're largely drudgery and paperwork. Somehow, Oldenburg's staff failed to inform their boss that much needed to be done before Young officially joined the Express. "He was upstairs in his office, and we're on another level with four lawyers and a lawyer I brought with me," said Steinberg. "Steve's off in a conference room studying for his microeconomics exam. And it's taking time, as it should, but Oldenburg has no patience. Little did we know it was his birthday, and he had people waiting for a birthday dinner. I'm prepared for a long evening, but Oldenburg's not. Because nobody told him this would take hours." At six o'clock, Oldenburg stormed into the room and screamed, "Why isn't this kid signing the fucking contract! What are you guys doing? What's the hang-up?"

One of the attorneys said, softly, "Well, we're talking through the guaranteed language" to which Oldenburg shouted, "Guarantees? Here's all the fucking guarantee you'll need!" He opened his wallet, which was stuffed with $100 bills, threw the money at the lawyer, and stormed out.

This was a first for Steinberg. "Did you explain to him that this is complicated?" he asked one of Oldenburg's employees.

"Well, not really," he said.

"This is going to be an interesting night," Steinberg said.

An hour passed, and Oldenburg returned to the room just as Steinberg was wrapping a phone call with LeGrande Young, Steve's father.

"Who are you on the phone with!" Oldenburg yelled.

"LeGrande Young," Steinberg said. "Steve's father."

"Well, you tell him to shove the Mormon temple up his ass!" Oldenburg railed, before once again exiting.

Steinberg asked that someone sedate Oldenburg, but was met with silence. The attorneys were intimidated, and aware of their employer's proclivity for firing people on a whim. Negotiations resumed, and at 11 o'clock Oldenburg was back. This time he was clearly intoxicated. His eyes were red, his speech slurred. "If you don't get this done in the next fucking hour," he yelled, "this deal is off the table!" *Bam!* — door slammed.

The talks continued. And continued. And continued. Steinberg pleaded with the attorneys to send Oldenburg out for dinner — "Tell him we won't sign anything without him" — but fear trumped logic. At 3:00 a.m., Oldenburg (via intercom) ordered, "Get Steve and Leigh in my office, right fucking now!" The two men took the elevator to Oldenburg's suite, where he sat atop his desk, scowling. The veins were bulging from his neck. His cheeks were puffy and fruit-punch red. "I've offered you the most money any athlete in history has gotten, and you won't take it!" he yelled. "You may not be good enough to play for my team!" By now, Oldenburg was jabbing Young in the chest with his doughy, stumpy finger. He called him a "fucking Mormon."

Young was the nicest athlete Steinberg had ever represented. He was polite, humble, genuine. At that moment, however, he snapped. "Mr. Oldenburg," he said, "with all due respect, if you do that again I'm going to belt you."

Oldenburg was, for the first time, momentarily speechless.

"You're snubbing me," he said to Young. "You're turning this money down."

Young shook his head. "I'm not doing anything. I'm studying. I don't know what's happening."

Oldenburg's office included a long glass bar, covered with dozens of jars and bottles. The owner drew back his right arm and smashed everything in sight. Snowflakes of glass seemed to hover through the air. He picked up a chair and ran toward one of the windows. "We're 30-something floors up," Steinberg said. "Steve grabbed him, and Oldenburg is swearing, cursing, screaming, 'Security! You guys are no fucking good! Security! Security!'"

It was 4:00 a.m., and Leigh Steinberg and Steve Young were escorted from the premises and left to fend for themselves on the dark streets of San Francisco. Klosterman, in town for what he presumed would be a celebratory announcement, received a call from Steinberg, asking for a ride to his apartment in Berkeley. He sped through the night, horrified. "I don't even know what to say," Klosterman moaned. "I'm so sorry."

Young was dumbfounded. "What just happened?" he asked.

"Steve," Steinberg said, "as awful as that was, try and look at it from Oldenburg's standpoint. His staff didn't prepare him, it was his birthday, he had people waiting, and he offered us the biggest contract in the history of sports. Plus, he was drunk."

Young and Steinberg spent the following day at the apartment. Word of the meltdown somehow leaked, and the phone never stopped ringing. Pete Rozelle called — urging Young to wait for the NFL. Staubach, Young's hero, called on behalf of the NFL, too. He encouraged the quarterback to go with the established league. Cosell reached out — "Sign with the USFL." Namath called again — "I'd stick with the USFL." Finally, in the early evening, Bill Oldenburg called. He had sobered up, and was unusually contrite. "I owe you guys an apology," he said. "There's no excuse for my behavior last night. I want you on the Express. We all do . . ."

The debate that followed was long. Young *did not* want to go to Cincinnati. He *did* want to work with Hadl and Gillman. He *loved* Klosterman, *disliked* Oldenburg. But, really, how often would he have to deal with the owner? Finally, Young called his father.

"Accept the offer," LeGrande Young said.

So he did.

On March 5, 1984, approximately 100 journalists jammed into a ballroom at the Beverly Hills Hotel to meet the new quarterback of the Los Angeles Express. Only the new quarterback of the Los Angeles Express didn't want to meet anyone.

The news conference was called for 11:00 a.m.

Ten minutes passed.

Then 20.

Then 30.

Young looked at Steinberg, his agent, as the two stood in a side room. "I'm not going out there," he said.

"Come on," Steinberg said. "We've been through this a million times."

"Sorry," he said, "but I'm not going through with it."

"Just sit tight," Steinberg said. "Don't go anywhere."

Steinberg exited the room and entered the ballroom, where Oldenburg and Klosterman were sitting at a long table — simultaneously stalling and bragging. "I can't remember another signing that would match this!" Klosterman told the assembled throng. "I guess you'd have to go back to Joe Namath's signing with the New York Jets. It's a tremendous step forward for the league."

Alone, Young paced nervously. He felt like vomiting. The money, the fame, the attention, the Broadway Joe comparisons — none of it was him. Young was the guy who kept a couple of dollars folded up in the pocket of his ripped blue jeans. He drove a 19-year-old Oldsmobile Dynamic (police in Provo, Utah, once tried to tow the vehicle because they presumed it had been abandoned), and if the muffler dragged, hey, the muffler dragged. He had yet to own a credit card, and still wasn't a college graduate. Now, with the bright lights awaiting, Young knew he was making a monumental mistake. *Maybe,* he thought to himself, *I should dump football and go to law school.* When Steinberg returned, Young shared the idea. "I can just be normal," he said. "Get married, attend school . . ."

"Steve, you've never been satisfied with that," Steinberg said. "You are a competitor. You've always dreamed of playing football on the big stage. Pro football is the biggest stage. You have to prove yourself there." By now, Young was sitting on a couch. Steinberg reached out his hand. "Come on," he said.

The quarterback rose, and walked toward the ballroom. He wore a beige blazer and a pained smile. After being introduced as "the newest member of the Los Angeles Express," Young was peppered with questions about money, and money, and money. No one seemed interested in his throwing motion, his footwork, his plans for USFL success. No, the dollars were the story, and

Young was horrified. When asked what he planned on doing with his new riches, he swallowed hard, rubbed his palms against his pants, and said, "I hope to fix up my car and take my girlfriend out to dinner for the first time in four years . . ."

There were chuckles. Not many, but a few.

"Look," he added, "the Express is a topnotch operation, and I felt like I would be well-coached. A quarterback couldn't ask for a better offense than this one."

Nobody bought it. This was about the money. Big money. Money Steve Young could not turn down. He knew it, Steinberg knew it, Oldenburg knew it, Klosterman knew it, and every member of the working media inside the room knew it.

"It was brutal — just brutal," Young said years later. "Leigh won't tell you this and I won't tell the whole story, but the idea that they would take this contract that, really, was an annuity that would take decades to pay off . . . the Express added it up and claimed some $40 million figure to make the world go crazy — it was such a disservice to a 22-year-old kid who didn't care about money."

The following day, Young's contract turned national joke. He was either called "the $40 million man" or "the $40 million quarterback" — neither of which thrilled him. Jerry Greene, the *Orlando Sentinel* columnist, ridiculed the whole affair, noting, "The funniest part of Steve Young's debut as a self-made millionaire is the universal claim by everyone involved that money had nothing to do with Young's decision to become a member of the LA Express for $40 million. That's like Burt Reynolds claiming he was dating Loni Anderson because of her mind or Hitler saying he invaded Poland because he needed a summer home."

Ouch.

When the press conference came to a merciful end, Young and Steinberg found themselves standing alone outside the Beverly Hills Hotel. The media had left, Oldenburg and Klosterman had left. It was official — Young was the new Express quarterback. Before they parted, Steinberg had one last thing to do. He reached into his case and pulled out an envelope. Inside were a pair of checks from the Express — one for $1 million, the other for $1.5 million. Young stuck them in a pocket, caught a car to the airport, and a plane back to Provo.

America's highest-paid athlete still had to wrap up his classes.

Generally speaking, when a professional sports team signs a superstar, league officials go out of their way to heap praise upon the addition. The Young inking, however, sounded the alarms inside the USFL offices.

Ever since he was hired to serve as commissioner, Chet Simmons insisted the goal was to keep salaries in check and build slowly. Now, with the certification of Young's contract, everything was spinning out of control. "As commissioner, I don't like it," Simmons told Bruce Lowitt of the Associated Press. "I do not think it is in the best interest of professional football or the league, but what can be done? These are wealthy businessmen seeking to build and sell a product and to compete."

He was far from alone in his angst. In Tampa Bay, John Bassett was beginning to see that the league might not last for long. Back in the USFL's dark ages, a memorandum on salary levels suggested $1.6 million per club in 1983, then gradual growth to $1.75 million and $2.5 million over the ensuing seasons. The entire concept had been ignored, and troubles loomed. Bassett insisted upon restraint and reason, only no one was listening. The Young contract was a slap in the face of caution. "Sports history is full of the likes of this," Bassett said. "It has established that you end up with one league or you go out of business. History also shows that salaries never go back to an old level. All this reminds me of the old commentary that goes, 'When a man with money meets up with a man of experience, the man with experience usually ends up with the money and the man with the money ends up with experience.' I have no intention of going bankrupt signing wealthy players."

Added William Dunavant Jr., owner of the Showboats: "This doesn't make any economic sense to me. I simply cannot conceive of anything like this."

Were Simmons's only concern the spending habits of a few rogue owners, perhaps he could have figured out some sort of solution. There was, however, an even bigger problem brewing, and it involved Trump, a man Simmons quickly came to both dislike and mistrust.

When Donald Trump agreed to buy a team, he uttered nary a word about his desire to see the USFL move to the fall and challenge the NFL. It wasn't written on any application, expressed in any sit-downs with Simmons, stated from the podium of his introductory press conference. The goal, he made abundantly clear, was to build the Generals into a terrific USFL operation. That sounded just peachy to the commissioner. Wrote Jim Byrne in his book, *The $1 League*: "Simmons believed that Donald Trump would be a great addition to the USFL ownership group. Trump was bright, young, articulate, and a very rich real-estate developer who built a lucrative family-owned business into one of the best-known in the world . . . [Trump] had paid close attention to the USFL's development during the first year and apparently liked what he saw."

Yet behind closed doors, Trump explained to confidants that he did not aspire to build a great USFL team — *but a great NFL team*. Just recently the New York Jets relocated from Shea Stadium in Queens, New York, to the Meadowlands in East Rutherford, New Jersey. With that move, New York City — America's grand metropolis and No. 1 television market — no longer had a National Football League presence. As opposed to USFL franchises, with their $6 million average value, NFL teams were gold mines. That year, the Denver Broncos sold for $78 million and the San Diego Chargers for $70 million. In 1989 the Dallas Cowboys would sell for a record $150 million. The NFL, though, wasn't simply giving away clubs, and inside league offices Trump was dismissed as a hot air balloon.

The New Jersey owner, however, had a specific strategy. He would purchase the Generals, build them into the class of the USFL, watch the upstart league either fold or merge with the NFL, and eventually relocate to Shea Stadium until his new facility was constructed in Manhattan. The end result: the NFL's New York Generals, hosting 80,000 red-and-white-clad fans in the state-of-the-art Trump Stadium.

The motive first became apparent to Simmons a week after the Generals' sale, when he received a call from William N. Wallace of the *New York Times*. The reporter wanted to know how the commissioner felt about Donald Trump's plan to have the USFL shift to the fall.

The what?

Read the ensuing article, headlined TRUMP WOULD LIKE TO TAKE ON NFL:

> Donald J. Trump, the new owner of the New Jersey Generals of the United States Football League, said yesterday that he favored the league playing its games eventually in the fall and winter rather than in spring and summer, thus competing directly with the National Football League.
>
> The 37-year-old real estate developer described what he called "a possible scenario" in an interview in his office on the 26th floor of Trump Tower on Manhattan's Fifth Avenue.
>
> The new league "in two or three years" will achieve parity with the N.F.L., he said, and the U.S.F.L. could then "perhaps go head to head with the N.F.L."
>
> Trump also suggested that the U.S.F.L. might "challenge the N.F.L. for a game" in some kind of a championship format. "In my opinion," he said, "some U.S.F.L. teams could beat N.F.L. teams now."

The piece was not well received by Simmons or the USFL owners. In a letter to a handful of his peers, Tad Taube of the Oakland Invaders warned that "it may be in Don Trump's best interests to pursue a strategy which gains him leverage, politically or otherwise, to move to Shea Stadium and become the NFL franchise that the City of New York is apparently ready to underwrite at any price. But Don's best strategy for the Generals could be devastating for the USFL as a whole." Still, Trump was given an early pass. He was new and eager, and sometimes a person's thoughts don't come across correctly. Then, on January 18, 1984, he showed up at the league meetings in New Orleans. Inside a conference room at the Hyatt Regency Hotel, Trump told his peers that the league had been "heading very rapidly downhill" . . . *until his arrival.* According to the book *Trump Revealed,* Trump alone, "by his own account, had orchestrated a complete flip in the perception of the USFL among fans, reporters, and the people who controlled the purse strings at major television networks." Trump added that he was tired of hearing other

USFL owners upset by his spending habits, and that they either needed to join him or get out of the way.

The words were troubling. What followed was nuclear. Trump told his colleagues that the USFL — a league he had barely been a part of — needed to move to fall and directly challenge the NFL. Not in 10 years, not in 5 years. *Now — as soon as possible.* "I guarantee you folks in this room that I will produce CBS and I will produce NBC and that I will produce ABC, guaranteed, and for a hell of a lot more money than the horseshit you're getting right now," he said. "Every team in this room suffers from one thing: people don't want to watch spring football . . . you watch what happens when you challenge the NFL . . . I don't want to be a loser. I've never been a loser before, and if we're losers in this, fellas, I will tell you what, it's going to haunt us . . . Every time there's an article written about you, it's going to be you owned this goddamn team which failed . . . and I'm not going to be a failure."

A small handful of owners found themselves nodding in agreement. The vast majority did not. First, because the plan from the beginning had been spring football. Second, because the plan from the beginning seemed to be largely working. Third, because . . . what the hell did Donald Trump know? At 37, he was still a kid; one who became wealthy off of an inheritance from his father; one whose sports background consisted of playing a little squash as an undergrad at Fordham University. "He understood nothing about football," said Mike Tollin, who produced the league's weekly highlight show. "Far less than your average fan."

Anyone who presumed a new owner would tiptoe cautiously was sadly mistaken. Over the months leading up to the kickoff of the second USFL season, Trump was everywhere — talking up a move to fall, talking down Simmons, signing NFL free agents, egging on owners who didn't share his goals and ideals.

"He was great entertainment," said Charley Steiner, the Generals' announcer. "But he was also poisonous. He didn't care about the league, about his players. He cared about one person, and one person only.

"He cared only about Donald J. Trump."

9

WILD FIELDS

If you've seen *North Dallas Forty,* you might understand the USFL.

— Scott Boucher, offensive lineman, Houston Gamblers

ENTERING ITS SECOND season, the USFL was strange.

That sounds like a blanket statement. And, indeed, it is a blanket statement. But of all the adjectives one could use to describe an entity that swapped a franchise for a franchise, expanded by six after one year, paid a college quarterback $40 million, ran a fixed draft, and featured a crazy, power-hungry New Yorker and a crazy, power-hungry Californian as owners . . . "strange" works quite well.

Yet for all the oddities in all the places, nothing — and no one — touched the man known as Big Paper.

Or Paper, for short.

His name, officially, was Gregory Keith Fields, and only the most die-hard of football fans knew of his existence. But for the 17 people who paid close attention to the 1983 Los Angeles Express, Fields — a six-foot-six, 265-pound defensive end — was something to behold. On an otherwise mediocre unit that ranked eighth of 12 teams in total defense, Fields tallied 12½ sacks, coupled with 63 tackles. He was big, strong, fast, durable, mean — and, according to Danny Rich, the team's best linebacker, "absolutely nuts. Like, crazier than batshit crazy."

Fields's path to Los Angeles *oozed* United States Football League. An escapee from the drug-infested San Francisco neighborhood of Bernal Heights (the center of which is, appropriately, a psychiatric hospital), Fields spent two years at Hartnell College before earning an athletic scholarship to Grambling University. A good — not great — pass rusher for the Tigers, in 1979 Fields signed a rookie free-agent deal with the Baltimore Colts, and made $25,000 to play special teams and occasionally give the starters a breather. One day, he was sitting in the locker room when Barry Krauss, the franchise's first-round pick out of Alabama, walked by with Steve Heimkreiter, a rookie linebacker from Notre Dame. The two were heading for a meeting, and Fields looked at Krauss — who was being paid $1 million over six years — and said, "*Wellllllll*, fuck — there's the big paper boy!"

Krauss laughed. "Big paper! Big paper!" he barked back toward his new teammate.

"Then I started calling Barry and Steve 'the big paper boys,'" Fields recalled years later. "That just meant they got big bonuses while I was making shit. Well, once two-a-days started they reversed it and started calling *me* Big Paper. The rest of the dudes didn't know what it meant, just the three of us. But they were being ironic, because I was the lowest-paid player on the team."

With that, a nickname was born. Greg Fields was Big Paper. But he was also Big Nutso. Fields was loud and brash and often irrational. He talked trash that made little sense and picked fights for no good reason. "Paper was a nice guy, but out there," said Randy Burke, a Colts wide receiver. "He would cash his check every week and carry around the bills like he was loaded."

"He would go to the A&P and bounce checks," said Bruce Laird, Baltimore's strong safety. "He just wasn't the brightest bulb in the box."

The Colts let Big Paper go after the 1980 season, and he signed as a free agent with the Atlanta Falcons. According to Fields (who leans toward exaggeration when it comes to his football prowess), "I owned that training camp. In Atlanta nobody could smell my jock. I destroyed everyone in my path." Yet on the morning of August 12, 1982, Fields was eating breakfast in the team's Suwanee, Georgia, facility when Jim Stanley, the defensive line coach, told him to wrap his meal, grab his playbook, and go see Leeman Bennett, the head coach.

"They were cutting me," he said. "Total bullshit."

Fields refused to comply, instead returning to his room inside the Falcon Inn. Ten minutes turned to 20, then 20 to an hour. Finally, Stanley knocked on Fields's door. The lineman would not open, which resulted in the following between-a-closed-door dialogue.

Stanley: "Greg, we're cutting you."

Fields: "I'm not leaving."

Stanley: "Really, you need to see Coach Bennett."

Fields: "Fuck off."

Stanley informed Tom Braatz, the general manager, who called the local police department and asked for the assistance of an armed officer. The next knock on Fields's door came via Braatz. He stood alongside a cop, and held Fields's plane ticket in hand. "I have the police with me," he said. "You might want to open up . . ."

Greg Fields's NFL career was over. He was blacklisted from the league, never to receive another shot. The USFL, though, wasn't the NFL. It enlisted your tired, your poor, your huddled masses, your one-armed and chain-smoking and half-blind and clinically insane. So when the league began, the Express signed Fields, then enjoyed as he played and behaved well. "He was a good guy," said Al Burleson, a veteran defensive back with the 1983 Express. "He was just this big gentle giant. All love."

"Greg was something," said Tom Ramsey, the Express quarterback. "Before one game the doctor was going around giving players B12 shots. Fields yells out, 'Hey, Doc, I need a double dose! Give me a B24!'"

The second training camp was different. First, with all of Bill Oldenburg's signings, the Express were a far deeper team. Second, Hugh Campbell had been replaced as head coach by John Hadl, the former NFL quarterback who employed a more hands-on approach to running a club. He watched the defense with a close eye, and conferred with Pokey Allen, the unit's coordinator, on individual players. At 29, Fields seemed a step slow and a tad dumb. He botched assignments, performed sloppily against the run, talked ceaseless nonsense. Finally, as they began to pare down the roster, the coaches decided Fields needed to go.

On the morning of Wednesday, February 22, Hadl arrived at the team's Manhattan Beach facility and reviewed the list of players who would be dismissed. When he broached Fields's name, an executive with the team warned, "Paper isn't going to take this very well."

"Really?" Hadl asked.

"Yup," the employee replied. "I'd have someone in the office with you."

At six foot one and 230 pounds, Hadl was no shrinking violet. He'd absorbed many crippling hits over a 16-year quarterbacking run with the Chargers, Rams, Packers, and Oilers, and feared not a journeyman lineman. Plus, at 44 he was young enough to hold his own in a scuffle. "I told two guys to stand *outside* the door," Hadl said. "Just in case anything happened and we needed to get him out."

Hadl summoned Fields, who sensed what was coming. A day earlier the Express had released Burleson, which enraged the lineman. "Al was my best friend on the team," Fields said. "It was bullshit. I wasn't in the best mood." Moments before entering Hadl's office, Fields spoke with Danny Rich, the linebacker. "If they dump me," he told him, "they're gonna have to get the National Guard to pull me out of here."

There is a robotic application to the typical player cut, one followed for decades by nearly all engaged parties in all professional team sports. The coach invites the soon-to-be departed to sit down. The soon-to-be departed gulps. The coach explains how impressed he was by the soon-to-be departed, but that it comes down to numbers. The soon-to-be departed thanks the coach for the opportunity, shakes his hand, leaves the room, packs his belongings, and (oftentimes) cries. "I've cut many players," said Hadl. "You pretty much know how it goes."

This time, Greg Fields entered scowling. He was wearing a gray Express T-shirt and blue shorts. "Have a seat," Hadl said.

"I'd rather not," Fields replied.

"OK," the coach said. "Well, we're letting you go. It's nothing you did wrong. We just . . ."

Fields turned toward the door and slammed it shut. He charged Hadl, drew back his right fist — bulky gold ring wrapped around the thumb — and swung it toward the coach's face, connecting enough to open skin and draw blood. "He got me on the cheekbone — that's all," said Hadl. "It wasn't so bad." The two men stationed outside the door burst in at the same time Hadl was popping Fields in the mouth with a counter left. "It was craziness," said Derek Kennard, the rookie offensive lineman. "That was my first day with the team, and security is everywhere, scrambling around. It's like, 'Oh my God, what have I walked into here?'"

"[Moments later] Hadl enters the room for the quarterbacks meeting and he has a rabbit-sized mark on his face," said Ramsey. "We were all shocked. He said, 'I cut Fields, and Fields didn't like it.'"

"It was," said Ricky Ellis, the Express tight end, "the weirdest thing ever."

Fields was escorted from the facility and ordered never to return. As he marched toward his car, he promised he would be back to "kill" Hadl and "beat down" Keith Gilbertson, the defensive line coach. Later that day Klosterman had someone in the organization reach out to Nelson Mercado, a well-known local security professional who, at the time, was working as a private bodyguard for Liberace, the flamboyant pianist. "They called and told me a disgruntled ex-player wanted to assassinate the head coach," said Mercado. "I'd heard a lot of things in my career. But that was a new one." Mercado said he advised the Express to contact the police, and that a team representative replied, "We did — but the cops won't handle this the way you would."

Mercado requested a leave of absence from Liberace, who was living and performing in Las Vegas, then made the four-hour return drive to his Southern California home. Although Fields later denied the claim, Mercado said the former lineman called in a threat to come to the facility on a random day armed with a six-inch-barreled Smith & Wesson .357 Magnum revolver and end Hadl's life. "It was no joke," said Mercado. "We followed him. I would report to the Express every day on Greg Fields's whereabouts. I had certain things put on his vehicle — tracers. He also had a beeper, and we put a tracer on his beeper. I had a crew of about six people working just to know his whereabouts. He would park his car two or three blocks away from the facility. He called me once and made the point of telling me he had a long rifle and was able to kill me and kill anybody on the practice field. He was not stable psychologically. The guy was definitely 5150."*

Greg Fields became Nelson Mercado's 24-hour-a-day obsession. Wherever the ex-Express lineman went, the security guard followed. He couldn't get his mind off of Fields, and came to the dark realization that this was his lot in life — to track and trace a pass-rushing whack-job with a gun fetish.

Then, in a this-can-only-happen-in-the-USFL moment, Mercado was granted a gift from the football gods.

On March 4, the San Antonio Gunslingers, desperate for warm bodies, signed Greg Fields to a free-agent contract. When he flew to Texas and entered the team's facility at Alamo Stadium, Fields was greeted by members of the coaching staff dressed in pads and helmets. On the wall was a newspaper clipping with the headline PLAYER PUNCHES EXPRESS COACH.

Fields grimaced. "Aw man, you got my shit on the wall," he said. "That's fucking cold."

Greg Fields lived to see another day.

Unlike its first campaign, which went straight from training camps into opening day, in 1984 the USFL decided a preseason would help clean the sloppiness. Therefore, each team would engage in two preseason games before kicking off the 18-game slog. This was good news for many franchises, but an absolute necessity for the league's most experimentally fascinating operation, the Houston Gamblers.

From the very beginning, the Gamblers defied convention. In Jerry Argovitz, Houston had the world's most outlandish dentist turned agent turned owner, and in Jim Kelly, the league's cockiest quarterback. But it was one of the organization's first hires that set Houston off on a different path.

Back in 1980, when he was still working as an agent, Argovitz sent a colleague to Portland State to watch a Division I-AA quarterback named Neil

---

* Section 5150 of the California Welfare and Institutions Code authorizes a qualified officer or clinician to involuntarily confine a person suspected to have a mental disorder that makes them a danger to themselves or to others.

Lomax. The kid was impressive enough to go to the St. Louis Cardinals in the second round of the 1981 Draft, but it was the Vikings' offensive numbers that caught Argovitz's eye. Under the guidance of an obscure young head coach named Darrel "Mouse" Davis, Portland State lit up opposing defenses with a system that felt downright incendiary. It was revolutionary stuff — four- and five-wide receiver sets, no tight end, no fullback, oftentimes no huddle. In one particular game, against Northern Colorado, Lomax completed 44 of 77 passes for 499 yards. As a senior Lomax led the Vikings to 93–7 and 105–0 triumphs. "My partner brought back tape, and, holy cow, what was I seeing?" Argovitz said. "It was this thing called the Run and Shoot offense — no one had ever heard of it. But one thing I knew was fans would rather see a team win 51–42 than 6–3. So as soon as the Gamblers began, I decided I would chase down Mouse Davis."

By this time Davis was working as the offensive coordinator of the Canadian Football League's Toronto Argonauts, and Argovitz brought him to town for an interview. The men spent two full days together, reviewing the Xs and Os of the Run and Shoot. The offense depended upon quick drops, speedy routes, and overmatched defenses lacking the personnel to keep up. "I was convinced, with the right personnel, that offense couldn't be stopped," said Argovitz. "Spread the field out, the defense is lost."

Davis had loved the USFL's first season, and asked Argovitz whether he could serve as the Gamblers' head coach.

"I'm sorry," the owner said. "I can't do that. I want you as offensive coordinator."

"Well, who's the head coach?" Davis asked.

Argovitz shook his head. The Gamblers had yet to name one.

"Wait, wait," Davis replied. "You're going to hire an offensive coordinator before you hire a head coach? That makes no sense."

"Mouse, here's my promise," Argovitz said. "You come here, it's 100 percent your offense."

Davis agreed to a contract. In the name of secrecy, there would be no press conference or news release. He was allowed to pick a wide receivers coach (June Jones, who was the quarterbacks coach at the University of Hawaii), but had no say in the head position. That's why he was flabbergasted when the Gamblers hired Jack Pardee, the former Chicago Bears and Washington Redskins head coach who enjoyed long passes and high-powered offenses the way he enjoyed crabgrass and iodine (he cared for neither). Pardee and Davis comprised the odd couple of coaching tandems. Not only had they never met — Pardee was unaware of Davis's existence. "We gave Pardee $100,000 to be our coach," said Argovitz. "But before he signed I said, 'Jack, there's one thing I have to tell you — you already have your offensive coordinator.'"

The news was not well received. Pardee liked offenses that featured a power running back, à la Walter Payton, and a dominant offensive line. Instead, he was being told the Gamblers planned on employing a fleet of tiny yet quick wide receivers. "One day Jack came to me and said, 'You don't have a tight end,'" recalled Davis. "I told him he should sign a tight end if he wanted one. So we did, and I never ran a play using the guy. Not one. After three games Jack cut him. He deserves a lot of credit — he adjusted, and it couldn't have been easy."

The key components to Davis's system were a strong-armed quarterback and a boatload of pass catchers. The first had been taken care of. The second turned into a riveting search-and-rescue program that Argovitz likened to a nationwide game of hide-and-seek. At the time, the NFL was going through a long-ball revolution, and every team seemed to be adding the type of receivers the Gamblers coveted. In Miami, for example, the Dolphins lit up the league with the gnat-like duo of Mark Clayton (5 foot 9, 170 pounds) and Mark Duper (5 foot 9, 185 pounds). Roy Green, the Cardinals' go-to burner, was 5 foot 11, as were the Jets' equally fast Wesley Walker and Atlanta's Stacey Bailey. Every franchise craved its own Mighty Mouse, which led to the belief that the cupboard was mostly bare.

So the Gamblers thought creatively.

Ricky Sanders was an obscure 5-foot-11, 180-pound halfback at Division II Southwest Texas State. He ran for an unspectacular 665 yards as a senior, and few NFL scouts knew of his existence. Only one, the Buccaneers' Bill Groman, watched him on multiple occasions. "He was too small to be a running back in pro football," said Groman. "But I always thought of him as a wide receiver." When the Gamblers hired Groman as their director of player personnel, he urged Argovitz to pounce. Sanders ran a 4.3 40. Surely, that was worth something. "I'm a small-school guy," Groman said. "I like taking guys who are overlooked and making them something."

"Rick twisted his knee one time and he was out for several weeks," said Todd Dillon, Houston's backup quarterback. "He wasn't sure if he was ready to return, and he ran a 4.3 40 . . . *limping*. I swear — he was limping."

Like Sanders, Richard Johnson had been a little-known and undersized college halfback at the University of Colorado. He was, at five foot seven and 182 pounds, shaped like a legged lumpy pear. The NFL could not have been less interested, so he spent 1983 on the Denver Gold bench, smoking joints and appearing in a single game. When Houston selected him in the 10th (and final) round of the expansion draft (each of the six new teams could take 24 players left exposed from the existing franchises), he presumed his undistinguished rushing career would continue. "I went to my first meeting and Mouse had the depth chart on the board and I didn't see my name at running back," Johnson said. "He said, 'We have you at wing now.' Wing? What the hell is a wing? But all of a sudden, instead of running into linebackers and linemen, I'm going up against guys my size. I liked it."

Clarence Verdin was listed at five foot eight and 160 pounds, but the figures were exaggerated. Like Sanders, his obscure-school background (he attended Southwestern Louisiana State) made him a long shot. Like Johnson, his size made him a no-shot. "I thought the Dallas Cowboys might draft me, because their scout came to our practice for a long time," Verdin said. "But it never happened." The Gamblers grabbed him in the 17th round of the collegiate draft — a flier on a nobody. "He wasn't as big as a popcorn fart," said Ray Alborn, Houston's defensive line coach. "But, boy, was he super fast."

Not quite as fast as Vince Courville, who played football at Rice but was better known as one of the world's elite sprinters. Courville clocked a blinding 4.2 40. He also had hands of cement. "In college they'd say, 'Don't throw a football at Vince, throw him a baton,'" said Verdin. "It's the only thing he can catch." That's why Courville was cut by the Canadian League's Montreal Concordes, then cut again by the Los Angeles Raiders. The Gamblers signed him up on a lark — "and then I ran a 4.31 for Pardee, and I was in," Courville said. "They weren't looking for a wide receiving unit. They were looking for a 4x100 relay team. And to tell you the truth, we had a 4x100 unit with the Gamblers that could have challenged any country in the world."

The two most unlikely members of Houston's pass-catching unit were Scott McGhee of Eastern Illinois University and Gerald McNeil of Baylor. McGhee had spent two unremarkable years doing little with Toronto and Hamilton of the CFL, but that's not what made him stand out. "He was a tough white boy who could fly," said Scott Boucher, a Gamblers offensive lineman. "I mean, he was tiny and quick and fit right in." McNeil, too, had no business playing professional football. He stood five foot six, weighed 137 pounds, and was primarily known for returning kicks. His nickname: "the Ice Cube."

"I'd never, ever seen anything like Gerald," said Dillon. "Up, down, left, right—he would buzz all over the field and nobody could find him. Just electric."

Houston was so infatuated with speed that it auditioned Stanley Floyd, the two-time NCAA 100-meter sprint champion at Auburn and the University of Houston. Though Floyd had never played a down of college football, he recently raced a horse on the television show *That's Incredible* . . . and won. That was good enough for Argovitz. "He wasn't a very good player," said Dennis DeVaughn, a Gamblers defensive back. "But, Lord, Stanley Floyd could run."

The wide receivers became the league's tightest unit, and their melding with Kelly was something to behold. Equally unfamiliar with the Run and Shoot, both the quarterback and his pass catchers struggled to adjust. During his college career at Miami, Kelly dropped back, sat in the pocket, and threw. Now, as Davis hounded him to roll out and make the quick read, the quarterback resisted. Lomax, excelling as a quarterback with the St. Louis Cardinals, traveled to Houston's camp on several occasions to serve as a covert translator. It did not help. "Mouse came to me one day, really upset," said Argovitz. "He said, 'Jim's not listening to me. He won't give the offense a chance.'"

"Don't worry," Argovitz replied. "I'll take care of it."

The next day the Gamblers traded with Los Angeles for the rights to Dillon, who had been a record-setting quarterback at Long Beach State. The two sides quickly agreed to terms, and Dillon arrived, unpacked his bags, and grasped the offense in a matter of hours. A furious Kelly demanded a meeting with Argovitz. "This is bullshit," he told the owner. "You signed me to be the starter."

"Look," Argovitz replied, "if this whole thing doesn't work, you and I are gonna be the laughingstocks of professional football. I'm gonna be the stupid schmuck who paid you $3.5 million to sit on the bench and you're gonna be the stupid schmuck who couldn't beat out a kid making $30,000."

Kelly pounded the table and yelled, "No one is taking my job!"

"Well, then, you better get used to this offense," Argovitz said, "because we're married to it."

A week later, a beaming Davis knocked on Argovitz's door. "What the hell did you say to that kid?" he said. "Kelly's unbelievable. The big kid has got it. He's dancing, rolling, his footwork is unbelievable, his accuracy is great, the leadership. Doc, what did you say to him?"

Argovitz smiled. "Mouse, it's Psychology 101," he said. "Jim's a competitor. He needed to feel threatened."

The increased comfort led to increased happiness. Boucher had recently earned his real estate license, and he sold Kelly a 4,529-square-foot mini-mansion on Springhill Lane in the Houston suburb of Sugar Land. It became the Gamblers' unofficial bonding headquarters, a place where football players of all ages, shapes, sizes, and races would congregate by the pool and drink and smoke and have sex. "We had parties there every single week," said Boucher. "The offensive line had a deal, where if we gave up less than two sacks a game we each got seven rib eyes. So we had more steak at those gatherings than you've ever seen . . ."

"It was craziness," said Courville. "One day we're all at Jim's house and a couple of guys are taking shots of Glenlivet scotch. I join in, and we get into a car. Richard [Johnson] is in there, and he has two chicks. And I'm in the backseat trying to get my mack on with this girl. Suddenly I have to barf, and I barf out the door on the ground. I close it, and now I'm trying to get my mack on with vomit breath. To this day, if anyone offers Glenlivet scotch I'm a definite no."

Once, after a game against Arizona, the entire Wranglers cheerleading squad came to Kelly's house — sans the Wrangler players.* "It was the best place on earth," said Johnson. "We'd spend the entire night at Jim's. We usually couldn't go home, so we'd all take a room."

Kelly didn't hold back. He dated a starting lineup of Lone Star State beauties, including Laura Shaw, the reigning Miss Texas. Thanks to an endorsement agreement with a local car dealership, Kelly also received a new Corvette every three months or 5,000 miles. He rarely paid for a meal, and lived the life of a swashbuckling 24-year-old quarterback on top of the world. "Jim was a wild man," said Kiki DeAyala, a Houston linebacker. "He went at it *hard*."

The Gamblers, as a whole, fed off the mojo. It was perfect timing — throughout the late 1970s the NFL's Houston Oilers enjoyed a magnificent stretch under coach Bum Phillips, but he was fired after the 1980 season, and everything felt stale. The Oilers won just two games in 1983, and fans craved a return to electrifying entertainment.

The Gamblers were electrifying.

No team talked more smack. No team fired off more shots after the whistle blew. No team ran more trick plays or employed more renegades. No team was faster, quicker, tougher. Houston's snazzy uniforms — black jerseys, silver pants — reflected its approach. There was a large mirror placed in the Astrodome locker room, and players took turns admiring themselves in their duds before jogging onto the field. "We'd pose," said Courville. "The sweet uniforms — oh, man."

Todd Fowler, the starting running back, wore the same T-shirt every day to practice. It read KILL 'EM ALL. LET GOD SORT IT OUT. Boucher, the offensive guard, purchased a black Trans Am off the showroom floor and had Gambler patches sewn into the seats. "It was the Gamblermobile," he said. "Badass."

"After games ended a couple of us would do a sweep of the room to make sure no one forgot anything," said D. J. Mackovets, the team's media relations director. "So there's this one time I'm walking out of the locker room with Jack and I hear this player yell, 'Oh, no! Here comes coach!' Well, there were

---

* In the USFL, cheerleading units often traveled to away games.

a group of players beneath the stands with a hooker, and she was giving all of them blow jobs before they got on the bus."

Most of the players carried Gamblers tickets in their pockets, currency to hand police officers should they be pulled over doing 90 in a 55, or to draw long-legged, big-breasted knockouts away from boyfriends. "We lived Studio 54 lives," said Verdin. "On many nights I'd go to the Rock, a club in Galveston, and sleep on the beach, wake up, and come to practice with sand all over me. It's just how we did it."

The Gamblers' offensive line coach was Bob Young, a 19th-round pick of the St. Louis Cardinals in the 1964 NFL Draft who lasted 16 seasons at guard thanks to determination, heart, and gargantuan quantities of steroids. Young twice participated in the CBS Sports World's Strongest Man competition, and believed there were no excuses when it came to the pursuit of greatness. "I mean, Houston had a six-foot-tall plastic syringe on display in the middle of the locker room," said Jay Bequette, an offensive lineman from Arkansas who attended training camp with the team. "It was nuts." The unit held its weekly meetings on Tuesday nights at a topless club on Highway 59, and those who missed out were fined $200. "Try telling your wife you have to go to a strip bar for work," said Boucher. Unlike other teams, where the use of performance-enhancing drugs was subtly encouraged, Young pulled no punches. As naked women writhed on nearby poles, the coach made himself clear. "You use steroids or you're not on my squad," he told his men. "It's that simple."

"I had several coaches almost demand I take steroids," said Joe Bock, a center from the University of Virginia. "I steadfastly refused, and they moved me from first-string center to third. [Young] pushed them on me four or five times, and I refused. The last time I said, 'If you keep forcing this on me, we will fight.' He wanted linemen to be at least 280 pounds. I was 252 pounds, and I'd eat five peanut-butter-and-jelly sandwiches a day to gain weight. I'd drink protein shakes. But I was not taking steroids, and that cost me."

"Drugs were all over that team," said Boucher. "A ton of us were on steroids. To be honest, I supplied them. I hate to say that, but I had a connection, and I'd get the drugs and distribute them. Also, after away games, we'd drink for free on the flights back. They walked down the aisle with Vicodin, with Toradol. You'd be holding three beers and they'd give you a bushel of pills. We were living large and partying large, and we needed the drugs to survive."

Although the elfin receiver corps didn't turn to steroids, Johnson — its star and biggest character — combined his love of catching footballs with his love of marijuana. In the hours before kickoff, as he drove to the stadium, Johnson would light up a joint and smoke it behind the wheel. "Nobody on the team could believe it at first," he said. "They were like, 'How do you smoke weed before games?' But it relaxed me." As the season progressed, Pat Thomas, the Gamblers' defensive backs coach, would tap Johnson on the shoulder in the locker room and ask, softly, "Have you had your medicine yet?"

If Johnson nodded, all was good in the world. "There was one game when Richard somehow didn't smoke his doobie," recalled Boucher. "He played terribly and we lost. The next week I walked past his hotel room and smelled the dope. I thought, *Richard's ready to play.*"

Courville lacked Johnson's abilities to find open spaces and catch flying prolate spheroids. In 1984, for example, Johnson led the USFL with 115 catches. Courville finished with 4. The two were roommates, and one day Courville thought, *If toking works for him, maybe it'll work for me.* To Johnson's shocked delight, en route to the Astrodome Courville placed a joint between his lips, lit a match, and took a deep pull. "Well, I go out there and I'm completely paranoid of everything," Courville said. "I didn't catch a single ball and I sprained my thumb. It was the last time I ever played high."

Much like the Gamblers, the new Chicago Blitz and *New* Arizona Wranglers were happy to have a little extra preseason preparatory time. Both organizations were going through the dizzying transitional phases of having been swapped for each other, but the Illinois-based outlet was at a distinct disadvantage for one big reason: George Allen.

The former Blitz/current Wrangler coach was known to be many things, but ethical and empathetic were not two of them. As soon as the swap was official, Allen made certain the old Blitz headquarters, at Maine North High School in Des Plaines, Illinois, was picked to the bone. Weight equipment

— gone. Desks and chairs — gone. Typewriters, staplers, doorstops, mirrors — all gone. Even though the old Wranglers were becoming the new Blitz, the old Blitz refused to leave behind stationery, pens, folders with the team logo. Anything that said "Chicago Blitz" was deposited into a dumpster. In Allen's overly competitive world, if the new Blitz needed supplies, they'd have to buy them. "The place was completely empty," said Michael Murphy, the new Blitz defensive line coach. "Not a crumb left on the floor."

James Hoffman, Chicago's owner, was chief of the cardiovascular surgery unit at Good Samaritan Medical Center in Milwaukee. Because the USFL desperately needed someone to take the Blitz, Hoffman was enlisted without a detailed vetting of his financials. Which (league honchos convinced themselves) would surely work itself out, because Hoffman, according to a *Chicago Tribune* profile, "drives race cars, raises show horses and has extensive holdings in investments in real estate, gas and oil." (What holdings? No one was quite certain.) He also had been a quarterback with the Denver Broncos — according to the Blitz media guide. "I called the Hall of Fame in Canton, Ohio, because I knew some people there and found out the guy had never played for the Broncos," said Jim Foster, the team's executive vice president. "As near as we could figure out, he went to a free agent open tryout camp, threw a few passes and said he was on the team." Everything was optical illusion. Hoffman had paid a mere $500,000 up front for the Blitz, and $200,000 in initial operating expenses. The rest of the $7.2 million purchase price was due in installments. "I met him one time and he was extraordinarily unpleasant," said Bill McSherry, the USFL's general counsel. "He showed up to our meeting a little bit drunk. The guy was way over his head." At an introductory press conference Hoffman promised Chicago would be competitive. It was either naiveté or a lie.

Were one to compile a STUFF THAT WENT WELL FOR THE 1984 CHICAGO BLITZ list, it would include a single entry: eats.

The Maine North facility was simultaneously being used by the director John Hughes, who was filming his latest movie, *The Breakfast Club*, inside a converted gymnasium. "They would cater meals for the cast and crew, and we'd get the leftovers," said Don Kojich, the team's media relations assistant. "That was pretty terrific."

Otherwise, it was football hell. When the team was still the Wranglers, then head coach Doug Shively noted that his men "look like we're a lost ball in high grass." The words perfectly surmised a talent-light roster, and the relocation failed to improve things. Shortly after purchasing the club, Hoffman made a push to sign Walter Payton, the free-agent Chicago Bears halfback who was closing in on Jim Brown's all-time NFL rushing mark. The Blitz roster was slowly being filled with the decayed carcasses of dilapidated ex-Bears, and the thinking was, with the presentation of a $2-million-a-year contract proposal, Sweetness would jump and join his old pals. So the Blitz sent Bud Holmes, Payton's agent, the paperwork, then grew frustrated when he took the figures back to the Bears. "That offer is not going to stay on the table indefinitely," Ron Potocnik, the Blitz GM, told the *Chicago Tribune*. "We need to know."

Within 24 hours of those words, Potocnik knew: no.

Hoffman quickly realized he had neither the money nor patience to run the Blitz, and discarded the franchise as one would a dented muffler along Emerald Lane. During the second week of the preseason, the Blitz faced the Michigan Panthers in Arizona and lost 21–20. Afterward, Hoffman was walking off the field with Dan Jiggetts, an offensive lineman who had played seven seasons with the Bears.

"You know, I'm out of here," Hoffman said.

"You're leaving town?" Jiggetts replied.

"No," he said, "I'm done with this."

"What do you mean?" Jiggetts said.

"I'm finished," Hoffman said.

"He just . . . left," said Steve Ehrhart, the USFL's executive director. "Came and went like a ghost." The horrified league took over organizational operations, committed the bare-minimal number of dollars to keep things running — and putridity ensued. Potocnik and nearly all of his front office coworkers were fired. Yet instead of slinking off quietly into the night, the exiled GM locked himself inside his office and began calling every member of the media he could think of. "I'm this 23-year-old kid with no idea what to do," said Kojich. "And Potocnik is telling radio stations he's not leaving the complex."

He eventually departed, but the problems did not. For the majority of

USFL franchises, the cover of the annual media guide featured a player, a coach, or a montage of players *and* coaches. The whole idea was to set an exciting tone for the coming season. In the absence of talent or big names, the Blitz publicity team turned toward a theme — "Puttin' on the Blitz." It was the brainstorm of Foster, the recently hired Chicago executive vice president, and the resulting image brought forth a cascade of mockery. The media guide cover displayed quarterback Vince Evans in a black tuxedo, standing alongside a woman's leg encased in fishnet stocking, à la the lamp from *A Christmas Story*. The ensuing pages offered not a word of explanation, leaving one to presume the new Blitz kicker was either a hooker or a Rockette. "It was the worst thing I've ever seen," said Kathy MacLachlan, an administrative assistant with the club. "Foster wanted to sell the Blitz as this luxury experience. Instead we looked like an escort provider."

Chicago's new coach, Marv Levy, and new player personnel director, Bill Polian, were exemplary football minds who would later team up to lead the Buffalo Bills to repeated Super Bowl glory. With the Blitz, however, they possessed no talent and no money. ("They couldn't even buy toilet paper," said Levy. "So at Christmas Bill and I gift-wrapped toilet paper and distributed it to everybody in the organization.") The backup quarterback, Dennis Shaw, was 37 and last threw a professional pass with the St. Louis Cardinals . . . *in 1975.* The three biggest names were Evans ("God, he just wasn't good," said Mark Middlesworth, the assistant trainer), halfback Larry Canada ("He chain-smoked!" said Middlesworth. "Our top ball carrier chain-smoked on the sidelines!"), and safety Doug Plank, best known for his eight years with the Bears as the NFL's dirtiest player. Plank sat out 1983 to open a Burger King franchise, but missed the thrill of crushing spines with the bridge of his helmet. "I was a big Doug Plank fan," said Tim Ehlebracht, a Blitz wide receiver. "Then my first day of practice I come over the middle, he hits me, and I end up in the hospital after landing on my head."

The Blitz would go on to win five games while surrendering a league-record 466 points. Chicago averaged 7,455 fans (in 65,077-capacity Soldier Field), and failed to provide coaches with contractually guaranteed team vehicles. ("Actually, we got them," said Tom Beck, an assistant coach, "with two games left in the season.") The players even suffered the indignity of incorrectly presuming a large throng outside a Memphis hotel was waiting for them. "We pulled up to this crowd, and thought all these fans were cheering us," said Beck. "Turns out the band Chicago was staying there, too. Nobody gave a shit that we were there."

On March 17, the Blitz hosted the Oklahoma Outlaws in what was billed in advance as "the World's Largest St. Patrick's Day Party." This was the team's biggest promotional push of the young season — *Come out, wear your green hats and drink your green beer and party with the Blitz like you've never partied before!*

That morning, a blizzard blanketed the city, and approximately 350 spectators showed up. A cameraman from one of the local TV stations asked the few fans in attendance why they bothered. One boy, probably 12 or 13 and dressed in a ski cap and full-body plastic bag, replied dryly, "It's unknown."

What's unknown?

"Why we're here."

The Blitz lost, 17–14.

Unlike Chicago, the old Blitz/new Wranglers knew exactly what they were doing.

Just as he had done in the Windy City a year earlier, George Allen came to Phoenix with an acute — albeit frantic and somewhat demented — vision of how to run a franchise. First, any reminders of the previous season's four-win Arizona monstrosity needed to be expunged ASAP. The majority of the 1983 employees were fired, and that club's results were stricken from all Wrangler archives. One of the few returnees to the organization was Steve Des Georges, the director of public relations. On an early January morning inside the team's Wrangler Ranch office, Allen walked up to Des Georges with a 1983 Wranglers' media guide in hand.

"Did you do this?" Allen asked.

"Yes," Des Georges replied.

"Well," said Allen, "you're a loser."

Des Georges was speechless.

"This is not a winning representation of who we are," Allen said. "It's a losing representation. That means you're a loser."

One by one Allen interviewed holdovers from the Wranglers. The dialogues were quite predictable:

Allen: "How do you think you did last year?"

Employee: "We sold a good number of tickets."

Allen: "You went 4-14. You were losers. You're fired."

Next.

The head coach had all of the high-caliber and high-technical equipment from Blitz headquarters shipped to Arizona. Once again he loaded the team with NFL veterans, and set high expectations. "We were going to win," said Sel Drain, a Wranglers safety. "It wasn't negotiable." In a moment that forever lives in USFL lore, one morning, while holding his daily six o'clock breakfast briefing with the local media at the Phoenix Ramada Inn, Allen went on a lengthy diatribe about success. "Everyone here plays a role in winning," he said. "From the secretaries to the equipment men to the coaches to the players, even to the writers . . . if you have one person in this chain who won't help the team win, it won't win." A waitress arrived with Allen's bowl of oatmeal. A creature of habit, the coach always ordered it the same way — piping hot, no sugar added, plenty of raisins evenly dispersed throughout the mixture. This time something was off.

"Ma'am," he said to the waitress, "where are the raisins?"

"I'm sorry," she said. "We're out of raisins."

Allen scowled and clenched his fists. "What do you mean you're out of raisins?" he growled.

"I'm sorry," she said. "We're out."

"Is this your responsibility?" Allen asked.

"No," she said sheepishly. "There's a guy in the kitchen and . . ."

Allen pounded his fists on the table. "The raisin guy!" he yelled. "That man is not doing his job — and that's the point of my story!"

The waitress was humiliated. Des Georges, sitting to Allen's side, was humiliated. The media members in attendance were humiliated. The only person not humiliated was Allen, who believed life was solely about conquering all challenges. Once, while preparing for a preseason game, Allen was taking his regular midafternoon run in the 100-degree Arizona heat when he spotted a decomposing dog carcass on some railroad tracks. "You see," he told a companion. "Had that mutt taken better care of itself, it makes it across the tracks."

It was madness. But it was madness the players bought into. Allen thought himself the Douglas MacArthur of football, chock-full of motivational lectures. *The New Arizona Wranglers,* he told his players, *were destined to be champions.* "I will accept nothing less," he often said. "Nothing."

It was the cult of George Allen, and the parallels between, say, Jonestown and Wrangler Ranch were not so far-fetched. You either believed what the old coach said, or you were gone. He insisted on 100 percent loyalty and adherence to the rules. "It was George's way, not yours," said Greg Landry, the veteran quarterback. "He was a great coach, but you had to understand that."

One man who failed to receive the memo was Tim Wrightman, the tight end who, in 1983, joined the Blitz out of UCLA and became the first NFL Draft pick to sign with the league. Injuries limited Wrightman to nine games and six catches his rookie season, and while he was hurt he began to silently question Allen's ways. "I was originally supposed to be out nine weeks, but I could have come back much earlier," Wrightman said. "I was healthy. But the team was winning, and George was so superstitious he wouldn't let me return. It was weird." The following January Wrightman reported to Wranglers' camp with high spirits and a clean bill of health. In the lead-up to two-a-days, Wrightman was playing basketball at the complex when he landed awkwardly and tore a tendon in his right quadriceps. He was rushed to the hospital, where a doctor had to reattach the tendon. While lying in bed he received a call from his agent, Steve Linett.

"Tim," he said, "I have bad news. They're not going to pay you."

"What?" Wrightman said. "But I hurt it *at* the facility."

"I don't know what to tell you," Linett replied.

Not only did Wrightman never play a game for the Wranglers — he wound up suing the team as a whole and (because part of his original deal was a personal services contract with the head coach) George Allen as an individual.

"It's the craziest ending to a story ever," said Wrightman. "We went to court and the jury was pulled because they ended up believing me, not George Allen. Ultimately, to get the money we had to put a lien on George Allen's house."

Wrightman laughed at the memory.

"That," he said, "was a *very* USFL type of thing to happen."

10

GERBILS AND SNOWBALLS

There was a really nice restaurant in Pittsburgh, so we all decided we'd eat there to celebrate every win. But we never won. So we changed plans and ate there after every game. When you go 3-15, you have to adjust on the fly.

— Mike Fox, equipment manager, Pittsburgh Maulers

THE USFL OPENED the 1984 season on Sunday, February 26, and — as was the case one year earlier — everywhere Chet Simmons looked, he found reason to be terrified.

For example, in Los Angeles the Gold were to visit the Express. Despite Bill Oldenburg's promise that his team would fill the Coliseum on the dual engines of a pregame Beach Boys concert and tremendous football, ticket sales lagged. With a week to go there were less than 20,000 purchased seats, and the commissioner knew how empty the building would appear. The Express wound up losing, 27–10, before a reported 32,082 spectators (the figure was greatly exaggerated).

The Generals were kicking things off in Birmingham, and while a USFL-record 62,300 fans attended New Jersey's 17–6 win at Legion Field, Simmons couldn't separate the expensive team from the insufferable owner. The friction between Trump and Simmons grew by the day; contrasting visions of one man who believed in slow and steady versus another who seemed to think money, greed, and narcissism trumped all.

In Jacksonville, the brand-new Bulls pulled in 49,392 to the Gator Bowl, then demolished the rotten Federals, 53–14. Despite the promises of Ray Jauch, Washington's coach, that his team would be vastly improved, the rout — at the hands of an expansion club, no less — spoke differently. "I will never get the hours back in my life that I spent covering the Federals," said David Remnick, the team's *Washington Post* beat writer. "I'll be on my deathbed thinking about that." How bad were the Federals? The team's new quarterback, Reggie Collier, was a blooming cocaine addict,* and his backup, Lou Pagley, was a long-ago Notre Dame wide receiver who was recently plucked from a Miami flag football league. "A friend told them I could throw a little," said Pagley. "That got me a job." Wrote Ron Cook of the *Pittsburgh Press:* "The good news for the Washington Federals is they do not have a quarterback controversy. The bad news is they do not have a quarterback."

Simmons had hoped he could somehow keep the Federals' dreadfulness concealed, almost like a crack beneath a throw rug. But such was impossible. During the game at the Gator Bowl, Gordon Davenport, the Federals' director of marketing, sat in the press box alongside Joe Morrison, the head football coach at the University of South Carolina. A former New York Giants wide receiver who had been around the sport for decades, Morrison never saw anything quite like the Federals. "Gordon, your defense literally doesn't know how to line up," he told Davenport. "It's not my place to critique, but it's asinine what they're doing out there."

Morrison scribbled his observations on a napkin, and on the return flight to Washington Davenport handed it over to Berl Bernhard, the team's owner. Already seething over both the defeat and the indignity of the USFL experience, Bernhard digested Morrison's take as if it were rotted sushi. The next day, when asked to assess his team's performance, Bernhard went ballistic. The Federals, he said, "played like a group of untrained gerbils . . . no fight, no aggressiveness, no spirit, and a lot of moping around."

---

* "I understood Reggie quite a bit," said Cornelius Quarles, a running back who played with him in Birmingham. "He came from small-town Mississippi, from a shack, and they gave him $1 million and told him he's the Man. He had a lot of hangers-on and he wasn't studying film like he should have. That catches up with you."

The quotation went viral in a time when quotations rarely went viral. Never mind that, technically, it is excruciatingly difficult to train a gerbil. The Federals were a mess, and there was no hiding it. Three days after the game, Jauch was fired and replaced by Dick Bielski, the foulmouthed offensive coordinator. "I wanted that job like I want a disease," Bielski said years later. "Fuck, the team couldn't put out for a decent pair of sweat socks. I'd been in the NFL for 21 years, and now I was in the minors. It was horseshit."

Unlike the Federals, Simmons had genuine optimism for the Pittsburgh Maulers, who were blessed with a terrific owner (Edward DeBartolo Sr.), a franchise running back (Mike Rozier, the reigning Heisman Trophy winner), an NFL refugee at quarterback (former Cowboys backup Glenn Carano), a market that loved its football, and one story in particular with the potential to grab fans.

Mark Raugh was a six-foot-three, 215-pound tight end who had been a star at West Virginia before being selected by the Pittsburgh Steelers in the 11th round of the 1983 NFL Draft. He was released, but spent the ensuing months back near campus, staying in shape and working at a car dealership. In the early-morning hours of Sunday, November 6, Raugh was driving with two friends through Morgantown, West Virginia, when his vehicle plunged down an embankment and flipped. One of the passengers, Tina Belt, went uninjured. The other died. His name was Andre Gist. He was 25, a former member of the Mountaineers, Raugh's best friend, and, in 1983, a Tampa Bay Bandits offensive guard. A medical examiner ruled that Gist's heart stopped after the collision, a result of his 260-pound body putting excessive pressure on his neck.

Around the time of the accident Raugh — who admitted to having consumed three mixed drinks, part of a fourth, and a beer before driving — was signed by the Maulers. Although he was later found innocent of negligent homicide, his presence in the USFL was shocking. West Virginia's stadium was an hour drive from Pittsburgh and Gist's death had been big news. "I never got over it," Raugh said years later. "Andre was the best — his laugh, his demeanor, his loyalty. To know that I was involved in the end of his life . . . you don't get past that. Ever."

It would have been easy for local football fans and members of the Maulers to reject Raugh. Instead, he was embraced for his willingness to talk about the tragedy. In the weeks following Gist's death Raugh lost 20 pounds, and wore the heartbreak on his face. Yet his fellow Maulers loved him. He was approachable and gritty and obsessed with football. He practiced hard and seemed more grateful than any teammate to be on the field. In short, he was a survivor.

Though Raugh's USFL debut in Tulsa wasn't the reason Simmons chose to attend Maulers-Outlaws in Week 1, it was an appealing sight to behold. Lacking any sort of history, the USFL had to sell gripping narratives. Connections needed to be made with fan bases. The NFL peddled tradition. You were a Steelers fan because of the franchise's four Super Bowl titles. You were a Cowboys fan because they had forever been America's Team. Yes, stars mattered to the NFL. But the brand power was organizational. The USFL, on the other hand, urged player-to-fan connections. The league wanted you to learn the storylines of its competitors, à la Raugh's return from tragedy. The commissioner was excited to see Rozier make his first start for Pittsburgh (the second Heisman winner joins the USFL!), as well as Doug Williams (jilted Buccaneer quarterback!) stand behind center for Oklahoma. So Simmons flew in to Tulsa, drove out to Skelly Stadium, and . . .

He got wet.

The weather was atrocious — a steady downpour with gusty winds and a windchill near 20. "There was a tornado warning," recalled Carano. "We didn't play in an actual tornado, but it definitely sucked." The official attendance was reported as 11,638, but that included 4,300 no-shows. The action (or lack thereof) was even worse than the conditions. Rozier's heavily hyped debut was a dud, as he ran for 27 yards on 16 carries. It was his lowest yardage total since being limited to 23 yards on 5 carries as a sophomore against Penn State, and afterward Ed Bouchette of the *Pittsburgh Post-Gazette* opined that "the only holes he found all day were the ones his line put him in." Williams was similarly dreadful — he completed 9 of 22 passes for 57 yards — and Carano could not have been less effective. He tossed 22 passes and completed 10. Two were intercepted. "Say what you want," he said years later, "but I outplayed Doug Williams." So did the stadium's 28,597 empty seats. "When your quarterback's voice quivers in the huddle, you know you have problems," said William Miller, a Pittsburgh running back. "Glenn was terrified." The final score, 7–3 Outlaws, was a relief only in that there would be no overtime.

Of all the expansion outlets, the USFL thrust most of its hopes upon the Maulers. The general belief was the team was too talented to continue to play with such putridity. Rozier was thought to be a marquee-level superstar, and two days after the Oklahoma loss his two new cars were delivered to the Maulers' practice facility. One was a red Mercedes-Benz, the other an otherwise identical white Mercedes-Benz. "Mike wanted his vehicles in the Nebraska colors," said Mike Fox, the team's equipment manager. "He was young, rich, talented, and making a statement. Or something. I guess."

Pittsburgh lost again in Week 2, 27–24, to the defending champions in Michigan, and focus turned to the March 11 home opener against Birmingham.

It would be an afternoon to remember.

Two months earlier the Stallions signed Cliff Stoudt, the Pittsburgh Steelers' starting quarterback for much of the 1983 season, to a three-year contract worth $1.2 million. Stoudt had filled in for an injured Terry Bradshaw, and his statistics were pitiful (12 touchdowns, 21 interceptions in 15 starts). He became, according to Russ Franke of the *Pittsburgh Press,* "public enemy No. 1 in Pittsburgh when he failed to live up to the fans' expectations."

"I tried to commit suicide," Stoudt once joked of his tough times in Pittsburgh, "but the bullet got intercepted."

The metropolis also never fully forgave him for what ranks as perhaps the dumbest self-inflicted injury in NFL history — in 1981, during a visit to a Seattle country music lounge named Montana's, Stoudt snapped the ulna bone in his right forearm while punching one of those machines that measure a person's strength. "I felt, 'What the heck. I'm not playing. Why not punch it?'" he explained to the media days later. "How can I be so stupid?"

Still, Stoudt was a tough kid who could get up from a hit, and the Steelers wanted him back. To the game's most passionate die-hards Stoudt's defection was treasonous. It was one thing to play poorly. It was another to jump ship. Stoudt was now the third ex-Steeler to turn Stallion, following the leads of Rollie Dotsch (the offensive line coach who became Birmingham's head coach) and wide receiver Jimmy Smith. As soon as Stoudt joined the USFL, he was dead to Pittsburgh.

But now he was returning, and the Steel City prepared itself. A sellout crowd of 53,771 spectators paid between $12 and $19 a ticket to brave blizzard conditions, enter Three Rivers Stadium, and mock, berate, intimidate, scorn, and physically assault Stoudt. T-shirts with the quarterback's image and the words DEATH WISH were peddled for $5 in the parking lot. When he arrived in town for the game, an unmarked van surreptitiously picked him up from Greater Pittsburgh International Airport, and he registered at the hotel without using his name. The Stallions practiced at Three Rivers in the days before the game and — in an effort to ready the quarterback — a team employee barked expletives at Stoudt through the stadium's PA system. Yes, the Maulers were making their home debut. But this wasn't about Carano or Rozier or dreams of a USFL championship. It was, 100 percent, about revenge. "We were second-class citizens in Pittsburgh, and we knew it," said Tony Lee, the Maulers kicker. "They weren't paying to watch us. They were paying to punish Cliff Stoudt."

To the attendees' dismay, the Maulers were ghastly. Rozier was once again held in check, rushing for a scant 52 yards behind increasingly ineffective blockers. "There was definitely some resentment from his linemen over the money," said Lee. "They were getting paid peanuts and he was a millionaire." Carano was proving himself the league's worst starting quarterback (Joe Pendry, the Maulers coach, used to complain in private that Carano had "a heart the size of a walnut"). To the attendees' glee, though, Stoudt the Stallion was even worse than Stoudt the Steeler. He finished with two completions in 16 attempts for 21 yards. His passes either bounced a solid five feet in front of the intended wide receivers or soared wildly against the arctic winds.

Whenever the Stallions were within 20 yards of the end zone, balls of ice and sludge were rocketed from the stands. Many hit the quarterback in the back, in the legs, in the helmet. At one point, while Stoudt was on the sideline, two full 16-ounce beer cans smashed against the wall near his head. Late in the fourth quarter, with the Stallions holding a comfortable lead, ice balls were replaced as projectiles by ice blocks and frozen oranges. The early anger was sober anger. The late anger was drunken anger. "We had to send security to clear out a section of the stadium," said George Heddleston, the Maulers' general manager. "The fans felt betrayed by Cliff, but it got mean and vicious."

When David Dixon thought up the USFL, then sold myriad wealthy men on the idea, one thing he didn't consider was that the window to grab fans would open and close in the blink of an eye. The Maulers wound up losing to Birmingham, 30–18, in a game as uninspired as it was cold. Afterward, Pendry walked to his car in the parking lot, only to find it stripped of its tires. "That was an omen," said Fox.

"We just weren't good," said Ike Griffin, a defensive end. "Mike Rozier sort of symbolized it all. He had swagger and he had talent, but he loved the nightlife more than football. You'd walk into practice and Mike would be asleep in a laundry hamper. He was a kid figuring it all out, just like 70 percent of our roster."

Before long, the 1984 Maulers were the 1983 Federals, minus the co-caine-addicted, halfway house–residing quarterback. The team's facility was located inside what was once the Wallace Intermediate School in Baldwin, Pennsylvania, and they shared the space with Lots of Tots, a day care and preschool center. "And the team stopped doing our laundry," said Griffin. "If you wanted your uniform cleaned, it was on you."

When the Maulers generated headlines, it was usually the result of unde-sirable circumstances. Save for Rozier, Carano, and ex–New York Jets cor-nerback Jerry Holmes, the club's biggest name was probably Bruce Huther, the former Dallas Cowboys middle linebacker. In the early-morning hours of April 13, Mark Raugh and Huther — close friends and neighbors at the Baldwin Commons apartment complex — engaged in a heated sidewalk al-tercation over which of the two men was tougher. "Bruce was a little jeal-ous [of Mark]," said Tom Rozantz, the backup quarterback. "He thought he was pretty tough." Huther refused to let the topic die, egging Raugh on and punching him in the chest. Raugh later explained to teammates that he felt reduced to two choices: keep getting hit or take a swing.

"We had a disagreement," Huther said, "and settled it man to man." Trans-lation: Raugh fired his fist into Huther's face, breaking his nose and fractur-ing his orbital bone, causing blurred vision and two black eyes. "I fell and struck the left side of my face against a stone ledge," Huther said. "That's how the other bone in my face was fractured."

The teammate-versus-teammate altercation was national news. Heddleston went to see Huther at Jefferson Center Hospital and, three decades later, vividly remembered a face that "looked like the elephant man. He couldn't talk. One side of his head was completely purple and swollen beyond belief. They were worried they'd have to perform brain surgery. Luckily, it didn't swell."

Despite the horror-flick visuals, the organization immediately stopped payment on Huther's contract, then released him. Raugh not only stayed, but missed nary a game. It was heartless. "[The Maulers] came to visit him in the hospital and told him to take all the time he needed to get well," said Spencer Kopf, Huther's agent. "Then three days later [Pendry] told him not to come back at all."

Pendry was, according to Bouchette, a "milquetoast guy." A black cloud hovered above his head, and lightning could strike at any moment. He found himself in a particularly dark mood following a 17–7 loss at Memphis that dropped the Maulers to 2-8. The night before the game, at approximately 3:00 a.m., somebody pulled the fire alarm at the Holiday Inn–Overton Square, and the entire Maulers organization had to shuffle out the front door and stand in the cold for 40 minutes. It was a harbinger of things to come. DeBartolo, the owner and native Ohioan, attended few games, but he happened to sit in the Liberty Bowl press box for the Showboats' shellacking. During the third quarter, Heddleston felt a tap on his shoulder. It was DeBartolo. "Tell your coach to replace Carano with Rozantz," he said. "This has to stop."

The order was relayed to the sideline — and ignored. It was relayed twice — and ignored twice. A third time — ignored again. Making matters worse, Carano dislocated his right arm during the first half, and all but begged the staff to take him out. Everyone in the stadium could see the quarterback's dangling appendage. The Showboat pass rushers, led by a rookie named Reggie White, were salivating. "I couldn't throw the ball 20 yards without suffering tremendous pain," Carano said. "They didn't care."

DeBartolo was irate. He paced back and forth in the press box. "I want to see you in my office in Youngstown tomorrow at 7:00 a.m.," he told Heddleston. "And bring that *fucking* coach with you."

Afterward, as the Mauler players changed in the locker room, Heddleston informed his head coach about the order.

"I'm not going," Pendry said.

"Bullshit," Heddleston said. "You're going."

"No," Pendry replied. "I most certainly am not."

Heddleston was astonished. He didn't even want Pendry as a coach — and now he was begging him to meet the owner who paid his $100,000 annual salary. "Joe, if you don't come I'm going to have to fire you," he said. "You don't want to insult Eddie."

The following morning, Heddleston arrived in Youngstown minus Pendry. DeBartolo told him to can the coach, and a phone call was made.

"You can't fire me," Pendry said, "because I quit."

Click.

He was replaced by Ellis Rainsberger, the 51-year-old offensive line coach, who immediately instituted two-a-day practices. "Here we are, late in the season, battered and beaten up, and we're practicing twice every day," said Carano. "It was insanity."

Actually, insanity arrived on June 22, when the Maulers and Jacksonville Bulls concluded their seasons by facing off before 31,843 fans in a Gator Bowl besieged by torrential rains. The contest was postponed for 80 minutes. Because one sideline was under water, the teams were forced to stand on the same side of the field. There were puddles everywhere, and waterfalls rolled down the cement stairways. Still, 31,843 people came so the Bulls could break the USFL's single-season attendance record (which, with an average turnstile count of more than 46,000, they did). "I remember the water flowing in under the doors in the locker room," said Lindy Infante, the Bulls coach. "Some people came just to go through the turnstile and left for home, just so we could lead the league." The Maulers lost 26–2, and wrapped the interminable season with three wins, 792 Rozier rushing yards, and an inflated average home attendance of 22,858. Just in case the franchise needed a bit more misery, on the return to Pittsburgh the chartered plane hit an air pocket and fell into a 1,000-foot nosedive. Food and beverages soared across the cabin. Enormous men screamed like babies. "Trays, cups, cans, and even people flew straight up to the ceiling," said Rozantz. "No one spoke for a minute or two afterward. I thought it was the end."

The next morning Hank Bullough, former Green Bay Packers defensive coordinator, was named the new head coach of the Pittsburgh Maulers. He celebrated the occasion by firing every assistant on his staff and promising great things for the future of the USFL in the Steel City.

The team never played another game.

Something needs to be said here. It's important.

Despite the sorrowful likes of the Federals and Blitz and Maulers, the USFL — as a whole — was improving. Whereas the 1983 season was plagued by a large handful of starting quarterbacks who were closer to the grave than to the Super Bowl, the USFL now featured legitimate signal callers in the primes (or at least near primes) of their careers. In New Jersey, Brian Sipe was playing similarly to how he had in Cleveland, when he won the 1980 NFL MVP. In Michigan, Bobby Hebert was a blooming superstar, as were Jim Kelly in Houston and Rick Neuheisel (to a lesser degree) in San Antonio. Steve Young had the formerly frightful Los Angeles Express looking downright explosive.

Although their roster remained largely unchanged, the Philadelphia Stars were somehow better than a year earlier, and many — including Carl Peterson, the general manager — thought them good enough to hang in the NFL. Houston's Run and Shoot offense, meanwhile, was dizzying. After being held to 17 points in an opening-week loss at Tampa Bay, Jim Kelly and the Gamblers hung 35 on San Antonio, 45 on Chicago, 32 on New Jersey, and 34 on Michigan. Before every game Kelly would throw up into a sideline garbage pail, then throw for 400 yards. "It never felt like you had enough people on the field to defend it," said John Bungartz, a Denver Gold linebacker. "We tried playing them with a two-man rush and Kelly just picked us apart. I hated that offense."

"We ran four wide receivers 70 percent of the time," said June Jones, the Gamblers' wide receivers coach. "Nobody knew what to make of us."

The overall USFL output was sharper, the officiating was more on point, the players were upgraded. An influx of NFL and college stars brought positive attention to an entity in desperate need. "I thought we were soaring," said Young. "The quality of play was really high. The future was bright."

One of the most fascinating early successes was brewing in New Orleans, home of the transient Breakers. The team spent the 1983 season in Boston University's Nickerson Field, averaging 12,817 fans to a low-level bandbox. Joe Canizaro, the new owner, looked over his team's roster and saw a lot of mediocrity, but no superstar for locals to connect with. Throughout the 1970s, the New Orleans Saints survived their standard three- and four-win seasons by featuring a marquee quarterback, Archie Manning, who played his college ball at nearby Ole Miss. New Orleans fans counted Manning as one of their own, and that kept them returning to the Superdome week after week.

Now, 13 years after Manning's arrival, the Breakers had their eyes on another regional son. For anyone who followed the sport, halfback Marcus Dupree was the embodiment of potential greatness being stunted by painful reality. Raised in poverty by a single mother in nearby Philadelphia, Mississippi, Dupree arrived at Philadelphia High in 1978 as a six-foot-two, 235-pound ball of potential. He ran the 40 in 4.3 seconds and deadlifted 700 pounds. He was a two-time Parade All-American who finished his career with 7,355 rushing yards and 87 touchdowns, a national high school record. "By the time he was an eighth grader he was as good as any high school football player in Mississippi," said Sid Salter, editor of the *Scott County Times* in Forest, Mississippi. "When he was in the ninth grade they were talking college. By the 10th grade they were talking phenomenon. In the 11th and 12th grades it was a matter of how many superlatives can they use."

Dupree was the most heavily recruited prep player in football history, and his freshman year at the University of Oklahoma showed why. Despite not starting until the seventh game, Dupree rushed for 1,144 yards and 13 touchdowns. "He was the best player on the field," said Barry Switzer, the Sooners coach. "Earl Campbell was the only other guy I ever saw who was like that — physically ready, as a true freshman, to be the best player on a great college team. Maybe even ready for the NFL at that age." Oklahoma's final game of the season would be the January 1, 1983, Fiesta Bowl against Arizona State. Dupree returned from Christmas break with 15 extra pounds, and Switzer ripped him to the media. For a shy, sensitive kid from a small town, it was hard to digest. Shortly after Dupree ran for 239 yards in the 32–21 loss to the Sun Devils, Switzer pulled him aside and said, "If you'd have been in shape, you'd have rushed for 400 yards and we'd have won the game."

The words cut, and the Dupree who came back for his sophomore season was not the same person. He missed the deadline for arriving on campus, skipped the official team photograph, looked overweight and disinterested. Switzer ripped and slammed his sensitive star, and dismissed him as "lazy."

Hampered by various injuries, the 19-year-old played in four of the first five games, then went AWOL. To the shock of a nation of college-football fanatics, Dupree enrolled in the University of Southern Mississippi, only to learn from the NCAA that he would have to skip an entire season due to transfer rules. With that, his collegiate career — a grand total of 16 games — was complete. "I figured," he said, "that I might as well try [to go pro]."

In the aftermath of the Herschel Walker signing, the USFL decided (in the name of positive PR) no other underclassmen would be allowed to join the league. Then a University of Arizona punter named Bob Boris sued, contending that the eligibility rule, as it was applied to him, constituted an unreasonable restraint of trade in violation of federal antitrust laws. A U.S. district court ruled in favor of Boris, and a few days later the Breakers worked a trade with the Generals (Oklahoma was, curiously, among New Jersey's territorial schools), swapping a first-round pick in exchange for the rights to Dupree. The running back had hired Ken Fairley, a Hattiesburg, Mississippi, pastor, to serve as his agent, and soon they received a call from the Breakers, itching for a meeting. That's how Dupree, Fairley, and Randy Vataha wound up sitting at a booth inside Mr. Gatti's Pizza in Picayune, Mississippi. "We're in the back of the parlor, hardly anyone is in the place, nobody notices us," said Vataha. "I handwrote a contract right there, and he signed it right there. We had to move it over to a league-wide contract, on official paper, but everything was agreed upon." New Orleans would give Dupree $6 million over five years, including a $1.1 million bonus to be paid over 10 years.

On March 3, Dupree was ferried to an outdoor press conference via horse-drawn carriage. "I went up to the guy with the carriage and said, 'How much do I have to pay you to drive Marcus and his family up here?'" recalled Steve Weaver, the team's director of marketing. "All he wanted was tickets. So Marcus and his family get on, but suddenly the mule can't pull them up the cement ramp. Me and Jack Galmiche [the team's VP of marketing] get behind the mule and start pushing it by the ass." Dupree wore black boots and a banker's blue suit, and was formally introduced to New Orleans in a ceremony that included the Heart Breakers sideline dance team, a sky filled with blue and silver balloons, and a Dixieland jazz band playing "Way Down Yonder in New Orleans." When asked what the team's fans could expect, Dupree gazed downward. "I don't talk that much," he mumbled.

"I call him my $6 million man!" barked Canizaro. "He's *got to* perform."

Dupree's arrival was a jolt. He joined New Orleans for the team's Week 3 home opener against Memphis, and 45,269 fans filled the stadium to see Archie II's professional debut.

The Breakers made certain the afternoon was an event. A man dressed in a full Breakers uniform parachuted from the ceiling. As the starting lineups were introduced, Dupree — resplendent in his blue No. 22 New Orleans jersey — trotted out to a standing ovation, his name blared with extra oomph. With 11:24 remaining in the opening quarter, the Breakers took over on the Memphis 23 and marched down the field. When the ball was on the 1, Dupree entered the game and lined up deep in the I-formation. He took a pitch from quarterback Johnnie Walton, rolled untouched into the end zone, and handed the ball to Dan Hurley, the burly offensive tackle, who spiked it into the ground. Dupree scored again, on a 2-yard run, moments later.

That he generated a mere 31 yards on 11 carries mattered not. New Orleans won 37–14, and the excitement was unmistakable. "He really ran a lot better than we expected he would," raved Dick Coury, the head coach. "For the first time out Marcus took to coaching real well."

With Dupree in tow, the Breakers jumped out to a 5-0 start and New Orleans responded with enormous crowds and unrivaled enthusiasm. The team was the talk of the USFL, and Dupree's reappearance from nowhere brought forth reporters from all corners of the nation. "I played against O. J. Simpson and Walter Payton in the NFL," said Tim Mazzetti, the Breakers' kicker. "Marcus was just as good an athlete."

Yet behind the scenes the rookie, only 19 years old, showed himself to be immature. Dupree was often late to practices, and took weight room work as a day at the beach. Shortly after signing his contract, he bought a motorcycle, then returned it after realizing he was contractually prohibited from having one. "So the next day he comes to practice with a green Mercedes with gold rims," said Marcus Marek, a Breakers linebacker. "He was a phenomenal athlete who didn't work hard enough and didn't understand money. But, boy, he could play football."

The Breakers eventually lost their luster, and finished 8-10 and third in the Southern Division. According to Jeff Gaylord, a defensive lineman, the city decimated any real chances of success. "Drugs, drugs, drugs — accessible everywhere in New Orleans, and we had so many guys who used," he said. "Most of us were poor, and once the checks started coming, so did the coke. The white guys snorted it, the black guys smoked it. We fell apart." Still, the positives outweighed the negatives. New Orleans bought into the USFL, and the Superdome felt like home. "Overall, we gained a lot of respect," said Buford Jordan, a running back. "We felt like we were on the rise."

The same, truly, could be said for much of the USFL.

If only Donald Trump and Bill Oldenburg weren't doing their all to ruin it.

When former Oklahoma halfback Marcus Dupree signed with the New Orleans Breakers, he was introduced to fans as the next Herschel Walker. Instead, immaturity and injuries make Dupree little more than a league footnote. AP PHOTO/ERIC RISBERG

The San Antonio Gunslingers featured some of the USFL's strangest uniforms, as well as one of pro football's most unlikely starting quarterbacks, former UCLA star Rick Neuheisel. AP PHOTO/PAUL SPINELLI

Kelvin Bryant, the Philadelphia Stars' transcendent halfback, raises the USFL's championship trophy after his team demolished the Arizona Wranglers in 1984. Though less acclaimed than New Jersey Generals' Herschel Walker, Bryant was the league's best offensive player. AP PHOTO/PAUL SPINELLI

Before engaging upon a legendary 15-year NFL career, Reggie White was a star with the Memphis Showboats. His claims to fame? Sacks, tackles, and once spending $199.50 on 57 pairs of socks.
AP PHOTO/PAUL SPINELLI

Pete Rozelle, the NFL's commissioner, was all smiles outside the federal courthouse in lower Manhattan after learning his league — technically a loser in court — would only be forced to pay the USFL $3 in damages. He had Donald Trump to thank for the good news. AP PHOTO/SUSAN RAGAN

Steve Young loved playing quarterback for John Hadl, but he wasn't around when his head coach was punched in the face by an angry defensive lineman who had just been released by the Los Angeles Express. AP PHOTO/NFL PHOTOS

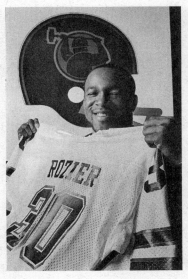

No professional sports league ever perfected the selling of sex quite like the USFL. Here, the big-haired Chicago Blitz cheerleading squad invites fans to the Sky Suites. CHICAGO BLITZ

The USFL fixed its draft in 1984 to make sure the Pittsburgh Maulers would select Nebraska halfback Mike Rozier with the No. 1 pick. The second-straight Heisman Trophy winner to join the league, Rozier rushed for just 792 yards on a club that won but three games. AP PHOTO/GENE PUSKAR

Craig James, halfback for the putrid Washington Federals, soars over the Philadelphia Stars' defense in a 1983 matchup. "We had a lot of hoodlums," recalled Mike Hohensee, Washington's quarterback, of his 4-14 team. "We out-drank everyone, we out-snorted everyone, we out-smoked everyone. We probably had more women than any other team. We had lots of fun off the field, but we were miserable on it." AP PHOTO/RON EDMONDS

The San Antonio Gunslingers' ever-changing cast of characters included a pair of former Grambling defensive linemen: Mike St. Clair (78) and Greg Fields (99). St. Clair was a solid professional. On the other hand, Fields, a.k.a. Big Paper, arrived in Texas after having punched John Hadl, his head coach in Los Angeles. The Express hired Liberace's bodyguard to keep tabs on him. SAN ANTONIO GUNSLINGERS ARCHIVES

In its two-year history, the Chicago Blitz went through thousands upon thousands of bodies, seeking out capable football players. Here, equipment manager Tom McVean talks on an amazing 1984 innovation: the portable telephone. CHICAGO BLITZ

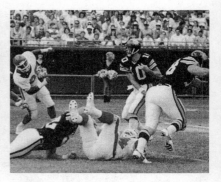

Denver Gold quarterback Bob Gagliano fires off a pass in his team's 28–24 win over the New Jersey Generals at Mile High Stadium. The Gold were one of the USFL's model franchises in the first two seasons, but once plans were announced for a shift to fall, interest dwindled. COURTESY OF THE AUTHOR

Denver's Darryl Hemphill leaps through the air in an effort to block the punt of the Oakland Invaders' Stan Talley. Once the USFL folded, Hemphill never again played professional football. Talley, meanwhile, punted in 12 games for the 1987 Los Angeles Raiders. RON RIESTERER

Before he emerged as one of college football's all-time great coaches at the University of Florida, Steve Spurrier was leading a USFL powerhouse in Tampa Bay. Here, he chats with quarterback John Reaves. JOHN D. HANLON/*SPORTS ILLUSTRATED*/GETTY IMAGES

Chet Simmons, the USFL's first commissioner, loved baseball far more than football. But he thought the young league had great potential, especially with its early television deals. Then Donald Trump came along.
BETTMANN/GETTY IMAGES

Donald Trump brought money, ambition, and ego to the USFL. His purchase of the New Jersey Generals was initially considered a boom to the league. Over time, however, he led the head-first charge toward demolition.
BERNARD GOTTFRYD/GETTY IMAGES

Featuring Jim Kelly at quarterback, and a fleet of speedsters at wide receiver, the nattily uniformed Houston Gamblers put on the best show in the USFL.
SCOTT CUNNINGHAM/GETTY IMAGES

George Allen, legendary coach of the Chicago Blitz, would do whatever it took to win. Leading up to the team's first-ever game, he had a pair of coaches spy on the Washington Federals' practices.
OWEN C. SHAW/GETTY IMAGES

Herschel Walker remains the most important player in USFL history. He was the league's first headliner, he broke the pro football single-season rushing record, and he gave the New York–New Jersey market the superstar it coveted. ANDREW D. BERNSTEIN/GETTY IMAGES

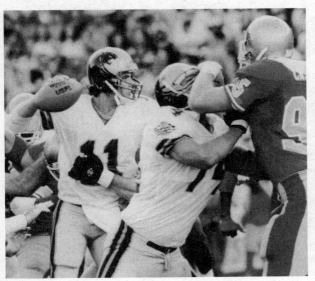

On July 17, 1983, the Michigan Panthers captured the USFL's first championship with a 24–22 win over the Atlantic Division champion Philadelphia Stars. Quarterback Bobby Hebert, whose thick Cajun accent rendered 50 percent of his words indecipherable, hit Anthony Carter on a 48-yard touchdown strike in the fourth quarter for what proved to be the deciding score.
DAMIAN STROHMEYER/
THE DENVER POST VIA
GETTY IMAGES

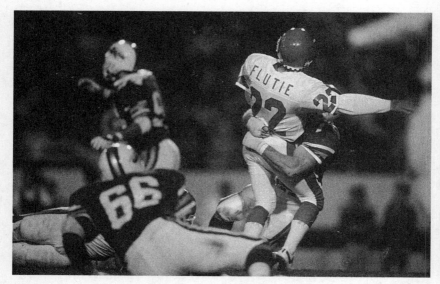

New Jersey Generals quarterback Doug Flutie gets hammered after a throw during a 1985 game against the Orlando Renegades. Although he arrived with tremendous hype, Flutie was merely an average player for New Jersey, throwing 13 touchdowns and 14 interceptions in his lone season. JOHN IACONO/*SPORTS ILLUSTRATED*/GETTY IMAGES

Brigham Young University quarterback Steve Young signed with the Los Angeles Express for a reported $40 million, but continued to wear old jeans and drive his even older (19 years, to be exact) Oldsmobile Dynamic. ANDY HAYT/*SPORTS ILLUSTRATED*/GETTY IMAGES

11

FALLING

I think without Donald, the league, in his words . . . it would have been small potatoes. But you know what? Nothing is wrong with small potatoes. They're very tasty and they can be successful food.

— Bill McSherry, former USFL executive director

BECAUSE EDWARD DeBARTOLO was a busy man with bigger things to worry about than a football team, the owner of the Pittsburgh Maulers often sent his general manager to the league meetings.

At but 35 years old, George Heddleston generally stood out as the youngest person in the room. He had spent the past four years as the public relations director for the San Francisco 49ers, and before that served as an assistant with the Dallas Cowboys, as well as the sports information director at both Mount Union College and Capital University. Such positions require a person to observe before speaking; to take mental notes not merely on rushing yards and touchdown passes, but mannerisms, decorum, methodology. A public relations chief is not part of the story, so much as a chronicler and preserver *of* the story.

"I was an untraditional choice to be general manager," he admitted years later. "But Mr. DeBartolo saw something in me."

Upon being hired, Heddleston steeled himself for irksome contract negotiations, cumbersome player transactions, nonstop roster shuffling. These are the things one must do, he rightly presumed, with a new franchise in a new league. What Heddleston could not have foreseen, however, was the absolute nuttiness that awaited him.

Three decades after the fact, Heddleston still vividly recalled his first encounter with Donald J. Trump, owner of the New Jersey Generals, at a USFL meeting in New York City. "Mr. DeBartolo stayed in Youngstown, Ohio, to work, so I was there, representing the Maulers," he said. "It was craziness. You had the owner of the Michigan Panthers [Alfred Taubman] and the owner of the Philadelphia Stars [Myles Tanenbaum] screaming at each other in Yiddish over some matter of unimportance.

"And then Donald Trump walks in. And he's bombastic from the start. He's loud, he clearly wants to be noticed. Just a jerk, and a jerk on purpose. He sits down, and the meeting starts and he's reading the *New York Times*. We're meeting, voting on things, and he's reading the newspaper. Finally, we get ready to hold a vote and Donald holds open the *New York Times,* stands to get attention, talks over whoever's speaking, and says, 'Look at this! Look at this! I build a skyscraper and nobody cares! I sign some obscure defensive back and I get three paragraphs in the *Times. That's* why I bought the Generals!'"

Later that day, the gathering broke off into an executive session, and Trump continued his verbal roller coaster. "I don't know about the rest of you people and I don't know how much money you guys have, but I have the money to get into the NFL," he said. "And that's where I plan on being. And I have more money than anyone in this room, except for maybe that guy right there . . ." He pointed directly at Heddleston, whom — despite a nearly 40-year age difference, a $33,000 salary, and no physical similarities — Trump somehow confused him with DeBartolo.

"I just froze," Heddleston recalled. "Here's this loudmouth guy . . . this outrageous guy, pointing at me. And, looking back, I believe he started to single-handedly take the league down that day. Nobody in that room wanted to move the USFL to fall. Nobody. Not one person. But there was something about Donald Trump . . ."

Indeed, there was *something* about Donald Trump. Beginning shortly after he purchased the Generals, Trump made it his objective to force the USFL's hand and convince his peers — often against their own best interests

— to change directions and switch seasons. Initially, the words were thought by some to be mere chatter. But as the second campaign headed from March into April and May, Trump shifted his movement to the next level and it became clear that the Generals' general would somehow scratch, claw, and shove the USFL into the fall and, as a result, his team into the NFL. "He was not an honorable man," said Jerry Argovitz, owner of the Houston Gamblers. "The truth wasn't his thing. But I've always said one thing about Donald Trump. You don't ever underestimate Donald. He can charm you out of your pants. And he's like getting involved with a rainbow or a tornado or a hurricane or a zombie — all at the same time depending on his mood."

"Trump was an intimidating, domineering force in a meeting," said Ted Diethrich, the Wranglers' owner. "Even though the owners were very powerful businessmen, when Trump came into the room he dominated."

Trump worked every angle. He devoted a large amount of time to undercutting Simmons to his fellow owners; trying to convince everyone within earshot that the commissioner — known first and foremost as a television guru — had damned the league with pitiful ABC and ESPN deals (that, in truth, were not Simmons's doing). Trump's words were poison, and they found the intended target. Simmons was one of the few USFL executives who saw Trump for what he was. Though loath to bad-mouth an owner behind his back, Simmons made an exception for Donald Trump. The man was a huckster. Expensive suits and cheesy hair and a way of walking past people without looking. Yes, Trump owned the New Jersey Generals. But Simmons wouldn't have trusted him to hold his wallet.

The optimism that carried the league through tough times was slowly being replaced by the whispers of a charlatan, dead set on getting his way. It wasn't that Trump disliked Simmons on a personal level, or even thought him to be a bad commissioner. Trump's brain didn't work in such a way. *Like? Dislike?* Neither mattered. He simply realized that the collective goal of the USFL was long-term viability, while his goal was to use the USFL as a quick ticket to NFL riches. "I was the chairman of our long-range-planning committee," said Joe Canizaro, owner of the Breakers. "And Donald would call me every night at 6:30 to tell me how important it was for us to move to fall. If nothing else, he was persistent."

Against Simmons's wishes, the league's executive committee (comprised of Trump, Bassett, Taube, and Taubman) took Trump at his I-can-get-things-done word, and gave him permission to negotiate with the various networks on their behalf. But it was all a sham. Trump met with Neal Pilson, the president of CBS Sports, and Michael Weisman, executive producer of NBC Sports, and focused upon one hypothetical: *If the USFL played in the fall, would they carry the games?* Both men presumed Trump was asking on behalf of the league, and said, sure, they would *consider* the idea. This, of course, was not relayed by Trump to his USFL peers. Instead, he told them that the networks *preferred* that the USFL exist as a fall entity. As far as many of the owners knew, he was doing his altruistic best to help the league survive and thrive. Wrote Jim Byrne in *The $1 League*: "[Donald Trump] would keep after Pilson and Weisman and try to get one or both of them to make a commitment to the USFL in the fall. There was no reason to inform Simmons or the executive committee of the gist of his conversations or that they had even taken place. It was premature."

Although he was now one of professional sports' most high-profile owners, Trump mastered the shadows. In public, he loved the USFL and believed in its splendor. In private, he considered the USFL — in his future words — "small potatoes." Those wooed by his big talk and promises of riches defended the man as a visionary. Trump was the greatest salesman anyone had ever seen. His thinking was grand, huge, enormous, spectacular — hypnotic. He held sway, even when his sway made no sense. Those who didn't buy a bit of it considered him a Svengali of the selectively blind. "I think he became a Pied Piper," Simmons said years later. "He spun a web of stories to the rest of the owners that all this could be accomplished, and it could be accomplished if you followed what he believed in and the people that he knew and his ability to get to people who we couldn't get in order to raise the level of income from the networks, to get the interest of Pete Rozelle and the rest of the National Football League in terms of the accommodation of a few teams. And away he'd go being the savior of what was the United States Football League."

John Bassett, owner of the Tampa Bay Bandits, was particularly horrified by Trump's ability to woo and coerce. Though far from the league's wealthiest man, Bassett grasped the perils of a sports entity moving too fast, too hard.

He believed in the USFL because it committed itself to a slow, steady pace. Trump was anything but slow and steady. "My overall observation was that [Trump] was kind of like a self-appointed savior of the league," Bill McSherry, the USFL's executive director, told the *Washington Post*. "That self-appointment was accepted by many of the owners, if not all of them. Maybe the only one that didn't was John Bassett. John Bassett was the smartest guy in the room. And Trump always thought he was the smartest guy in the room, so you can imagine how well that went with anybody."

Bassett was particularly agitated when, on April 15, 1984, the *New York Times* ran a piece headlined USFL ENVISIONS FALL SCHEDULE BEGINNING IN 1987. The article featured insights from "two prominent USFL executives," and included the nuclear quote, "It was an executive decision made by the people who control the league. It is the only logical way for the league to continue. There's virtually no chance that it's not going to happen." The words, he knew, belonged to Donald Trump — and they undercut the very mission of the USFL. Trump actually put his name behind a few passages in the story, including this: "If I thought this league would not have gone for a fall schedule, I wouldn't have come into this league."

When asked about the article by Paul Domowitch of the *Sporting News,* Bassett refused to hold back. "The *New York Times* is supposed to be the most respected newspaper in the world and all they listen to is Donald Trump, who has duped them," he said. "[The story] is absolute nonsense. I hate to see [the *Times*] used by a conman. It upsets the hell out of me. Donald Trump is trying to manage the news and browbeat the rest of the owners in this league. In the end his philosophical view may be correct, but his tactics stink. At this point there is absolutely no basis to [the story] whatsoever."

Bassett's outrage was palpable, and he didn't even know the worst part. Shortly after the *Times* piece went to print, Trump arranged a secret meeting with (of all people) Pete Rozelle, the NFL's commissioner. It was held inside a room of New York City's glitzy Pierre Hotel and *paid for* by Trump. For the next three decades, the reason for and content of the sit-down would be debated. Was Trump trying to get into the NFL? Was the NFL trying to get Trump? Was this about killing the USFL? Merging the USFL? All of the above? None of the above?

When asked, Trump and Rozelle offered differing accounts. Here is what we know:

1. The meeting was arranged so Trump could see what he had to do to wind up in the NFL.
2. Trump made clear to Rozelle that the USFL was relatively unimportant.
3. Trump explained to Rozelle that the other USFL owners mattered not to him.

Three additional men were invited to the otherwise private meeting. Two never discussed it. A third was Leslie Schupak, cofounder and senior managing partner at KCSA Worldwide, the USFL's marketing and public relations firm. Trump formed a tight bond with Schupak, and 30 years later, the memory remained vivid. Schupak stayed for the early portion of the session, then left. In the immediate aftermath, however, he was briefed by one of the two men. "He told me that before Pete could start talking on a casual basis, Donald started his diatribe on how great he would be for the NFL, and what it would mean to the NFL to have him as a franchise owner," Schupak said. "Donald was apparently going on in his typical style, telling Pete Rozelle what he believes in, why he would be wonderful. It was typical Donald, in that Rozelle couldn't get a word in." According to Schupak's colleague, at the conclusion of the meeting Rozelle looked at Trump and said, "Mr. Trump, as long as I or my heirs are involved in the NFL, you will *never* be a franchise owner in the league."

"And that was that," said Schupak. "It was over."

Trump — gifted with the ability to hear and ignore information as he saw fit — remained undeterred. He was building something big, and he knew it. The sorrowful New Jersey Generals finished the 1983 season with a 6-12 record, and now — following a 31–21 Week 10 thumping of Michigan before 50,908 fans at the Meadowlands — his team was 8-2 and one game behind Philadelphia in the Atlantic Division. That the Generals' $5 million payroll was the league's second highest (behind the estimated $13.1 million spent by Los Angeles) was of no matter to Trump. *He was a winner, dammit, and winners spent whatever it took.*

The truth of the matter was the original USFL model actually worked — if anyone dared pay attention. In Jacksonville, the first-year Bulls were averaging a league-best 46,730 fans per game with not even a single household name or $1 million player on the roster. Bassett's Bandits, composed mostly of former Florida Gators and Florida State Seminoles (and with only one huge salary — running back Gary Anderson's $1.375 million over four years), were averaging 46,158 — largely based on quirky marketing and fun promotional giveaways. Yes, the majority of organizations were in the red. And, yes, TV numbers rose and dipped like an ocean wave. But the franchises losing the most were their own worst enemies. The Michigan Panthers, for example, dropped $6 million in 1983 — but spent outlandishly on three NFL offensive linemen. The Chicago Blitz (now Arizona Wranglers) reported losses of $3 million. Much of that, though, was due to George Allen's willingness to throw cash everywhere. Trump's Generals were hemorrhaging dollars. But whose fault was that? Who decided to hurl around money as if it were confetti?

Under David Dixon's initial plan, franchises were *supposed to be* losing money. It was an understood by-product of beginning a football league. Trump, though, was a master of, in one colleague's words, "banging the bruise." Instead of reminding people that most investments (including his own myriad real estate holdings) struggle before they surge, Trump mocked his peers for watching their dollars blow away. Tellingly, in all his the-USFL-is-a-disaster spewings, Trump never (not once) referred to its ultimate success story and (in a sense) model franchise.

In Denver, the Gold played in relative obscurity. Their black-and-gold uniforms were uninspired; their helmet design (an exploding gold starburst) among the league's dullest. In 1983 the club was owned by Rob Blanding, a notoriously thrifty construction executive. He sold the organization to Doug Spedding, an even thriftier head of a chain of used-car dealerships. "I don't want to frame the Denver Gold as cheaping out," said Jim Van Someren, the team's media relations director. "But for a year or two the idea was to keep expenses low, don't go crazy with contracts, and work within the framework of the USFL."

Over its first two seasons, Denver's roster was home to (count them) zero household names. The coaches — first Red Miller, then Craig Morton — were ex-Broncos with high regional appeal. Otherwise, it was a whole lot of Darryl Hemphill and Tom Davis; of Greg Gerken and Pat Ogrin. Their star, running back Harry Sydney out of the University of Kansas, had been cut in camp by the Seahawks and Bengals. He would not have been recognized had he walked through the city's downtown wearing his own jersey. "That's probably true, but the fans loved the Gold," said Sydney. "The spring weather in Denver was great, the people were passionate about their football. It wasn't fancy or especially electrifying, but the Gold was working."

For the 1983 season, Denver turned a small profit. For the 1984 season, the team lost a reasonable $1 million. They ranked first and fifth in the USFL in attendance, respectively. On the one hand, it wasn't always an ideal spot to be a player. "Spedding was a very bad guy . . . a crazy man," said Morton. "He started reading the players' mail, marking notes in the letters. It was bizarre. I actually told the team in meeting, 'I've always told you the truth, and this guy will not be a blast to play for.'"

And yet, Spedding did the organization well. Fred Mortensen, one of the quarterbacks, was simultaneously playing football and pursuing his MBA from Arizona State. He viewed the league with the intrigue of a player and the curiosity of a financier. "You could see how the Gold — for their flaws — resisted the temptations that killed other organizations," Mortensen said. "The original owners knew what the plan was. But too many got lost. Denver did not." It bothered no one that the Gold lacked a Herschel Walker or Steve Young. They played sound football and kept the payroll down. "The Gold were run exactly as the USFL model dictated it to be run," said Craig Penrose, a Denver quarterback. "We didn't have problems getting paid. Expenses were taken care of. Guys like Donald Trump wanted to be in the NFL. They didn't appreciate what we had in the USFL. They never gave it a chance."

As Trump was pushing for a move to fall, Oldenburg — his USFL brother from another mother — was heading in a different (but equally destructive) direction. On the field, the Los Angeles Express were exciting, daring, inventive, and living up (if that was even possible) to the $13 million Oldenburg was paying his players and coaches. Steve Young, in particular, was a show unto himself. His first start came on April 1 at home against the New Jersey

Generals, and 19,853 entered the Coliseum to watch him complete 19 of 29 passes for 163 yards in a 26–10 setback. The reviews were solid — "Well, maybe the guy *is* worth all the money," wrote Chris Dufresne of the *Los Angeles Times* — and improved three weeks later, when Young became the first player in the history of professional football to rush for 100 yards and pass for 300 yards in a single game. The 49–29 thrashing at the hands of the Chicago Blitz was played before only 11,713 souls at Soldier Field, but those who attended witnessed a merging of Fran Tarkenton and Harry Houdini. Teammates nicknamed him Hobbs, after Roy Hobbs from the film *The Natural*. Young spun, twirled, dipped his way to 120 yards on the ground (a team record), while hitting 25 of 37 passes for 302 yards. "Steve was amazing," said Jo-Jo Townsell, an Express wide receiver. "I'd never seen a quarterback like him. He did 1,000 things at the highest level."

After starting the season 2-5, the Express went on a roll, at one point winning five straight behind Young and his wealthy teammates. The quarterback spent much time sitting alongside assistant coach Sid Gillman, the legendary passing guru (who, it can be noted, referred to anyone who made a mistake as "slapdick"). On one memorable afternoon, Gillman looked over a practice field, spun to face Young, and said, "Steve, this is not a game. It's a canvas and you are Michelangelo."

"I had never heard anyone talk about football in those terms," Young recalled. "It is a game of brute force, an unforgiving contest of wills that leaves men bloodied, bruised, and broken. The concept of the quarterback as an artist changed my approach to passing."

Yet no matter what magic Young pulled out of his hat, it couldn't save the Express. The attendance figures were embarrassing — 10,049 for a visit from Memphis, 10,193 for the defending-champion Panthers. There were simply too many competing entertainment opportunities in the city. Young felt the pressure of $40 million, and it followed him like a bad scent. "The kid wore ripped jeans and he had holes in his socks," said Dufresne. "The team had to call him and tell him to deposit his checks, because he was just sticking them in a drawer and it was messing up payroll. I assure you, Steve was about anything *but* the money." When, on June 3, the Express trekked to Washington to play in front of 5,263 in a win over the Federals, Young's parents, LeGrande and Sherry, drove down from Connecticut. At one point the Los Angeles quarterback lost his grip on a moist ball and fumbled. The crowd started a chant: "Forty million down the drain! Forty million down the drain!" Sherry couldn't take it. She stood, faced her boy's tormentors, and screamed, "That's my son! He's a good boy! And by the way, it's . . . an . . . annuity!" Her words were drowned out by the repeated mantra, "Fuck you! Fuck you! Fuck . . ." "My mom told me she would never attend another USFL game," Young said. "I told her that was probably a good idea."

Oldenburg could not believe what was happening. He purchased the Express with the understanding the franchise would own Los Angeles. A week before the opening of the 1984 season, Oldenburg hosted a dinner for team management at the Bistro restaurant in Beverly Hills. Midway through a seemingly joyful meal, he barreled toward one of his marketing executives, jabbed a finger in the man's face, and screamed, "Are we going to get 100,000 people for the first game? Are we?"

"Um, no," the employee replied. An inebriated Oldenburg began kicking and throwing chairs, then fired his entire marketing staff and ejected them from the restaurant. A few weeks later, Oldenburg threw a plate of spaghetti at John Hadl, the head coach. During a game against the Gold, Oldenburg grabbed Dick Daniels, the Express personnel director, shoved him in the chest, and screamed, "It isn't fucking young players! This is shit coaching!"

Some of the behavior could be dismissed as Oldenburg's trademark erraticism. Much of it was surely due to the crumbling of his empire. Well before the rest of the USFL knew Bill Oldenburg was a con man, the FBI knew (or strongly suspected) Bill Oldenburg was a con man. That April it began investigating an Investment Mortgage International Utah land deal, then started looking into Oldenburg's full business portfolio. By June, his image graced the front page of the *Wall Street Journal,* beneath the headline LOAN BANKER IS SAID TO LURE WEAK S&LS INTO MANY SHAKY DEALS. According to the piece, Oldenburg bought a piece of property for $800,000 and sold it to a savings and loan under his control for $55 million. Wrote G. Christian Hill and Victor F. Zonana: "IMI apparently is, indeed, kind of magical. Its aura of success is mostly an illusion spun out by Mr. Oldenburg, according to regulators, former employees, developers and lenders across the country, and court documents." Oldenburg's biography as a business genius was fiction. He was "nearly broke" in 1970 and settled a civil suit in 1974. He was named in 17 fraud lawsuits. One S&L regulator interviewed for the piece said many of Oldenburg's successes were "largely a figment of his imagination."

Here was a man who shopped for designer clothing on Rodeo Drive, who traveled via a Rolls-Royce and chartered jets, who dropped thousands upon thousands of dollars on single meals — *and he had no money.* "There have always been people with ambition and dreams who thought they could make 2 plus 2 equal 5," explained John French, head of mortgage banking for Grubb and Ellis, a San Francisco–based firm. "Mr. Oldenburg is trying to make 2 plus 2 equal 50."

When he purchased the Express, Oldenburg provided the USFL with a financial statement that showed his net worth to be $100 million. Because the league needed a team in Los Angeles, and fell for the flash and panache of men like Trump and Oldenburg, a detailed, look-over-absolutely-everything background check was never commissioned. "It may be that we didn't do our homework," Tad Taube, owner of the Invaders, said (correctly) at the time. On June 3, 1984, the *New York Times* ran an examination of Oldenburg's finances. Within a week, the owner's attorneys informed the USFL that the Express could no longer afford to pay their bills, and that Mr. Dynamite had imploded. "He disappeared," said Paul Sandrock, the Express controller. "Just one day, he was gone. Poof . . ."

The USFL was already trying to figure out how to handle Chicago, what with a vanishing owner leaving the franchise rudderless. Now it had Los Angeles. The league didn't waste time. It drew upon Oldenburg's $1.5 million letter of credit and, in the words of Jim Byrne, "arranged for additional financing to cover the team's expenses." Simmons fired off a threatening letter to Oldenburg, suggesting that should word leak of the Express possibly folding, the USFL would see him in court.

"It was a disaster," said Young. "We went from having all this money to having nothing."

Some 3,000 miles away, Trump smelled blood. The league's three big-market operations were New Jersey, Chicago, and Los Angeles, and two were on life support. Maybe, just maybe, the USFL could survive without the Blitz and Express (though its deal with ABC required franchises in all three of the markets). But it needed the Generals and, therefore, needed Donald Trump. The plan to move from spring to fall (and, ultimately, into the NFL) was time-consuming and onerous, but Trump could see the mechanisms slowly working. With each passing day and every nugget of bad news, Simmons's power eroded. He was, Trump told all who would listen, impotent and weak. The league required a leader, he insisted, not a lemming. In Chicago, meanwhile, Eddie Einhorn, part-owner of Major League Baseball's Chicago White Sox, had emerged as a potential buyer of the Blitz (or, if not the Blitz, a new Chicago franchise to replace the Blitz and start anew), which thrilled Simmons until it turned out he, like Trump, wanted the league to switch to the fall.

Everything seemed to be spiraling out of control — especially if one listened to Trump. Yet the doom and gloom was somewhat exaggerated. The USFL was *one and a half years old.* "We were an infant," said Jim Mora, coach of the Stars. "You have to expect a lot of growing pains." ABC's telecasts of USFL games fell from a 6.0 rating in 1983 to a 5.5 in 1984, and the figure was cited as a reason for dread. Yet upon closer examination, the drop — according to William Oscar Johnson of *Sports Illustrated* — was "roughly the same as the decrease all three networks have experienced of late in virtually all sports programing." Naysayers liked to point to the Washington Federals (average attendance: 7,694) as exhibit A of the USFL's disaster. Yet when Dixon introduced the USFL idea, he explained there would be a number of franchises that might struggle and, in all likelihood, have to relocate. Sure, Washington was a catastrophe. But Sherwood Weiser, a wealthy Miami hotel developer, agreed to purchase the franchise for between $5 million and $6 million and bring it to his hometown for the 1985 season. Putting league before self, Bassett surrendered the Bandits' South Florida territorial rights, and applauded as Weiser offered Howard Schnellenberger, the legendary University of Miami coach, a five-year contract worth $2.5 million. The new organization, to be named the Spirit of Miami, opened offices and hired a savvy NFL veteran named Ken Herock to serve as personnel director. "It was very promising," said Schnellenberger. "The Dolphins had a new stadium so we'd play in the Orange Bowl. We put the spirit of the city into our team at the University of Miami, and I was committed to doing that in the USFL. It was going to work — I had little doubt."

"It really had a chance," said Herock. "But it had to be smart."

The biggest indicator of league-wide problems wasn't TV numbers, but revenue losses. The Express dropped $15 million in 1984, the Maulers $10 million, and the Blitz $6 million. The Gamblers, Wranglers, Showboats, Panthers, and Generals (yes, Generals) all took approximately $5–6 million hits. Only the Bandits reported profits, and those were slight. But, as Bassett said repeatedly (and often loudly), who told the Maulers to spend a fortune on Mike Rozier? Who told the Express to drop $40 million on Young? Who told Trump to turn his roster into a museum of NFL veterans? The USFL's struggles were self-inflicted, but — Bassett insisted — didn't have to be that way.

If Bassett required a game to back up his point, all he needed to do was focus on a Week 17 matchup between the Birmingham Stallions and Memphis Showboats at the Liberty Bowl. In a metropolis that craved professional football, inside a stadium that was crumbling and antiquated, a sellout crowd of 50,079 witnessed a clash that featured everything the USFL tried to offer its fans. First, Cliff Stoudt, the Stallions quarterback, hit on 13 of 19 throws for 282 yards, including a thrilling 44-yard pass-lateral-run to Joe Cribbs, the team's splendid halfback. Second, the Showboats' Derrick Crawford, a 5-foot-10, 185-pound speedster from nearby Memphis State, made a series of electrifying, how-did-he-do-that catches, resulting in six grabs for 52 yards and two touchdowns. Third, in a play that was later watched and rewound and watched and rewound in both teams' facilities, Reggie White, Memphis's 6-foot-5, 290-pound defensive end, chased Cribbs down the sideline and captured him after a 50-yard pursuit. Fourth, while the home team wound up losing 35–20, the city's denizens were in love. The Showboats were not particularly good (they would finish 7-11 in their debut season), but the franchise was going about the process wisely. It spent big money on two regional college stars, White from Tennessee and quarterback Walter Lewis of Alabama, and forked over $440,000 over two years for linebacker Mike Whittington, the onetime New York Giants starter. As its coach, Memphis hired Pepper Rodgers, the former UCLA and Georgia Tech head man and a walking, talking PR machine. Though far from football's most adroit Xs and Os practitioner, Rodgers devoted himself to pimping the league and the team. No event was too dull, no audience too small. "He always knew where the camera was," said Art Kuehn, Memphis's center and a former Seattle Seahawk. "For a new team, that was an important skill."

Rodgers had a line for every occasion, a chuckle for every minute, a sales pitch for every potential customer. Asked how he wooed a player to Memphis over the NFL, Rodgers said, "I take him out for a big spare ribs dinner, show him Memphis and take him to my big ol' house overlooking the beautiful Mississippi. Then I ask him, 'Boy, do you really want to play in Buffalo?'" Rodgers's assistant coaches were required to schedule time for naps. He never wore socks. He devised oddball offensive alignments off the top of his head and demanded the team give them a try. "He'd split the offensive line in half—three guys on one side of the field, two on the other," said Billy White, a halfback. "I thought it was the craziest thing ever." Rodgers was weirdly lovable. Once, on a team flight to New York, he wore an Air Force bomber jacket and stood at attention the entire trip. "I remember he came to practice in a tuxedo and football shoes," said Jairo Penaranda, a Memphis running back. "And he took every rep that entire week as the scout quarterback in the tux and cleats." One afternoon, following a long practice, the Showboat players punished Gary Shirk, a rookie tight end who refused to bring the veterans donuts, by taping his shoes, pants, shirt, watch, and car keys to the goalpost. Rodgers thought it was the funniest thing he'd ever seen. So the following week players grabbed *Rodgers* and taped him to a goalpost. "I was never around another coach who would allow that," said Bill "Brother" Oliver, the defensive backs coach. "The managers had to go back there and let him down."

"Pepper took me and another player out to a shot and beer place in Memphis," said John Banaszak, the defensive lineman. "We walk in and everyone yells out, 'Pepper! Pepper!' We're pounding beers, and after a couple of pitchers I said, 'We should probably get out of here. We have practice tomorrow morning.' Pepper says, 'No, no! Another round!' He finished that round then left us for a party."

Rodgers wasn't merely a drinker and night owl. Though probably not the only USFL coach to smoke marijuana, he was the most blatant about it. "Oh," said Oliver, "it wasn't hidden."

"One time I had to ask Pepper a question, so I knocked on his hotel room door," said Penaranda. "A huge cloud of smoke comes out, and he's totally naked, smoking a joint and standing there."

Rodgers was the face of the Showboats, but the talent belonged to Reggie White. Because he played defense, and because he was deeply religious and a non-self-promoter, White's signing with Memphis generated few national headlines. His contract ($4 million over five years) befitted his size and skill set, and it took but a month for opposing offensive linemen to see that the USFL had its own Lawrence Taylor–esque pass rusher. "Reggie didn't know how strong he was," said Alan Reid, a Memphis running back. "And fast, too. The guy ran a 4.65 40, he could slam a basketball. Off the field, as mellow as can be. On the field, a beast."

"I played with Lawrence," said Whittington. "Lawrence was faster and meaner, but Reggie had 60 pounds on him. You could double-team and contain L.T. a bit. But Reggie — no. Lawrence revolutionized the game, but Reggie changed offenses. Whatever they had planned was based upon his existence."

"We were playing Memphis, and our starting left tackle got hurt and left the game," said Fred Mortensen, a quarterback with Denver. "The backup comes in and he doesn't even slow Reggie down. Not a second's delay. He finally came to the huddle and said, 'I can't do this.'"

Thirty years later, teammates recalled a game against the Stallions when Birmingham's center cut the back of White's legs on a block. "That ain't right, man!" White yelled. "You're trying to ruin my career."

"Fuck you," the center replied.

White was playing to the left of nose tackle Brett "the Toaster" Williams. In the huddle he told Williams they were switching positions. It was not a request, but an order. White lined up across from the center on the ensuing play, and as the two men settled into their stances. White said, "Hey, you ever meet Jesus?"

Hutt . . .

Hutt . . .

"Reggie picked that guy up where only his toes were touching the ground," said Sam Clancy, a Memphis lineman. "He pushed him back 10 yards and slammed him into [quarterback] Cliff Stoudt for the sack. Nobody could believe it."

White was perfect for the USFL, in that he recognized his duty to not merely play in Memphis, but become one with the city. On off days he'd walk the streets, preaching the gospel. He'd shake hands and invite people to the Liberty Bowl. He was warm and friendly and the new franchise's greatest (non-Pepper) ambassador. He was also, at age 22, painfully naive. Midway through his rookie season, White decided to visit Laskey's Big and Tall, an upscale Memphis men's clothing store, to purchase a suit. He and Lewis shared an apartment, and that evening White returned and told his teammate, "Man, I had a big day today! I needed a wardrobe so I went shopping."

"Oh, yeah," said Lewis. "What did you buy?"

"I got some suits, some warm-ups," White replied.

"How much you spend?" Lewis asked.

"Oh," said White, "about $5,000."

The precise total was $5,262.

"What?" said Lewis. "On clothing?"

"Well, you know," White said. "Dressing isn't cheap."

The following morning, the *Memphis Commercial Appeal* ran a small piece about White having purchased 57 pairs of socks worth $199.50. Someone at Laskey's leaked the information, thinking it funny. Only White was not amused. The Showboats were participating in a local parade that began at noon, and when the defensive lineman arrived to board the bus he was greeted by a pair of veteran teammates, Myke Horton and Billy Roe. They pounced. "Reggie," said Horton, "what the hell does a man need 57 pairs of socks for?"

"Yeah," said Roe, "you gonna wear different socks every day of the year?"

The crestfallen White was near tears. He retreated to the rear of the bus and held his head in his hands. Rodgers pulled him aside and insisted it was no big deal. The next week, in a game against the Maulers at the Liberty Bowl, hundreds of Memphis fans arrived with socks atop their heads. They called themselves the Sock Exchange, and cheered every White move. What was once a humiliation turned triumph. "That was Reggie White," said Rodgers. "Things always worked out for him."

The Showboats were a model USFL franchise. They lost what ownership considered to be a reasonable amount of money, but gained a following. They relied on a roster made up primarily of regional college stars to establish fan loyalty, played hard, and got involved in the community. "I loved Memphis and Memphis loved us," said Reid. "I don't care what people like Donald Trump were saying. We had something that was working. The naysayers just weren't paying attention."

The regular season concluded on the night of Monday, June 25, as the Gamblers dominated the overwhelmed Showboats, 37–3, at the Astrodome. With that victory Houston secured a 13-5 mark and the top seed in the Western Conference. This was fantastic news for the USFL, which leaned heavily on Jim Kelly, the franchise's dynamic quarterback, for marketing prestige.

In a rookie season that rivaled anything the NFL had ever seen, Kelly passed for a pro-football-record 5,219 yards and 44 touchdowns. The Run and Shoot offense that Jerry Argovitz insisted upon, and Mouse Davis installed, was a thing of beauty. The league's top-two pass catchers were Gamblers (Richard Johnson had a pro-football-record 115 grabs for 1,455 yards and 15 touchdowns, and Ricky Sanders added 101 for 1,378 yards and 11 scores), and the team's 618 total points were the most ever scored on the planet. The league's weekly highlight show, *This Is the USFL*, served as a regular ode to the black-and-red Gambler uniforms streaking down the field, dancing in the end zone, speeding here and there at 400 mph. The USFL desperately sought to portray the NFL as the "No Fun League," and Houston provided the perfect contrast. "We were the best entertainment in the league," said Argovitz. "No close second." Players from the NFL's Houston Oilers used to come to the Astrodome to jealously watch the Gamblers. The Oilers offense was, by and large, Earl Campbell plowing into the line for 3 yards, then Earl Campbell plowing into the line for another 3 yards. "We were a rocket," said Johnson. "*Boom! Pow!*"

Inside the USFL offices, the hope was for the Gamblers to reach the championship game and square off with Philadelphia, who at 16-2 once again finished with the USFL's best mark. The Stars were so good and so consistent and so workmanlike that people who followed the league tended to take them for granted. Quarterback Chuck Fusina led the USFL with a 104.7 passer rating (he tossed 31 touchdowns and only nine interceptions), and halfback Kelvin Bryant was again the USFL's greatest all-around weapon, rushing for 1,406 yards and 13 touchdowns and catching 48 balls for 453 yards and a score. The defense, led by middle linebacker Sam Mills, was the USFL's best. "We could have hung with most of the NFL teams," said Dave Opfar, a Stars defensive lineman. "That was pretty clear." The lone thorn in the Stars' side seemed to be Trump's Generals, who handed Philadelphia its only two regular-season losses. The Philadelphia–New Jersey rivalry was a fierce one, in large part because Myles Tanenbaum, the Stars' owner, thought Trump to be a self-serving ignoramus intent on ruining the league. As opposed to the man who ran the Generals, Tanenbaum believed in patience, in slow growth, in thriving within Dixon's confines. Unlike Trump, whose fortune had been inherited from a wealthy father, Tanenbaum worked his way through Wharton School and Penn Law by selling fruit and life insurance and doing tax returns. A decade before Trump would receive five military deferments, Tanenbaum was serving two years in the Air Force. Trump, to him, symbolized everything wrong with privilege. The Generals and their buy-everyone approach felt cheap and easy. The Stars' rise through the USFL was due in large part to scouting and professionalism. Yes, Bryant and offensive linemen Irv Eatman and Bart Oates had been highly coveted collegiate standouts. But the vast majority of the team's slots were filled by hungry and overlooked men like Fusina, like Mills, like Sean Landeta, the punter out of Towson State. "Our bond was our drive to make it and prove people wrong," Landeta said. "That was some powerful shit."

The playoffs opened on the afternoon of Saturday, June 30, and while two of the games went as Simmons and his lieutenants had hoped — the Stars crushed the Generals, 28–7, before a scant 19,038 at Philadelphia's Franklin Field (Veterans Stadium was occupied by baseball's Phillies) and the Stallions outlasted the Bandits, 36–17, in front of 32,000 in Alabama — the weekend was, by and large, a magnificent disaster.

In Los Angeles, where Oldenburg's organization (it still, technically, belonged to him) was losing money by the minute, the Express and visiting Panthers played one of the most remarkable football contests in the sport's history. Trailing 21–13 with less than nine minutes remaining, the Express took over on their own 20. On the first play of the drive, Young dropped back, found no open receivers, then darted up the field. As Michigan linebacker Kyle Borland approached, the quarterback — a stubborn nonslider — rammed headfirst into a man who outweighed him by 20 pounds. "He hit me so hard I nearly lost consciousness," Young recalled. "I land out of bounds, right near Coach Hadl's feet. I wobble when I try to stand up."

Young refused to exit the game, and a few plays later scrambled for a first down on fourth and 1 while being drilled by Ron Osborn, the Panthers' free safety. "My right arm goes numb," Young recalled. "I hold it up with my left hand. I'm pretty sure my right hand is broken."

Again, the coaches demanded he leave. Again, he refused. A 22-yard pass to tight end Darren Long and a pass-interference call helped get Los Angeles to the Michigan 12. There were now 59 seconds left in regulation. It was third and goal. Young dropped back, pump-faked, ducked beneath a missed tackle, and sprinted wide and left toward the end zone. As he approached pay dirt Young lowered his shoulder and was pulverized, this time by cornerback Oliver Davis. He landed out of bounds at the 1, and on the next play — fourth and goal — halfback Kevin Nelson dove in for the score. It was now 21–19.

The USFL instituted the two-point conversion for this very reason. The ball was placed on the 3. The Express offense lined up. Young faked a handoff to halfback Tony Boddie, rolled to his left, pointed downfield, sprinted for the end zone, crossed the goal line, and was immediately hammered by John Corker, the ferocious linebacker. Both men writhed in pain on the ground, but the crowd went crazy.

Sports Illustrated later nicknamed Young "Little Stevie Wonderlust," and it fit. In completing 23 of 44 for 295 yards and rushing seven times for 44 yards, he was a dizzying whirling dervish. The game went into overtime, then another overtime, then a third overtime. It remains the longest professional football contest ever played. Young was brilliant and physically broken. Panthers kicker Novo Bojovic was merely broken — he missed two short field-goal tries that would have ended things. ("On the flight home, somewhere around the Rockies, John Banaszak threatened to throw Novo off the plane," recalled Ray Bentley. "I think he was kidding. Maybe.") Finally, on Los Angeles's 100th offensive play, rookie halfback Mel Gray took the handoff, cut to his right, and dashed 24 yards into the end zone. The Express won, 27–21. Wrote Ralph Wiley of *Sports Illustrated*: "Flat on his back and in obvious pain, but with the ball clutched in his left hand, [Gray] hazily saw an official signal touchdown — and then dropped the ball. He lay writhing in the end zone for 10 minutes; he had broken the humerus bone in his left arm."

It was amazing, startling, fabulous. "I spent four hours in the bath [afterward]," said Jo-Jo Townsell, the Los Angeles wide receiver. "Then I lay in bed all day and had someone nice take care of me."

So why was the USFL disappointed? Because nobody watched. ABC abruptly switched from the game to local programming after the completion of the first extra period. A pathetic 7,964 fans entered the Los Angeles Coliseum, some even holding tickets that were not giveaways. "You could hear the vendors screaming, 'Peanuts! Popcorn!'" recalled Danny Rich, the Express linebacker. "It was so disheartening." A few weeks earlier, the Express had appeared to hit an all-time low when a postgame Chuck Berry concert drew about 17 people. (Recalled Tom Vasich, an attendee: "Old Chuck's rhythm guitar echoed throughout the empty stadium.") This — playoffs, national television, $40 million quarterback — was significantly worse. Every time ABC's cameras panned the stadium, the plastic empty seats seemed to taunt the league's very existence. *Playoffs? This is your playoffs?*

The following day, Houston hosted Arizona, and 32,713 entered the Astrodome with the expectation that the Gamblers would light up George Allen's antiquated collection of geriatric NFL castoffs. The home team was favored by 2½ points, and many experts considered the contest's outcome a foregone conclusion. Yet in all the hype over the Run and Shoot, there was one systemic flaw: the defense spent *a lot* of time on the field. Were the Gamblers scoring 30 or 40 points, it was no problem. But against the Wranglers, they were held to a season-low 16. "Their defense was dog-tired," Arizona fullback Kevin Long said afterward. "Maybe their offense has always put them so far in front the defense hasn't had to play hard in the fourth quarter before."

Allen, a defensive mastermind dating back to his start in 1947 as an assistant on the long-disbanded University of Michigan 150-pound team, used to tell his players, "Defense wins championships, offense sells tickets, and special teams wins games." Against Houston, he employed a bend-don't-break game plan that was a departure from his usual aggressive man-to-man approach. By refusing to press the Gamblers' speedy wideouts, the Wranglers watched as Kelly settled for short completions and frustrated glares. Allen, wrote Bob Hurt of the *Arizona Republic,* "spent the week preaching poise and patience. He had warned the defenders not to get upset by a rash of short completions."

The close final score, 17–16, revealed a painfully dull contest, and also thrust the USFL into yet another humiliating quandary. The Wranglers were now expected to travel to Los Angeles for the following weekend's Western Conference championship match. Only, the Los Angeles Coliseum was readying for the Summer Olympics and unavailable. Therefore, instead of trying to hold the game in the regionally compatible Rose Bowl in Pasadena, or perhaps Jack Murphy Stadium in San Diego, the USFL decided it should be played at Sun Devils Stadium, the Wranglers' home turf. This was strange enough news, but the weather report for the scheduled 12:30 Saturday kickoff — 103 degrees, 28 percent humidity — was crippling. Two days before the game, the USFL changed the time to 8:30 p.m., and a local TV blackout was lifted. Amateur hour had arrived.

In the east, the Stars cruised past the Stallions, 20–10, to reach a second-straight title game. They would face the Wranglers, who capitalized on four Express turnovers to capture a 35–23 away-but-really-home triumph. "It was hot as hell, and too many of our guys used white crosses [a.k.a. speed] before games," said Sam Norris, an Express linebacker. "Speed works, but if it's 100 degrees out it drains you. We were drained, and they spanked us." Afterward, Allen, 62 but spry, gathered his players in a circle and led them in a cheer:

Hooray for Wranglers!
Hooray at last!
Hooray for Wranglers!
They're a horse's ass!

The whole evening would have been strange enough without the presence of Oldenburg, who somehow convinced a private jet outfit to ferry him from San Francisco to Phoenix to watch his team one final time. He arrived at the stadium with an entourage, and afterward told the scant collection of reporters that he had no interest in selling. "It would have to be a great offer," he said. "Building this team up from nothing to something is one of the most exciting things of my life."

Memo to Oldenburg: you have no money.

The Stars were in the championship game. That was good news. Otherwise, little was going as planned. The USFL had approved a budget of $1.35 million for player shares to cover the first two rounds of the playoffs, as well as $6,000 per member of the championship winner and $3,000 per championship loser. Those costs were supposed to be accounted for, in large part, by playoff gate receipts. Yet the attendance figures were disappointing and *highly* inflated. There were, reportedly, 33,188 people at the Express-Wranglers game. There was also no foreign substance ever placed on a Gaylord Perry baseball. "It was a joke," said Long, the Wranglers fullback. "Numbers rarely met reality."

Were Simmons merely trying to keep up a positive public image, it would be one thing. But he knew that, kicking back in a Trump Tower office, a certain New Jersey Generals' owner was observing all the mayhem and snickering. Every misstep and misdeed was yet another bullet to be used against the USFL's spring existence, and Trump took notes.

The second USFL championship clash, scheduled for Sunday, July 15, at Tampa Stadium, was an opportunity for Simmons and the league to steady themselves. Yet while the first title game felt bold and fresh, this time there was a sense of unenthusiastic staleness. With the exception of Philadelphia's Kelvin Bryant, who was as quiet as he was gifted, the matchup lacked marquee star power. A year earlier the USFL could sell Bobby Hebert and Anthony Carter and newness and inventiveness and a league on the fast rise. Now? *Pfft.*

Simmons flew into Tampa early in the week to handle the onslaught of media and promotional opportunities, only to find there were few to no media and promotional opportunities. "I remember Simmons was supposed to have a 'State of the USFL' press conference with the reporters, and he showed up wearing this big neck brace, and explained that he didn't have much time because of the pain [from a disk problem]," said Jim Lefko, covering the event for the *San Antonio Light*. "Later that night we see him at a Tampa steakhouse with all his cronies, laughing it up. And guess what? No neck brace." In his initial efforts to have Tampa host the game, Bassett guaranteed a packed stadium and $40 ticket prices. Now seats were selling at a snail's pace, and could be had for $5–$10. The only USFL-related story to make national news was an unfortunate one: three days before the game, three Wranglers employees (two coaches and a groundskeeper) were arrested for jumping into a $25,000 limousine parked in front of a nightclub, turning the key (left in the ignition), and driving it four blocks. Harry Burr, owner of the vehicle, pressed charges. "It makes us look bad that somebody's going around and pulling that kind of prank," Burr said. "It's not a laughing matter."

Ted Diethrich, the Wranglers' owner, demanded the employees immediately return to Phoenix. Allen was fine with Mark Whipple, an administrative aide/coach, and David Wood, the groundskeeper, not hanging around, but felt as if he needed Patrick Schmidt, Arizona's quality control specialist. "So we found Schmidt and we wouldn't let him leave his room," said Deek Pollard, the defensive backs coach. "That's not an exaggeration. We kept him hidden where he could prepare us, and we told the owner we had no idea where he was."

The kickoff was scheduled for 8:00 p.m., and fans filed into the stadium as dark clouds and streaks of lightning loomed in the distance. Simmons shook hands and smiled and promised a great game to all who asked, but he understood the reality. Winners of 16 of their last 17, the Stars were a juggernaut, while the Wranglers were merely good. Philadelphia was also playing with a distinct competitive advantage. A few days before the game, Steve Des Georges, Arizona's director of public relations, met in the lobby restaurant of the Tampa Marriott with the ABC television crew. Des Georges had been allowed to sit in on several of Allen's coaching sessions, and he knew the team's strategy inside and out. "I told them who was healthy, who was hurt, what we do on third down," recalled Des Georges. "And as I got up I saw Bob Moore, the Stars' PR guy, standing in the wings of the hotel restaurant."

"Hey, Steve," Moore said.

"Hey, Bob," Des Georges said.

"Thank you for the game plan," Moore said. "That was terrific." He turned, did a waving motion toward the restaurant, and a dozen Stars staffers and players rose to leave.

"I felt like the biggest fool on the planet," Des Georges said years later. "George Allen was the one usually taking advantage. And here I was, and I gave the Stars so much."

In 1983, the Philadelphia-Michigan championship game was such a fantastic battle that, afterward, the USFL went to great lengths to note the NFL's recent string of Super Bowl flops. This time, as the Stars jumped out to a 13–0 first-quarter lead, the flop belonged to Simmons. The Wranglers looked old and sluggish, the Stars fresh and explosive. Along the Arizona sideline, scores of players were wilting in the 100-degree heat (an on-field thermometer registered 130 degrees). Thirteen were hooked up to IVs. Kit Lathrop, the All-USFL defensive lineman, was pursuing Fusina on one play when he slid across the turf, felt his right hand burn, and came up with a nasty blister that stretched across his entire palm. Midway through the third quarter, defensive back Bruce Laird approached Hal Wyatt, the team's trainer, and said, "My feet are sliding in my shoes because they're so wet." Wyatt sat down, pulled off his socks, and handed them to Laird. "It didn't help," said Wyatt. "We got killed."

Ralph Wiley, the excellent *SI* scribe, described Philadelphia's 23–3 romp as "cheerless." He also wrote that 52,622 fans watched — and, in this, he was off, by, oh, 12,622. In the game's worst-kept secret, the league yet again inflated the attendance to save face. Not that anyone cared. "The game stunk," said Lathrop. "Horrible day. Why would anyone stay for the whole game?" Indeed, by the time the attendance was announced, a solid 80 percent of spectators had gone home.

The following afternoon the Stars were feted in Philadelphia with a tickertape parade down Broad Street. The players were loaded onto a fleet of flatbed trucks and ferried through the city. An estimated 250,000 locals lined up to salute a team many in town believed could beat the NFL's Eagles. Bottles of beers were tossed to the Stars. Confetti dropped from the sky. "It's July, so you have open buses and we're waving, slapping high fives," said Sean Landeta, the punter. "It was craziness. And the rings we were given rivaled the Super Bowl rings. For me, life as a football player was never better than it was in the USFL. It was the most fun a guy could have.

"I wanted it to last forever."

ACT THREE

I was at a Generals game with my brothers Dave and Jeff. We some-how started talking to two of the players, James Lockette and Tom Woodland. I told them that after the game we were gonna tailgate in the parking lot of the Meadowlands. You say that kind of thing, but you never expect a response. Well, the game ends, we're grilling, and who joins us? Lockette and Woodland. Think that happens in the NFL? Ever?

— John Kovach, New Jersey Generals fan

12

MOVEMENT

Why, they let people buy into the league that I wouldn't even let into my box.

— Alfred Taubman, owner, Michigan Panthers

THE REPORT COST $600,000.

The figure is important, because with the USFL losing money as if it were water rushing down a drain, big dough wasn't to be spent frivolously. That's why, when — in the spring of 1984 — the league commissioned McKinsey & Company, a management-consultant firm, to piece together a lengthy and detailed study on the demographics of switching to a fall schedule, it was with the idea that the findings would be taken *v-e-r-y* seriously.

At the time, owners were sparring with owners, and officials were sparring with officials. The one thing everyone seemed to agree on was the idea of the study. For those who wanted to move to fall, it would surely conclude the USFL needed to move to fall. For those who wanted to stay in spring, it would surely conclude the USFL needed to stay in spring. In a league that often seemed to be plagued by uncertainty, both sides were certain McKinsey & Company would support *their* perspective.

On the morning of August 21, 1984, the USFL's owners gathered inside a conference room at the Hyatt Regency Hotel in Chicago to finally hear the presentation of what was officially known as the Long-Range Planning Committee study, then make an official decision on the league's future. It was an awkward time with awkward emotions, made 1,000 times worse by a *New York Times* piece that ran one day earlier with the headline USFL SET FOR FALL PLAY IN '86. When Chet Simmons heard of the article, he felt like slamming his fist into a wall — or Donald Trump's head. Once again, the Generals' owner had manipulated the media and taken the league's future into his own hands. *How,* the commissioner wanted to know, *could a decision have been made if they hadn't even heard the results of the report?*

With approximately 25 men gathered around a conference table, Sharon Patrick — the McKinsey executive who managed the study — stepped forward. Over the past few months she had interviewed team owners, league honchos, TV executives. She learned everything there was to learn about the inner and outer workings of the United States Football League. And now, at this moment, she was . . . *pissed.*

The USFL, she made clear, was a good idea. Spring football. Managed costs. Untapped markets. Fun promotions. Slow growth. Plugging into regional allegiances. It was smart and inventive and borderline revolutionary. And yet, the men before her were screwing it all up. According to Patrick, the USFL lacked unity. There were few common goals and a nonstop barrage of disagreement, contention, hostility. Patrick turned to a public-opinion survey that had been conducted by Yankelovich, Skelly & White. Much of the information was positive — the USFL's name appeal was near universal, and people loved the idea of a rival league. But, thanks to the public sniping and conflicting headlines, nobody seemed to trust the USFL's long-term sustainability.

Finally, after speaking uninterrupted for nearly an hour, Patrick announced the bottom line: the USFL, she concluded, *needed* to stay in the spring for at least the 1985 and 1986 seasons, and perhaps even longer. Then, Patrick wrote in the report, "once the full ownership makes its decisions, strongly and publicly affirm collective and individual concerns concerning the League's future to underscore its permanence to the public and to reduce conjecturing about its future in the media."

With that, Sharon Patrick waited for the barrage of questions. She gazed out toward an ocean of suits and ties and blank stares and bored looks. "Patrick was thanked for her efforts and assured that her information would prove invaluable in their discussions," Jim Byrne wrote. "She was excused from the meeting and asked to remain available in the event there were any subsequent questions."

As soon as Patrick exited, Simmons opened the meeting up for comments. The first man to speak — *always* the first man to speak — was Donald Trump, who slammed the report as "bullshit." He said, should the USFL decide to stick with spring, it would likely have to do so without his presence. He then went on a lengthy rant about the "great potential" of the fall television market and asked Eddie Einhorn, the presumptive owner of the Chicago franchise (which was still controlled by the league), to elaborate. Einhorn rose, cleared his throat, and backed Trump's idea. Having once masterminded Major League Baseball's $1.2 billion TV contract, he was considered a guru in the medium. He now said his myriad conversations with television executives convinced him there would be plenty of fall USFL telecasting, and that the league could expect "approximately $90 million" off of the inevitable multi-network deals. When asked whether he could be more specific — which networks, which executives — Einhorn turned quiet. "That," he said, "will have to wait."

Enough was enough. Simmons adjourned the meeting, and afterward the owners held a press conference with their commissioner front and center. When he initially took the job, a large part of the appeal was being able to guide a league toward prominence. Now, however, Simmons was reduced to humiliated puppet. "A pretty good size group of writers came, and I was told to say that the move to the fall is very important and we're going to look into it and go eventually," he confessed years later to the writer Paul Reeths. "I said, 'I don't want to say that.' They said, 'Well, that's what it's going to say.' That's what I faced. It was not all of them, but a bunch of them. Donald was in the group that told me, 'To hell with what McKinsey said; we know better and we're going to move to fall.' My instinct at that point was to say the owners said this is what I should say. That was my instinct, but I knew if I did that it would not be fun and I didn't want to do that at that point, but I did it privately with every writer in the room."

Simmons had as little regard for Einhorn as Trump. The so-called "television genius" had no idea whereof he spoke. Roone Arledge, ABC's news and sports president, told *USA Today* that the network liked the USFL because it was a "small, modest programming venture." Neither ABC nor ESPN had any interest in televising non-NFL fall football games. Einhorn's promises were either exaggerations or lies.

In the end, it mattered not. On August 22, the league's owners — spooked in large part by Trump's doomsday promises — voted and officially reached the decision that the USFL would switch to fall beginning in 1986. Gun to head, Simmons announced the move was made unanimously. That was a lie. There needed to be a two-thirds majority approval for a move to fall — and there was. Trump had won in convincing his peers that riches awaited on the other side of the seasonal rainbow. The blunt weight of his thuggish insistence proved irresistible. Patrick had told the owners that taking on the NFL was likely death. They heard her, listened to her convincing arguments, and opted to line up behind their new king. Too many of the USFL's power players would kneel before Zod. Donald Trump was right. He *had to be* right. "The vote was eventually called unanimous, but it really wasn't," said Bill McSherry, the USFL's executive director. "I think on the final vote everybody voted yes, but there was a lot of discussion." Some owners thought it was pure stupidity. Others considered it the only path toward survival. Marvin Warner, owner of the Stallions, hated the idea of having to compete with Alabama's and Auburn's college seasons — but hated the current spring TV deal even more. He voted fall. Jerry Argovitz of the Houston Gamblers loved the Astrodome, loved the Run and Shoot, loved spring football. Like Gold versus Broncos, his Gamblers would surely be crushed by the Oilers. "But I was a guy who did it for the team," he said. "It felt like most of my peers wanted to move. So I was OK with fall." The Maulers, owned by Eddie DeBartolo, would disband if the league changed seasons. Bobby Taubman, the son of Panthers' owner Alfred Taubman, was sitting in on his father's behalf. He told those present that they were embarking on a suicide mission, that this would not work. The Panthers were one of the USFL's elite franchises. But there was no way they could survive in Detroit alongside the NFL's Lions.

Myles Tanenbaum of the Stars was less dramatic, yet he, too, could not support the spring move. His team was the best the USFL had to offer, and could likely hold its own in an on-field battle with the Philadelphia Eagles. But, in the world of stadium leases and attendance figures, the Stars would be financially steamrolled. He wanted to stick to spring.

Donald Trump's dream — born and nurtured from the seedlings and roots of narcissism — was set to come true. "It was Donald's sheer force of personality," said McSherry. "For good or bad, he was a leader of men."

John Bassett desperately desired to stay the course. He also knew a lost cause when he saw one. Trump was about to get exactly what he wanted, and there was nothing he could do about it. Bassett was neither the wealthiest nor most intelligent man in the room, but his skill set included being able to sniff out a fraud from miles away. "Donald thought John was a knucklehead," said D. J. Mackovets, the Bandits' director of media relations. "But John was one of the brightest people I worked for in 40 years." A mere six days earlier, Bassett had fired off a letter to Trump that was written on Tampa Bay Bandits stationery, and mailed to Trump Tower's 26th floor:

> Dear Donald:
>
> On a number of occasions over the past meetings, I have listened with astonishment at your personal abuse of the commissioner and various of your partners if they did not happen to espouse one of your causes or agree with one of your arguments. It is obvious from the record that you are a talented and successful young man. It is also a fact that I regard you as a friend and an owner who has made a contribution to the league in general and been a savior to New York/Jersey in particular.
>
> While others may be able to let your insensitive and denigrating comments pass, I no longer will.
>
> You are bigger, younger, and stronger than I, which means I'll have no regrets whatsoever punching you right in the mouth the next time an instance occurs where you personally scorn me, or anyone else, who does not happen to salute and dance to your tune.
>
> I really hope you don't know that you are doing it, but you are not only damaging yourself with your associates, but alienating them as well.

Think before you shoot and when you do fire, stick to the message without killing the messenger.

Kindest personal regards,

John F. Bassett

"That letter is so Dad," said John Bassett Jr., his son. "He just wasn't afraid to call people out. I don't know if he'd call Donald the enemy, but he was certainly a foe in the league."

Now, faced with the near-certain death of a league he believed in, Bassett meekly expressed his preference to remain in spring, coupled by the thin hope that fall would somehow work itself out.

This would not go smoothly.

This *could not* go smoothly.

Not with Donald Trump now calling most of the shots. Not with Chet Simmons diminishing in stature by the day. Not with more uncertainty than certainty. The ensuing months brought forth a bevy of shifts and slides that raised the question of whether the USFL was a football league or some sort of precursor to a sports-themed reality-TV show.

It was a little bit of both.

Wrote Howard Balzer of the *Sporting News*: "The USFL ceased being a league when owners who were committed to making a go of spring football gave in to the likes of Donald Trump. Scriptwriters already may be working on a spring release of 'Indiana Trump and the Temple of Doom' to play at a theater near you."

Indeed.

With the decision to go fall in 1986, a number of USFL owners and franchises found themselves lost and bewildered and shuffling aimlessly like blindfolded pigs atop an ice rink. The Maulers were the first clear casualty —not only was Eddie DeBartolo Sr., the club's dignified owner, unwilling to play in the same city at the same time as the legendary Pittsburgh Steelers, but he was also unwilling to rival his son, San Francisco 49ers' owner Eddie DeBartolo Jr. "[Team president] Paul Martha called me to his office one day," said George Heddleston, the Maulers' general manager. "We sat down at his big desk and he said, 'George, I think Senior is going to put the team out of business.' He's not going to do fall football. He didn't get into the USFL to fight the NFL.' It was one of the most crushing moments of my career. But I understood. The USFL was embarking on a suicide mission."

The Oklahoma Outlaws followed suit. Though the team made a mark by adding former Tampa Bay Buccaneers quarterback Doug Williams before its debut season, not much went well. Playing on a battered left knee, behind a Swiss cheese offensive line with nary a single go-to offensive weapon, Williams was sacked 30 times, while throwing 21 interceptions with just 15 touchdowns. "There was no talent around him," said Isaiah West, an Outlaws linebacker. "He tried his best. We all did. But . . ."

"We had a fan base in Tulsa," said Kelvin Middleton, the Oklahoma defensive back. "Just not a huge one."

Indeed, the Outlaws drew 21,038 per game, which would surely plummet playing in the fall against a lineup of regional college-football games. In an odd twist, Oklahoma planned on relocating to Phoenix and merging with the Arizona Wranglers, who in large part were considered one of the USFL's successes. Behind the scenes, however, Dr. Ted Diethrich was tired of losing money, and he decided to surrender ownership to the Outlaws' Bill Tatham (technically, the Wranglers folded and Outlaws took their place). In all the mayhem, the one thing Diethrich forgot to do was inform his general manager. Neither Bruce Allen nor his father, coach George Allen, were in the new organization's future plans, and they were not happy. On October 30, Bruce Allen held an impromptu press conference at a Phoenix bar, the Dreamy Draw. The setting — loud noise, drunk patrons, a jukebox filled with sad country songs — felt appropriate. "They think they know what they're doing, but they don't," Bruce Allen said of the Outlaws officials who would be taking over the franchise's operations. "This club doesn't need one Oklahoma player. Right now this team is one game away from the championship. We're right on target. Our 70th player is better than the Outlaws' first. The worst thing is that the people of Phoenix have lost."

It mattered not. In 1985 the *Arizona Outlaws* would be coming to town. And it wasn't even the most unexpected merger. Reluctant to challenge the Lions for fall football supremacy, Taubman's Michigan Panthers paid the

Pontiac Silverdome a $1.325 million release settlement, packed up their belongings, and folded into the Oakland Invaders, thereby ending one of the USFL's great success stories. The Panthers had won the league's first championship, then reached the playoffs again in 1984. However, football in Michigan was *expensive*. The Panthers had commissioned a study of conditions in Detroit versus other USFL cities, and the conclusion was that they were in the wrong metropolis. The Silverdome lease cost more than $1 million annually, advertising rates in local media were outrageous, and fans had little patience for anything but success. "Very simply," the report concluded, "the Michigan Panthers would be better off moving out of the Michigan market." Now they ceased to exist.

"You have to remember, Al Taubman was a Lions fan," said Vince Lombardi Jr., the Panthers' general manager. "He refused to even consider challenging them. So that was the end of the USFL in Detroit."

Because the Philadelphia Eagles spent their fall Sundays performing inside Veterans Stadium, the Stars, too, found themselves without a viable home. Hence, Myles Tanenbaum transformed the Philadelphia Stars into the Baltimore Stars. From afar it made perfect sense. On March 28, 1984, Robert Irsay relocated his Baltimore Colts to Indianapolis, and the city — a football hotbed — was desperate for a new NFL franchise. "Clearly the plan was either the USFL survives and we're the Baltimore Stars of the USFL, or the NFL adds us and we're the Baltimore Stars of the NFL," said Scott Fitzkee, the team's star receiver. "But it didn't work out that way."

Edward Bennett Williams, the owner of the Baltimore Orioles, cared little about football, and he wasn't about to split Memorial Stadium with the Stars. The Orioles had recently signed a three-year lease with the city, and the document banned any football team from sharing the facility. The Stars, therefore, maintained their Philadelphia-based offices (inside Veterans Stadium) and practice space, but would play home games two hours away at the University of Maryland's football stadium in College Park. "It was a crazy idea," said David Riley, a Stars running back. "But what were the other options?"

Everything was dizzying. The Wranglers were now the Outlaws, but still in Arizona. The Panthers were part of the Invaders, with Bobby Hebert at quarterback and Anthony Carter a star wide receiver. The Washington Federals had planned on becoming the Spirit of Miami, but that spirit died as soon as the fall transition became inevitable. Sherwood Weiser, the wealthy real estate developer who agreed to pay $6 million for the Federals, wasn't about to go head-to-head against Don Shula's Miami Dolphins. He wisely backed out, and the franchise was instead sold to an Orlando, Florida–based businessman named Donald R. Dizney.

This caused some initial national confusion. Orlando's Dizney was not Orlando's Disney. The USFL would have preferred Disney, in that, in 1984, the Walt Disney Corporation had accumulated assets valued at $3 billion. But Dizney, the owner of a chain of acute care and psychiatric hospitals throughout the southeastern United States (as well as a minority owner of the Bandits), was successful and excitable and willing to supply the $250,000 down payment on what was rumored to be a $5 million sales price. Unlike Miami, Orlando was a non-NFL city that could, perhaps, thrive in the fall. After the deal was confirmed, the *Orlando Sentinel* sponsored a naming contest, and 10,000 people submitted suggestions. The monikers were narrowed down to Lazers, Lightning, Lynx, Condors, Avengers, Thunders, Wildcats, Juggernauts, Tribe — and the ultimate winner, Renegades. The man who submitted the victorious concept was Blake Harper, a 29-year-old plumber from Christmas, Florida, who admitted the idea came to him as he was plunging a customer's toilet. Harper was rewarded with an all-expenses-paid trip for two to the 1985 USFL championship game in glamorous Detroit. Which, with the death of the Panthers, was changed to even more glamorous East Rutherford, New Jersey.

The last big change, franchise-wise, was another direct result of the seasonal switch. Although the New Orleans Breakers had drawn well in the Crescent City for much of the 1984 season, averaging a respectable 30,557 fans per game, the NFL told local officials that, were a USFL team to remain in the Superdome, the Saints would seek another location. "There's no way around it — the NFL forced us out," said Joe Canizaro, the Breakers' owner. "We pretty much broke even our first season, and we were excited about coming back. But it wouldn't happen."

The options were limited. Canizaro loved professional football, and wanted to stick with the USFL. But where could he go? There were brief talks about merging New Orleans and Birmingham, but the idea went nowhere. Sacramento, California, was a quick thought. Then Bud Clark, mayor-elect of Portland, Oregon, began speaking nationally about his city as a future sports mecca. The Portland Trail Blazers were one of the NBA's best-drawing franchises, so why not football?

Before long, the Boston Breakers turned New Orleans Breakers were moving out west and being rechristened as the Portland Breakers. Their home, the Portland Civic Stadium, was a 60-year-old monstrosity that seated only 32,500. Their TV market was the nation's 21st. A decade earlier the World Football League's Portland Storm (coached, coincidentally, by Breakers coach Dick Coury) played in the stadium, and the results were disastrous. So why would it work this time? Well, save for the $1-per-year stadium lease offered by the city, nobody seemed quite sure. "I find this is an extraordinarily important day for the United States Football League and the city of Portland," Simmons said at a press conference announcing the move. "I believe we're getting a terrific sports market and are opening the entire Pacific Northwest to the league, and that is something I have coveted."

The words were delivered on November 13, 1984, from behind a podium inside the Civic Stadium. Standing alongside Canizaro and Vic Atiyeh, the governor of Oregon, Simmons looked and sounded convincing. The USFL, his cheek-to-cheek smile conveyed, was here to stay and about to be better than ever.

Hmm . . .

Nothing ever came easily for the USFL, especially after the decision to move to fall. For example, Simmons was praised for finding a purchaser of the Los Angeles Express. His name was Jay Roulier, a Denver-based real estate developer who had been a minority owner of the Houston Gamblers. One of the USFL's standard (and historically flawed) background checks found the 39-year-old Roulier to be well-off and financially stable (his assets were valued to be between $100 million and $125 million), and shortly after being approved he sought to show his commitment to the team by suggesting a $25,000 documentary be produced on . . . *the Express's cheerleading squad*. It was a head-scratcher — the film would be commissioned at the same time the team faced a $1 million debt, and word began to spread that, just maybe, this Roulier knew not whereof he spoke. Shortly thereafter, Jay Roulier Interests, Inc., filed a $10 million lawsuit against its namesake, alleging fraud and misapplication of funds. He was never heard from in USFL circles again, and the league was forced to continue to run the Express. "Los Angeles never truly worked for the USFL," said Chris Dufresne, the team's *Los Angeles Times* beat writer. "And as the years passed, everything just got harder and harder and harder."

Over its first two seasons, the USFL's collegiate draft was an event the league cheered and the NFL feared. It was a celebration of what could be. No, the USFL did not add every player it wanted. Or even most of the players it wanted. But whenever a Craig James, a Tim Spencer, a Mike Rozier, a Steve Young picked the upstart over the establishment, it fed and bolstered the USFL brand.

The third draft was slated to take place on January 3, 1985, but instead of hope, the prevailing emotion among USFL teams was apprehension. There was simply too much chaos and confusion to expect America's most talented college players to place their futures in the hands of a question mark. So, when the Birmingham Stallions led things off by taking a Mississippi Valley State wide receiver named Jerry Rice, Rollie Dotsch, the team's coach, told reporters that "Rice has great speed and runs a 4.45 40-yard dash. And to go with his great speed, he has a great pair of hands." And Jerry Sklar, the Stallions' president, insisted the team would do everything it could to sign Rice. And Rice maintained he was "very interested in the USFL." And . . . and . . .

It was blarney.

Jerry Rice was not coming to the USFL. Neither was Tory Nixon, a San Diego State cornerback plucked second overall by the Arizona Wrang— er, Outlaws. Neither was Issiac Holt, the Alcorn State defensive back who went third to the Gunslingers. The fourth overall selection, linebacker John Nevens of Cal State–Fullerton, did indeed become a Denver Gold — but only because the NFL had minimal interest in his talents. The open draft lasted 12 rounds, and was coupled by the territorial draft that spanned 25.

While the relatively anonymous Rice went first in the general phase, the big name belonged to a 5-foot-9¾-inch, 176-pound Boston College quarterback named Doug Flutie, who was selected by Trump's Generals in the territorial phase. By all measures, Flutie was a questionable professional prospect. His arm was strong, his instincts sharp. Were it not for a miracle Hail Mary pass to beat the University of Miami, however, as well as the resulting Heisman Trophy that came with it, Flutie was a borderline NFL prospect, perhaps a candidate to switch to wide receiver or tailback. There simply wasn't a market for elfin quarterbacks. "He wasn't impressive," said Doug Woodward, a Breakers quarterback who played collegiately at Pace University. "He was a nice guy, he was a competitor. But I didn't think he had the tools needed to compete in the NFL or USFL. I'm pretty sure that was a common belief among those of us playing at the time."

The Generals were one of the USFL franchises *least* in need of a quarterback. Their starter in 1984, Brian Sipe, was a longtime NFL standout who guided New Jersey to a 14-4 mark while throwing 17 touchdowns and 15 interceptions. Sipe was a savvy veteran who cost Trump $1.9 million. "Brian was excellent, and you have to remember we didn't really have great pass catchers," said Kevin MacConnell, the Generals' media relations director. "He got hurt a little, and that impacted the numbers. But he was still one of the league's best quarterbacks."

Trump knew little about football but everything about headlines and eyeballs. During the 1984 season, for example, he dressed the Brig-a-Dears, New Jersey's cheerleading squad, in the USFL's skimpiest outfits, making them look, in one member's words, "like hookers. The outfits fitted poorly in the back and exposed too much." By most accounts, the uniforms were tasteless and prone to vaginal/breast exposure — and Trump *loved* it. "He was an attention whore," said Jerry Argovitz, owner of the Gamblers. "No spotlight was too bright." On the field, his team needed Doug Flutie like it needed a ham sandwich, but the whole package was simply too irresistible. Over the past two months, since the November 23, 1984, Miami win (forever nicknamed "the Hail Flutie Game"), Flutie emerged as America's most-talked-about athletic superstar. He appeared on the cover of *Sports Illustrated,* on an endless number of talk shows. He was the subject of a *New Yorker* cartoon and the grand marshal of a parade. He helped carry gymnast Mary Lou Retton onto the stage for Bob Hope's Christmas special. Flutie was Tom

Cruise–handsome and aw-shucks humble, and the nation ate it all up. "Doug is a better person than he is a player," said Jack Bicknell, the Boston College head coach. "I've never known anyone like him."

Four days after the draft, Jay Seltzer, the Generals' president, met with Bob Woolf, Flutie's attorney. As negotiations were progressing, Trump told a reporter, "I don't know if we'll be lucky enough to sign him. But I'd go as much [money] as I think is reasonably ridiculous." Before long the two sides agreed on a six-year deal worth $8.3 million — the largest contract in the history of pro football. The first three years were guaranteed, should the USFL die. The announcement was pure balderdash. Flutie was introduced as a General at a press conference inside the Trump Tower atrium, and thanks to a cascading waterfall, those in attendance could pick up every fourth word. "We were yelling, 'Hey, we can't hear! We can't hear!'" recalled Barry Stanton, the *Reporter Dispatch* beat writer. "I swear, Donald snaps his fingers and the waterfalls turn off."

In all the hubbub, Trump neglected to tell Walt Michaels, his head coach, that Flutie was about to be a General. This would have been fine had Michaels wanted the quarterback — which he did not (Michaels preferred to stick with Sipe and try to sign Randall Cunningham, the senior quarterback from UNLV*). "I'm just the coach," Michaels later deadpanned. The other little nugget Trump failed to initially share was that he expected the rest of the USFL's owners to help foot the Flutie bill. As Trump saw it, the quarterback wasn't merely a Generals player, but a league-wide marketing gem who would help enrich all involved. He purchased multiple ad spots, calling Flutie the Miracle Man, and referred to his new star as "the Joe Namath of the USFL."[†] Trump *might* have been able to recognize Joe Namath were he wearing a sign that read I AM JOE NAMATH.

---

* According to an excellent 2016 piece from SI.com's Tim Rohan, Michaels far preferred Cunningham to Flutie. Wrote Rohan: "Michaels tried explaining to Trump why Cunningham was the wiser choice, and he brought in two assistant coaches for a demonstration. He pointed to his defensive coordinator, who stood about 6'3", and told Trump he wanted a quarterback of that size. Not one the size of his wide receivers coach, 5'9'."" ("Donald Trump and the USFL: A 'Beautiful' Circus," SI.com, July 12, 2016.)

† Thereby making him the 324th "Joe Namath of the USFL."

That was the biggest joke of it all — the man raving about his players understood literally nothing about the game. If one asked Trump a football question, "he sort of went blank," said George Vecsey, the *New York Times* columnist. "We all knew more about football than him."

Tim Rohan, the excellent *Sports Illustrated* football writer, chronicled a particularly telling moment from a 1984 game between the Generals and Breakers.

Wrote Rohan:

> People other than the beat writers had wondered how much Trump knew about football. When making decisions regarding the team, he sometimes canvassed people until he reached a consensus. A few people joked that he even asked the janitor his opinion.
>
> On one kickoff, the Generals' kicker sent an awkward wobbler to about the five-yard line, and the returner was swarmed by tacklers as he struggled to pick the ball up. [General Manager Jim] Valek screamed during the play, while Trump watched quietly.
>
> He turned to Valek. "How do you like that kick, Jim? Effective, right? Powerful." He jabbed the air with his right fist twice to emphasize his point.
>
> "Effective if they can get him quickly," Valek replied.
>
> "That's what happens."
>
> "Yeah."
>
> Trump paused and then leaned over again. "Is that intentional, a kick like that?"
>
> "No, no, I think he missed it."

With Flutie's signing, Trump utilized John Barron, his pretend spokesperson (a.k.a. Trump doctoring his voice). Barron called a reporter from United Press International and said, "When a guy goes out and spends more money than a player is worth, he expects to get partial reimbursement from the other owners. Everybody asked Trump to go out and sign Flutie . . . for the good of the league."

None of the USFL owners took the request seriously. In fact, the idea was ridiculed as a preposterous piece of silliness. "There won't be any takers," said Tanenbaum. "The rest of us already gave at the office. And at home. And everywhere else." From afar, the NFL once again guffawed. If Trump thought the Flutie addition would impress and/or intimidate Pete Rozelle, he was badly mistaken. Herschel Walker was a Heisman Trophy winner the NFL coveted. Mike Rozier was a Heisman Trophy winner the NFL coveted. Doug Flutie? Meh. The contract Los Angeles had handed Steve Young was farcical, but at least he was going to be the No. 1 selection in the NFL Draft. Hype aside, Flutie was no Steve Young. Or Steve Dills. Or Steve Pisarkiewicz. "He's a perfect college player," said George Young, the Giants' general manager. "But that doesn't make him a great pro."

Because the addition of a third-straight Heisman Trophy winner was announced with such fanfare, another fresh USFL news item was somewhat overlooked. Yet standing on the stage alongside Trump and Flutie at the introductory press conference, appearing dapper if not somewhat diminished, was Harry Usher — the league's new commissioner. The news was officially announced a couple of days earlier, via a *New York Times* headline that read, simply, USHER NAMED U.S.F.L. HEAD. The story — only nine paragraphs long — was neither particularly interesting nor shocking. Trump and several of his peers had devoted months to undercutting Simmons, and now the fruits of their labor had paid off. Officially, Simmons "resigned." Unofficially, he was bludgeoned to death, and grateful to be done with the whole USFL fiasco. "Chet wanted to get out of there in the worst way," said Bill McSherry, the league's executive director. "He knew what a disaster looked like."

"It was kind of mutual," Simmons said. "I didn't want any more of them, they didn't want any more of me and that's lovely. I really wanted out of there. I wanted away from the owners. I wanted away from the stupidity. I wanted away from Donald Trump."

Unlike Simmons, who always appeared as if he'd escaped the innards of a vending machine, Usher was dapper, poised, corporate. At 45, he was fresh off a successful stint as the executive vice president of the Los Angeles Olympic Organizing Committee, and he was the only choice of Trump and Co. to replace Simmons. The first contact between the league and Usher actually took place in October 1984, when Don Klosterman, the Express president and a casual acquaintance, placed an out-of-the-blue call to gauge his interest in the position, should it open. A former Beverly Hills–based attorney with a focus on the entertainment business, Usher's Olympic run caused him to seek out a career path that felt . . . *exciting*. Usher told Klosterman that, yes, he would certainly be intrigued by the USFL. A few weeks later Trump reached out, and the two spoke at length. Again, Usher sounded interested — even with the switch to fall and the television drama and the out-of-control spending and the egos.

Even with the antitrust lawsuit.

That's right. The lawsuit. It was, of course, Trump's idea — a plan for the USFL to sue the NFL for (in short) monopolizing professional football. Rozelle identified the wounded animal that was the USFL. He knew it was coming, and actually mailed Simmons a letter shortly before anything was filed. Wrote the NFL commissioner: "It is . . . becoming clearer and clearer that a treble-damage lawsuit features at least as strongly in the USFL plans as does making your league a business and entertainment success."

Bingo.

Filed in the United States District Court for the Southern District of New York on October 18, 1984, the 39-page complaint focused specifically on the NFL having broadcast arrangements with ABC, NBC, and CBS — the three commercial television networks. The suit, which aimed to limit the NFL to *two* networks and charged Rozelle's league with monopoly practices making it impossible for rival football entities to exist, sought damages of $440 million (trebled to $1.32 billion under the Sherman Anti-Trust Act). Although Simmons was still, technically, commissioner when the lawsuit was filed, he was a figurehead. The man who submitted the suit was McSherry, the former league counsel now serving as its executive director. "I did the initial research and concluded the NFL was a monopoly," McSherry said. "The USFL was a mess, no question. But I honestly thought the case against the NFL was a strong one."

The morning after he filed the complaint, McSherry was asleep in his Larchmont, New York, home when he was stirred awake by his wife, Betty. "Bill," she said, "you have to see this!" The nearest television was tuned to the *Today* show, which was featuring the USFL's lawsuit against the NFL. On

the screen, at a Manhattan press conference, were two men: Trump and . . .
Roy Cohn?

"Of all the people he could hire as my cocounsel," said McSherry, "Donald
gets Roy Cohn."

Yes, Roy Cohn — the attorney who discovered initial fame in the mid-
1950s, when he represented Joseph McCarthy during the Wisconsin senator's infamous tarring and feathering of alleged American communists and
communist sympathizers. Cohn, wrote Michelle Dean of the *Guardian,* "was
a kind of stage director of the major events of the red scare . . . another man
would have let himself be an invisible functionary in those proceedings, but
not Cohn. He made himself visible. He wanted to be front and center . . .
shamelessness, in fact, was Cohn's defining trait." In this regard, the veteran
attorney found a sibling-like figure in Trump, who first sought Cohn's services in 1973, when he and his father were sued for discriminating against
African Americans in their real estate rental policies. Although the Trumps
had, indeed, violated the law, Cohn's advice was simple: "Tell them to go to
hell." He then filed a $100 million countersuit that, while immediately dis-
missed, generated (surprisingly positive) headlines. Four years later, when
Trump was about to marry a Czechoslovakian model named Ivana Zel-
níčková, Cohn was hired to draw up the prenuptial agreement — one that
gave his soon-to-be bride a mere $20,000 a year.

Now, at the press conference, Cohn was at his absolute best. *Life* magazine's Nicholas von Hoffman called him "a master performing on television,"
and the compliment fit. An odd-looking man with sleepy, almond-shaped
eyes, elephantine ears, and a nose the shape of a crooked ice pick, Cohn
stood before the gathered press and spoke persuasively of the NFL's monopoly, and insisted a secret committee existed, with the sole purpose of destroy-
ing the USFL. Though his refusal to cite any actual evidence raised eyebrows,
it was overshadowed by his bombast and masterful articulation. Cohn may
well have been one of America's most loathsome men, but he was the best
loathsome man to have on your side. "I can only make analogy to a criminal
case," he said at the press conference, a jabbing index finger accentuating
his words. "How do they catch anybody? They catch anybody by someone
on the inside coming along and saying, 'Hey!' You have a right to do a little

concluding that we didn't find this out and place it in the lawsuit by accident. Obviously, the information was supplied to us, and obviously, when [the NFL] created this USFL committee, they did not create it for the purpose of going over the airwaves or into newsprint. They didn't expect anybody to know about it. We allege it was done in a clandestine manner."

Trump, sitting alongside Cohn on a couch in the attorney's office, was unusually quiet, though at one point a reporter asked why he was serving as the USFL's mouthpiece. Before he could answer, Cohn stepped in and said that, as the owner of the New York–area franchise, Trump was the ideal representative.* In a statement, the NFL called the lawsuit "baseless." Which wouldn't have been so bad, were it not for the fact that, behind the scenes, many of the USFL's owners secretly agreed. Even those who supported Trump in most areas were puzzled by his rise from relatively new franchisee to the most important man in the universe. And, really, was the NFL the reason the USFL was struggling? Or was the USFL the reason the USFL was struggling? What would have happened had the owners adhered to David Dixon's original plan? What would have happened had they ignored Donald Trump? Wrote Hal Lancaster in the *Wall Street Journal*: "The USFL's own missteps make it hard to take the league seriously."

Never had a truer sentiment been penned. The press conference was organized by Ira Silverman, a young New York–based sports public relations guru who was warned beforehand that Cohn and Trump were notorious for failing to pay their tabs. When Cohn asked how much his services would cost, Silverman told him the bill was $2,000 — half due up front. "I got the check for $1,000, did the job, put together a media invite list, we had a great turnout . . . and they never gave me the second $1,000," he said. "I handed a bill to Cohn — nothing. Then I started calling Cohn and Trump, and they never picked up the phone. They just ignored me."

A day after the press conference, the league held its annual meetings at the

---

* A day after the press conference, Trump bragged to his fellow owners that some 400 people attended. An owner later checked with a reporter who covered the event. "Oh, maybe 35 people were there, maybe 40," he said. "Maybe a few more if you count TV technicians and hangers-on."

Amelia Island Plantation, a resort just north of Jacksonville, Florida. Trump, Cohn, and the lawsuit starred as the hot topic among team owners, and while some expressed optimism about a fight with the NFL, many were overcome by dread. A league that began with such promise now seemed rudderless. Simmons was being replaced by Usher (whose contract included a clause providing for a substantial bonus for every USFL franchise that merged with the NFL). Eighteen teams were about to be reduced to 14. Eddie Einhorn, thought to be the man who could negotiate a magnificent fall television deal, seemed smaller by the day. There would be no Chicago franchise in 1985, and his initial boasts of huge TV money and widespread fall interest were nowhere to be found. "I'm sure a lot of guys are disappointed with me that I didn't come back with a deal," Einhorn said. "But I never promised them a rose garden." Wrote William Taaffe of *Sports Illustrated*: "The networks have told the league to drop dead if it insists on moving its '86 season to the fall. Eddie Einhorn . . . was talking big a few months ago about landing a long-term network deal for a fall schedule, but [it didn't happen]." Put simply, none of the networks liked the idea of the USFL competing directly with the NFL. A growing number of owners even wondered what Einhorn was doing in the league and at the meetings. He didn't own a franchise. He'd put forth not a dime. The so-called Chicago savior was saving nothing. Taube, the Oakland owner and one of the league's wiser men, thought Einhorn a fool, and remained convinced Trump's actions were all in the name of landing himself an NFL franchise. Tampa's Bassett agreed—he saw Trump as a fraud, and little more.

Alas, it mattered not. The USFL's plan was clear: it would play one last spring season while pursuing action against the NFL.

Then, in 1986, it would move to fall.

13

WOBBLING

Negativity is contagious. There were lots of negatives. But the USFL
had something special going on. It just became harder and harder
to see.

— Eric Truvillion, wide receiver, Tampa Bay Bandits

IN THE LEAD-UP to the USFL's 1985 season, John Bassett was asked by *Time*
magazine whether his league could survive.

Under normal circumstances, the owner of the Tampa Bay Bandits would
have smiled, nodded, and spoke glowingly and hopefully of new teams, great
players, exciting opportunities, creative rules (the USFL would be imple-
menting an instant-replay challenge system for the upcoming season). He
was a born hype man, generally anxious to extoll the virtues of whatever
project sat before him. Yet there was nothing normal to what was transpir-
ing. Yes, 14 teams would compete for a championship. Yes, footballs would
fly through the air. Yes, stadiums would charge people money to observe
competitions between two teams. Beer and soda would be sold. Reporters
would interview players. Television cameras would show the games.

But survival? Long-term survival?

"I don't know," Bassett said. "It's like being in a canoe and rowing like hell
against a wind and tide moving 200 mph. You have to wonder."

In his new job as league commissioner, Harry Usher faced multiple un-
enviable tasks — first and foremost, convincing football fans that the USFL

would continue with business as usual, and that the games were worth attending. Sure, the lawsuit lingered like a grim reaper. And, sure, organizations like the Houston Gamblers and Denver Gold and Tampa Bay Bandits were almost certain to die once they had to compete with NFL franchises in the same cities. And, sure, in a private meeting with owners, Usher said, clumsily, "I wasn't hired to rearrange the chairs on the deck of the *Titanic.*" And, sure, the *Wall Street Journal*'s preview article was headlined USFL FACING FOURTH-AND-LONG AS IT STAGGERS INTO THIRD SEASON. And, sure, the league's plight was quickly emerging as fodder for parody. Inside the newsroom of the *Tampa Tribune,* for example, a sports columnist named Tom Ford borrowed from the recent hit song "We Are the World" to write "We Are the League."

> *There comes a time, when we heed a certain call.*
> *When the league must come together as one.*
> *There are teams folding; we're going to lose them all.*
> *Unless we change, the game from spring to fall.*
> *(Chorus) We are the league, we are the stepchild,*
> 	*we are the ones who make the fans all say they're sick of football.*
> *There's a choice we're making, we're saving our own hides.*
> *It's true we'll make a better league just wait and see.*
> *Oh, open your wallets and help support our greed.*
> *Even though the owners are hardly in need.*
> *Donald Trump has told us the league is here to stay.*
> *But will it be for more than just a day?*

The song was recorded by Tedd Webb, a local radio personality, and copies found their way to stations across the country. Though it never reached a "Super Bowl Shuffle" level of phenomenon, "We Are the League" had plenty of folks laughing at an increasingly laughable situation.

Approaching the start of the season, what the USFL needed — *desperately needed* — was quick credibility. There was an urgency to remind people that the football was genuinely good; that the USFL was packed with players who could excel in the NFL and bring forth repeated *ooh*s and *aah*s and thrills;

that the newest rule (coaches could throw a flag to challenge an official's ruling, and instant replay would be deployed) would only add to an inventive approach to football. That's why, after three weeks of exhibition games, all the attention of the opening weekend was centered around Birmingham, where Doug Flutie and the Generals were slated to visit the Stallions for a nationally televised matchup.

In the two months since the diminutive quarterback was signed out of Boston College, Flutie went from mere national phenomenon to the fledgling league's greatest (last?) hope. In fact, a smiling Flutie appeared on the cover of *Sports Illustrated* alongside the daunting headline CAN THIS MAN SAVE THE USFL?

It was a preposterously unfair question, but Donald Trump expected nothing less.

The rookie reported to the Generals' Orlando training camp and immediately endeared himself to teammates. The kid worked hard, minded his business, seemed humble and deferential. "You expected him to be a jerk," said James Lockette, a New Jersey defensive end. "But Doug was very down to earth." When asked about sharing a room with the $8.3 million man, Dwight Sullivan, a low-paid fullback from North Carolina State, noted, "That's fine. Between us, we'll be worth $7,025,000." After Sullivan was cut, Flutie was reassigned to live with Vince Stroth, a 267-pound offensive lineman. "It's great," Stroth cracked. "He doesn't take up any room."

In early February, in a transaction that had nothing to do with skill or wins and everything to do with ticket sales and TV ratings, Brian Sipe, the incumbent starter, was exiled to the Jacksonville Bulls for a draft pick. Naive to the ways of the Trump universe, Flutie figured he would be able to learn from Sipe as his backup. The trade, he recalled, "was a rude awakening to the realities of professional football." Flutie lacked Sipe's stature, composure, and skill set, but money was money and hype was hype. Trump rightly knew football fans didn't specifically pay to watch Brian Sipe throw passes. They *would* to see Flutie.

The New Jersey–Birmingham game sold 34,785 tickets. Which sounded quite impressive when compared to 23,622 in Oakland for Gold-Invaders or the (yikes) 18,828 for Gamblers-Express at the Los Angeles Memorial Coliseum. Yet the figure was a profound disappointment. One year earlier, the Generals and Stallions drew 62,300 to Legion Field, and that was without America's favorite little quarterback. Part of the problem was weather — the Alabama sky was gray, and the threat of a thunderstorm quite real. But the other issue was enthusiasm. Birmingham loved its football, and the denizens took to the Stallions. But now, with the lawsuit and the fall and Trump, everything felt shaky. There might be a league in 1986, there might not be a league in 1986. The Stallions could stay. The Stallions could disappear. Whatever sense of permanence the USFL provided fans had largely vanished, and there was a price to pay.

It hardly helped that Flutie stunk. With both Trump and Usher in attendance, the quarterback — wearing a red No. 22 jersey and white pants — bounded onto the field, stepped into the huddle, called a play, took the snap . . . and . . . *misfired. And misfired. And misfired. And misfired. And misfired. And misfired. And misfired. And misfired. And misfired.* He failed to complete his first nine passes, and two of those were intercepted. By the time Flutie ("The $7 million fluke," wrote Jerry Rutledge of the *Anniston Star*) finally connected with a receiver, there was 2:10 remaining in the third quarter. The Stallions led 31–7 heading into the fourth, and the small remaining handful of ABC viewers had changed the channel to watch either ESPN's coverage of World Team Volleyball or the *Silver Spoons* episode in which Alfonso becomes frustrated with his performance in school until Bruce Jenner, visiting the Stratton mansion, suggests the problem may well be dyslexia (really, it was quite excellent). Meanwhile, Birmingham's defensive players were spewing nonstop trash talk Flutie's way. He was an undersized Muppet who belonged in the circus. He was an overrated turd. Behind the Generals' bench, a flock of approximately 50 women squealed whenever Flutie walked to the sideline. There were homemade signs — WE LOVE DOUG FLUTIE and CINDY AND TANYA LOVE DOUG FLUTIE. The Stallion players found it irksome. Especially in their home stadium. "All we heard and read all week was Flutie and the New Jersey Generals," said Cliff Stoudt, the Stallions quarterback. "Even our own media was doing it. The city boys with their Heisman

Trophy winners were going to come to Alabama and show up these country boys from Alabama. Right now, down deep inside, I'm grinning."

The final score was 38–28. The Generals' game plan (Herschel Walker carried five times for 6 yards) was befuddling. Usher, who knew surprisingly little about football, could be spotted in his box, half-asleep. ABC scheduled a halftime interview with the commissioner, but canceled it at the last minute. As if the night could go no worse, Walt Michaels, the Generals coach, made a bit of USFL history by challenging a penalty call. The referee dropped the red flag to signify an instant replay — then realized after prolonged awkwardness that penalties could not be reviewed.

When the game ended, Flutie sat by his locker and took question after question. His statistics (12-of-27, 189 yards, two touchdowns, three interceptions) were better than his performance. He seemed upbeat and reasonable, and accepted responsibility for underwhelming. "No one said it was going to be easy," he said. "I played very poorly throughout the first half. In the second half I felt very confident."

Did he take any positives?

"No," he said. "Not when you lose a ballgame."

On the same day Flutie debuted as a General, Brian Sipe was making his first start for the Bulls. Were there one USFL city showing itself to be of true professional football caliber, it was Jacksonville. The Bulls averaged a league-best 46,730 fans in their premiere season, and that was with a 6-12 team that lacked skill and star power.

Although it remains an unofficial (and unmeasurable) designation, in 1984 the Bulls set a professional football record for quarterback ungodliness. The opening-day starter was ex-Jet Matt Robinson, who tossed seven touchdowns and 12 interceptions before being replaced by Robbie Mahfouz, a rookie from Southeast Louisiana and the finest quarterback to ever feature a last name with a side-by-side *h* and *f*. Mahfouz tossed 13 interceptions (with 11 touchdowns) and gave way to Ken Hobart (no touchdowns, three interceptions) and Ben Bennett, who was allowed to attempt but 13 passes.

The roster was a mess, and Fred Bullard, the owner and Jacksonville native, was committed to action. He agreed to take on much of the $800,000 owed Sipe, then grabbed Mike Rozier, the former Maulers halfback, who was available for Jacksonville after Pittsburgh folded. While Flutie was the USFL's biggest rookie name, the best newcomer was Keith Millard, a quarterback-eating defensive tackle out of Washington State. Suddenly, as other franchises seemed to be crumbling, the Bulls were the hottest operation of the preseason. Larry Csonka, the former Dolphins running back and Bulls senior vice president, even turned creative, signing free-agent halfback Archie Griffin, the two-time Heisman Trophy winner at Ohio State. Now Jacksonville could market its all-Heisman backfield of Rozier and Griffin — even though the ex-Buckeye was 30 and two years removed from his last professional season, with the Cincinnati Bengals. "I wanted to see if I had anything left," said Griffin, who carried 10 times for 11 yards before leaving midway through the campaign. "Well, I didn't."

The Sipe-Rozier pairing was a hot subject around the league, and 51,045 jammed the Gator Bowl to see Jacksonville take on the defending-champion Stars. On the bright side, it was a warm, sunny day, and beneath a clear blue sky kicker Brian Franco booted a league-record-tying five field goals in the Bulls' 22–14 triumph. On the down side, Sipe suffered a separated shoulder late in the first quarter, and threw only six passes before he was replaced by Mahfouz. "Brian was so frail," said Greg Blache, the defensive line coach. "He looked more like a tennis player than a quarterback. Getting hurt was only a matter of time." The injury kept him sidelined for much of the season, but set up one of the greatest financial success stories in pro football history.

With Sipe out and Mahfouz, well, Mahfouz, Jacksonville desperately needed an established quarterback. The pickings around the USFL were slim, as was the NFL availability list. Three free agents who stood out were Phil Simms of the New York Giants, Don Strock of the Miami Dolphins, and Ed Luther of the San Diego Chargers. The first two men showed no interest in crossing league lines. Luther was intrigued. A fourth-round draft pick in 1980 out of San Jose State, Luther, though, spent five years backing up the legendary Dan Fouts in San Diego. His lifetime numbers were awful (245 of 460, 3,187 yards, 12 touchdowns, 23 interceptions), but his arm was a rocket and he was well regarded around the sport. "It came down to two things," Luther said. "The first was finally getting an opportunity to start. And the second was finances."

Ah, finances. Although men like Herschel Walker and Doug Flutie and Mike Rozier generated larger headlines, in measures of money there is no more triumphant USFL saga than that of Ed Luther. With the Chargers, his annual salary was $160,000. The Bulls gave him a four-year, $2.6 million deal — all of the dollars guaranteed. By comparison, the Cowboys' Danny White and the Bengals' Ken Anderson, longtime NFL starters with Pro Bowl résumés, were making only $550,000 annually. With his new contract, Luther would have been the seventh-highest-paid quarterback in the NFL, and earned less than only three running backs and one wide receiver. "I had to do it," Luther said. "You don't get too many opportunities in football like that one."

As was the case when Walter Lewis signed with Memphis before the 1984 season, Luther's arrival in Jacksonville — a place he had never before visited — was greeted with the unjustified fanfare of a presidential visit. His contract was signed inside the office of Jacksonville mayor Jake Godbold, who presented the quarterback with the key to the city. Before taking questions, Luther pledged to donate $100 to charity for every successful touchdown pass. A giddy Godbold then vouched to match the generosity, as did the Bulls. Lindy Infante, Jacksonville's coach, smiled toward Luther and cracked, "I hope he's a poor man by season's end."

The concern was misplaced. The $2.6 million quarterback's 15 touchdown passes were coupled with 21 interceptions. "Ed was a No. 2 quarterback," said Buddy Geis, the Bulls' wide receivers coach. "He could throw all the routes, he was solid. But if Brian stays healthy we have a chance to win a championship. Not with Ed."

The Bulls wound up finishing a disappointing 9-9, yet the fans kept coming. There were 40,112 inside the Gator Bowl for a Week 2 setback to Memphis, 41,298 to watch a Week 10 win over Birmingham, and 60,100 for a Week 11 triumph over Flutie's Generals. Only twice did the USFL draw crowds in excess of 70,000 — both times at the Gator Bowl. Part of the appeal was Jacksonville's limited entertainment options. The city's 600,000 residents were not distracted by Broadway, or Beverly Hills, or Michigan Avenue, or the White House. In fact, of the five USFL franchises that averaged more than 30,000 fans in 1985, four (Tampa Bay, Jacksonville, Birmingham, and Memphis) played in smaller markets with slim entertainment options (the Generals, who averaged 41,268 fans but featured a league-high $7.3 million payroll, also cracked the list). "Being in Jacksonville was almost like when I played at Notre Dame," said Mansel Carter, a Bulls defensive end. "Players would go to restaurants and bars and never pay for a thing. We were kings. There were always commercials to film, important businesspeople to meet. Tremendous."

What also helped the Bulls was quirkiness. The team was named after its owner, and its uniform colors paid homage to three area college football powers — red for Georgia, orange for Florida, gold for Florida State. Smoking in the locker room wasn't merely tolerated, but encouraged (Don Bessillieu, a safety, would famously take a deep, exaggerated drag before running onto the field). Rozier liked to show up at practices wearing deflated footballs over his shoes. Millard, an excessively cocky rookie who thought he belonged in the NFL ("God, he was a real asshole back then," said Geis), used part of his four-year, $1.9 million salary to purchase a Mercedes-Benz convertible. "It's an awesome story," said Ike Griffin, a defensive end. "Keith enjoyed life, and every now and then he'd be a little late for practice. One time they fined him an entire game payment. I picked up my check, he picked up his. Then he crumpled the paper in a ball, because it was $0.00, and says, 'Big Griff, let's buy a Mercedes!'"

Griffin thought his teammate was joking.

"Come on," Millard said. "Let's do it."

The local dealership had one available car on the lot. It was the color of diarrhea, and screamed, "Hey ladies, I am a professional athlete!"

It was also far too small for a six-foot-six, 260-pound man.

"I was such a dummy," said Millard. "I drove the car to practice that next day and I could barely get out. I caught a lot of crap for that, and rightly so. My teammates wouldn't let me get away with something so stupid. But you know why? Because there was a genuine sense of unity. We felt like we were all fighting together."

• • •

At the same time the Bulls packed the Gator Bowl, the rechristened Arizona Outlaws exemplified an increasingly lost league trying to get out of its own way.

When the merger between the Wranglers and Outlaws was initially announced, the reaction was a muted groan. It was sort of like watching two cousins French kiss — unnecessary and uncomfortable. Though George Allen's Wranglers reached the 1984 title game, the team averaged only 25,568 fans at Sun Devil Stadium. It wasn't a tough one to figure out — "too damn hot," said Hal Wyatt, the team's trainer. "You'd have days of 112, 115 degrees. For the players, it's exhaustion, it's cramps. For the fans, it's sitting on metal bleachers in the Arizona sun." For their part, the Oklahoma Outlaws were largely invisible. Stupendous uniforms, good quarterback (Doug Williams), bad surrounding players, popular yet overmatched head coach (Woody Widenhofer), underwhelming market.

After the deal was consummated, one presumed the merged team would retain Allen or Widenhofer to run things. Instead, on December 13, 1984, Bill Tatham Jr., the general manager and owner's son, announced that the new Outlaws had hired Frank Kush.

Wait. *Frank Kush?*

Wrapping up his third-straight year of disastrous football with the NFL's Colts, Kush was better known for the 22 seasons he spent at the helm of Arizona State. He compiled a 176-54-1 record with the Sun Devils, and earned the status of Grand Canyon State gridiron icon. Yet behind the curtain, Kush bullied and beat his players into submission. During a game between Arizona State and Washington on October 28, 1978, Kush punched Kevin Rutledge, the Sun Devils punter, then urged his assistant coaches to cover for him. Rutledge quit the team, transferred, and filed a $1.1 million suit against the university and Kush, who was fired.

He went on to coach a season with Hamilton of the Canadian League before the Colts called. "I played for him with Baltimore, and he was ruthless," said Jeff Hart, an offensive tackle. "There's an unwritten rule you don't cut guys right before Christmas. He seemed to take joy in ruining people."

Kush's two NFL claims to fame are having Stanford quarterback John Elway refuse to play for Baltimore (in part because of Kush) and a wide receiver named Holden Smith dumping a Fanta root beer over his head after being cut. "I'd kind of had it with him," Smith said, "and I had to do something to the guy to vent steam."*

Those under contract with the Arizona Outlaws greeted news of Kush's hiring with pronounced gulps. Not a single player attended his introductory press conference, and several expressed their anonymous fears to the *Arizona Republic*. "If I've got to go through boot camp next month," one said, "I think I'll take a long siesta." The members of the team prepared for the absolute worst — a ruthless tyrant dictator ordering 10-mile runs in the 120-degree midday heat. Then training camp commenced at the team's Phoenix facility, and Kush was a plush throw toy. He complimented players and coaches, ran the bare minimum number of drills and practices, spoke politely with the media. He seemed to always be signing autographs, or posing for a photograph, or telling some story about that catch John Jefferson made in Arizona State's 1975 Fiesta Bowl victory. Kush was being paid $700,000 in guaranteed money to coach the Outlaws, and was quite content to enjoy the all-expenses-paid vacation. "He was the biggest disappointment in my life," said John Teerlinck, the team's defensive line coach. "Kush was bullshit. All of us heard about Frank, and Camp Arizona, and running the mountains. Well, he wasn't shit. One time he sat me down and said — to my face! — 'You care too much about winning. You wanna win too bad. We have the rights to the stadium. Don't make any waves and we'll get an NFL team and our owners will make a lot of money.'

"He was a complete sellout. All he cared about was getting paid. I never worked for a worse coach than Frank Kush, because he didn't care about any of it."

The Arizona Outlaws debuted at home with a Week 1 clash against the (also rebranded) Portland Breakers, and the scene at Sun Devil Stadium was dispiriting. In the days leading up to kickoff, the Outlaws' print advertising blitz included the sentence, "Matt Robinson and the rest of the Portland

---

* The Arizona Outlaws' 1985 media guide made Baltimore's 0-8 strike-shortened season sound like a Super Bowl championship. Kush, it read, "revitalized a struggling Colts' side in 1982 with a two-thirds turnover in player personnel. The team posted a 3-2 preseason slate."

Breakers take on the Outlaws!" — which led to two questions for the majority of American sports fans:

1. Who is Matt Robinson?
2. Portland has a football team?

That same advertisement also urged Arizonians to catch "all-time collegiate scoring champ Luis Zendejas!" marking the first time a professional sports franchise reached the desperate depths of begging fans to behold a kicker.

"It was rough," said Kit Lathrop, the Outlaws defensive lineman. "It didn't feel like people cared so much about us."

The announced paid attendance was an exaggerated 20,321. What those who stayed home missed was a 9–7 Outlaws win, made possible by three Zendejas field goals. Wrote Walt Jayroe in the following morning's *Arizona Republic*: "Frank Kush escaped with a victory in a game salvaged as entertainment only by the exciting conclusion."

It was the highlight of the season. The Outlaws featured one of the USFL's best defenses, and no one cared. The team's season-ticket base fell from 18,500 in 1984 to 12,000 in 1985. Attendance dropped by the week. A nationally televised Monday-night meeting with the visiting Bulls brought 13,025 to the stadium. There were constant rumors from the NFL about a soon-to-be Phoenix franchise, thereby marginalizing the Outlaws' existence. Leonard Tose, owner of the Philadelphia Eagles, hinted his club might wind up in the Valley of the Sun. Then Bill Bidwell, the St. Louis owner, said the Cardinals would love to head west. Each NFL statement made the Outlaws feel increasingly small.

It did not help that Kush's team was a mess. Karl Lorch, All-USFL defensive end, was drinking between 25 and 30 beers per day after practices. Trumaine Johnson, the Arizona Wranglers' top receiver in 1984, walked out of camp over a contract dispute and never returned. Kevin Long, the beloved veteran fullback, was waived to save money. Tim Spencer, one of the league's best halfbacks, was allowed to flee to Memphis. John Lee, the reigning sack champion, was nearly traded to Jacksonville following a dispute with a coach. Doug Williams's knees were soggy paper, and he was briefly benched for Rick Johnson, pride of Southern Illinois University. In occasional flashes to his old self, during film sessions Kush bestowed on players emasculating nicknames. Tight end Ron Wheeler was "Wheeler Around the Rosies." Zendejas, the rookie kicker, was "Louise." Linebacker Daryl Goodlow was "Shitlow." Defensive back Mike Fox was "My Cock." It was never funny. "He would embarrass you deliberately," said Wheeler. "Run the play you messed up back and forth, back and forth for everyone in the room to see 100 times."

On May 26 the Outlaws traveled to Houston to face Jim Kelly and the high-flying Gamblers. The game was, by most standards, of little remark. Houston won 41–20 behind Kelly's three first-half touchdown passes. Afterward, the dejected Arizona players showered, changed into street clothes, and caught the team buses for Houston Intercontinental Airport. With non-existent revenue, the organization no longer flew chartered jets, and instead booked commercial flights. This time, the Outlaws boarded a Continental Airlines 727 for a direct three-hour journey to Phoenix. "It was probably my genius idea," said Tatham Jr. "Why pay $100,000 on a charter when we can spend $35,000 to go commercial? I mean, what could possibly go wrong?"

As they waited in the gate area, many players and coaches found the nearest bar to booze away their troubles. On the plane, the imbibing continued. "We didn't control their alcohol intake," Tatham Jr. said. "Another brilliant move by me." Among Arizona's biggest personalities — and drinkers — was Teerlinck, the six-foot-five, 270-pound defensive line coach. A onetime All-American at Western Illinois University, Teerlinck spent two years as a grinder with the San Diego Chargers before recurring knee problems ended his career. He jumped into coaching, and following six years in the collegiate ranks joined the Chicago Blitz in 1983, then the Wranglers and Outlaws. Teerlinck was a unique breed — 24/7 intensity, combined with 24/7 furor; a mouth that fired off 200 curses per minute and a temper that knew no limit.

"God, John was a tremendous coach," said Dave Tipton, a veteran Arizona defensive tackle. "That year in training camp he got ahold of me and said, 'I know you. I know what you can do. If you wanna take one snap in the preseason, that's fine with me. If you wanna take 100, that's fine. Just be ready for the opener.' He trusted you as a player and as a man. But, if I'm being honest, he loved the alcohol, and drinking really hurt him and his reputation. I don't think that's entirely fair — but it's true."

"John would sit back, get drunk, and ridicule players," recalled Kelvin Middleton, a defensive back. "We did not like it." On the Houston-to-Phoenix flight, Donnie Hickman, the Outlaws' starting left offensive tackle, could be heard near the rear of the plane ripping on Tipton, who jumped off-sides four times against the Gamblers. Teerlinck was drunk and angry, and he rose from his seat to confront the six-foot-three, 275-pound Hickman as he stood in the aisle smoking a Marlboro.

"John told him to sit down, shut up, and put out the cigarette," said Vic Koenning, a linebacker. "John was badass. But he wasn't badass enough to get Donnie to stop smoking."

Hickman snickered — a no-no even were the coach sober. Teerlinck charged forward and Hickman drew back his right fist and popped him in the face. Somehow, Teerlinck continued forward and reached for Hickman's head. "The black man [Hickman] had a hold of the white man's [Teerlinck] mouth," said Connie Dreistadt, a passenger in seat 26C. "It looked as if he was trying to rip off his jaw."

The two behemoths grappled like a pair of sumo wrestlers, slamming into seats, rolling into passengers, barreling into the wall of the plane. "Donnie Hickman was a really tough guy," said Mike Westhoff, the offensive line coach. "He grew up just with his father in a lumberjack camp. He was as tough as nails." Other Outlaws jumped into the fray, and many of the 136 passengers found themselves screaming and cowering. "Women were yelling at the team members," said one witness. "A little boy wept all the way to Phoenix after the fight."

The men were separated, and when the plane landed it was greeted by federal agents. The following day Tatham Jr. handed Teerlinck his pink slip. "John took it well," Tatham Jr. said. "He accepted responsibility."

The defensive players, however, were heartbroken. Seventeen members of the unit paraded to Tatham Jr.'s office to express their displeasure with the firing. The following Sunday, Arizona won a 13–3 snoozer at home, and afterward the game ball was delivered to Teerlinck's house. "We loved John," said Sel Drain, the safety. "He always looked out for us."

Lost in the gesture was the fact that the victory came against a team only able to devote one and a half days of the past week to practicing; a team whose players hadn't been paid in weeks.

Yes, the Arizona Outlaws were a sad lot.

They had nothing on the San Antonio Gunslingers.

14

BLANK GUNS

My second practice with San Antonio, they had us doing full contact drills with nothing but helmets on. A guy tackled me and broke my collarbone in three places. A coach comes up to me minutes later and asks whether I can take the bone out and get back early. I went home to San Diego and never returned.

— Doug Banks, running back, San Antonio Gunslingers

YOU ARE READING a book about the United States Football League. Were there a market for it, however, you'd be reading a book about the San Antonio Gunslingers.

And it would run at least 5,000 pages.

"It'd be a tremendous movie," said Mike Schaefer, the team's ticket manager. "Only *nobody* would believe it's real."

There had never been a professional sports franchise like the San Antonio Gunslingers, and there will never again be a professional sports franchise like the San Antonio Gunslingers. It is an organization that had no real business existing, yet — in its very existence — brought joy and laughter and frustration and bewilderment and disbelief to large swaths of humanity. Think of your favorite *Looney Tunes* episode, inject it with a line of cocaine, then stir in some really hideous uniforms, a senile seven-fingered coach, a wackadoo owner, a decrepit stadium, bouncing checks, speeding cars, murderous defensive linemen, and a football.

Toss in bad judgment, ego, and clinical insanity, and you have the Gun-slingers.

When the USFL decided to expand for the 1984 season, its power players envisioned markets like Minneapolis and Seattle. Yet as several marquee spots fell through, the league turned its attention toward San Antonio, Texas, and, specifically, the application of a 60-year-old rancher and oilman named Clinton Manges. According to the USFL's requisite background research, Manges had loads of money. His official biography in the 1984 Gunslinger media guide described him as "controversial, colorful, opinionated and hard-nosed," then added, "Known throughout South Texas as a man who knows what he wants and how to obtain it, Manges prides himself on having as many loyal friends as detractors." The sentences were all true, but mere words camouflaging a marvelous saga.

Born in Cement, Oklahoma, in 1923, Manges — one of seven children — dropped out of school before fifth grade to pick cotton. He never returned, and when his family relocated to Junction, Texas, young Clinton served as an assistant to his grandfather, who traveled from ranch to ranch renovating mattresses. As he grew, Manges held a bevy of low-paying blue-collar jobs, ranging from shrimper to movie theater janitor. His dream was to become a watch repairman.

He served in the United States Coast Guard, then, in the late 1940s, managed a service station in Port Aransas, Texas. It was there, while pumping gas and checking beneath the hoods of automobiles, that Manges caught his break. Or, to be more accurate, *created* his break. One day he found himself working on the car of Lloyd Bentsen, the investor who amassed a fortune worth $100 million in real estate, oil, and cattle. As he fixed up the vehicle, Manges inquired as to where Bentsen was heading. "Farmer Jones up the road has some land for sale," the former congressman said.

"Hmm," said Manges. "That's interesting."

He excused himself, rushed inside the service station office, and placed a call to Farmer Jones, a man he knew well. He returned moments later with a smile plastered to his face. "Mr. Bentsen, you can deal with me directly," he said. "I just bought the land."

Bentsen could have been furious. Instead, he was dazzled. From that day forward, he took Manges beneath his wing, involving him in land deals and teaching him the tricks of the game. Wrote Paul Burka in *Texas Monthly*: "Clinton Manges learned from Lloyd Bentsen the fundamental lesson of South Texas: *land is power*." Within a half-decade, Manges was known as one of his state's elite property traders. "He had no peer when it came to understanding the intricacies of deals," Burka wrote. "He knew how to read people; he could be charming and expansive and incredibly persuasive, or he could be belligerent and ruthless."

As the years passed, Manges's sliminess was as much a calling card as his savvy. He was regularly sued for unpaid bills and skipped taxes; he wrote fake checks and was content bribing any available on-the-take public official. In 1963 Manges was indicted for making false statements on a loan application to the state of Texas. "There is a role for the bull in the pasture," Henry Cisneros, the former San Antonio mayor, once explained. "It's not to everyone's taste, but it's an essential role to be played."

Clinton and his wife, Ruth, adopted four children, and the family moved to San Antonio. He became heavily involved in the oil and gas business, and wide-ranging successes (via both drilling and lawsuits related to drilling) allowed the Mangeses to purchase a magnificent 10,000-acre ranch in Freer, Texas, that he named the Magic Kingdom.

Oh, one other thing: Clinton Manges loved sports.

When he heard of the USFL, he wanted in. So Manges applied for an expansion team, then talked the ears off of anyone with the league who would listen. No, San Antonio was not an ideal location. It was America's 45th-largest television market, and the only potential facility, Alamo Stadium, had been built in 1939 and seated just 24,000. "It was a high school stadium with a track around it," said Tim Griffin, the beat writer for the *San Antonio Express-News*. "It was known as the Rock Pile." The paint was peeling. Rust could be seen in every corner. The elevator that ferried people to the press box regularly stopped moving somewhere between the first and second levels. "You'd get stuck in it all the time," said Daryl Mueske, an offensive lineman. "Like a broken carnival ride." Nonetheless, Manges jabbered and jabbered and jabbered. He would pay to refurbish the structure. He would add seats. He had powerful friends in powerful places who would make certain the team received regional support. "At the time he was on the Forbes 400 list," said Steve Ehrhart, the league's executive director. "He was the king of the world. He flew up to meet us in this unique aircraft, he had some fancy wheels take him to the office. We did legal checking on him, and he had plenty of money. Then I went down to his ranch. He had a landing strip, a helicopter at the house, all these imported animals. I'm talking African antelopes. They just ran around, like the land was a safari. And he had big plans for the team and the league, and everything seemed great." So impressed was the USFL that it decided not to force Manges to make an initial capital investment.

"He seemed to be rolling in money and glamour," said Thomas Aiken, the Gunslingers' director of marketing. "He took me on a trip before the start of the season to visit the teams in Tampa and Los Angeles to see how they did things. Well, we went to the Bandits on a Learjet, and then, instead of going to California, he decided we should go hunting in Alaska. So we went there, killed some caribou, came home. Like it was no big deal."

It is hard, in hindsight, to pinpoint the first neon warning sign. Perhaps the USFL would have been wise to take note of the outlandishly cheesy refrigerator-sized four-leaf-clover diamond pendant dangling from Manges's neck ("It would choke a horse," David Knaus, a Gunslingers coach, recalled). Or maybe it should have alarmed some in league offices when, after Manges and the San Antonio Independent School District (which owned Alamo Stadium) reached agreement on a lease,* two lawsuits were filed—one by neighborhood residents, who argued traffic congestion and excessive crowd noise would result in plummeting property values; the other by the city, which cited a 1939 deed restriction that banned professional sports from Alamo Stadium. Or the USFL might have been scared off by a $100 million lawsuit filed against Manges by Seattle-First National Bank. Or by the time he called San Antonio "the most pessimistic city in the country." Or by his announced plan to have his daughter, MaLou, blessed with the vocal abilities of a drowning emu, sing as the regular halftime entertainment. Or by

---

* "I can tell you this—Clinton greased a lot of palms to get the stadium worked out," said Tim Griffin. "That was his thing."

his choice for the team's uniform colors (blue, green, silver — "What sort of fucking gunslinger wears green and blue?" asked Jim Bob Morris, a safety) and preposterous logo (a cartoonish cowboy drawing his gun — "like an anorexic Woody from *Toy Story*," said Mueske).

"There were some real red flags," said Ehrhart. "We probably should have paid closer attention."

The USFL finally came to see, with 100 percent certainty, that Clinton Manges wasn't all there when, on July 19, 1983, he hired Gil Steinke to be the first coach in team history.

At the time, the USFL was known for its coaching luminaries.

Chicago's George Allen was a legend.

New Jersey's Chuck Fairbanks was a legend.

Oakland's John Ralston was a legend.

Steinke had seven fingers.

It's true — just seven. Well, technically, seven and some leftover bacon bits. Three of the fingers on his right hand were sliced to the nubs from a long-ago lawn mower accident, leaving him with an alien-looking appendage. "Whenever Gil would talk he'd use that hand to gesture," said Knaus. "You'd try to pretend not to notice, but it was very distracting."

Steinke's hand was unusual, as was his résumé. Though he had enjoyed much success, first in four seasons as a defensive back with the Philadelphia Eagles, then in 23 years as an NAIA head coach at Texas A&I, his last season with the program came in *1976*. Back then, along with the fingers, Steinke was best known for spending the majority of games observing from the stands. "The sideline is the worst possible place to watch," he once explained to *Sports Illustrated*. "You can't tell what the devil is going on from the sideline. I try to sit about 12 rows up where it's high enough that I can see and low enough so it's not too far to run down to the field. At home I ask a student to save me a couple of seats on the aisle. I sit at about the 20 so I can stay away from the band, and I sit on the student side so I can stay away from the alumni."

The 64-year-old Steinke was far past his expiration date. He walked slowly and spoke slowly. "We didn't use the term 'dementia' back then," said Jim Bates, the defensive coordinator. "But the first meeting we had as a football staff, we all came in and sat around a table, and Gil spoke for five minutes, got up, and left the room. We waited. And waited. And waited. He just never returned. I think he forgot." Steinke often drove the wrong way in a city he knew well. An appointment for 10 o'clock would be pushed back until 2. He was awful at remembering names. And plays. And play calls. Once, in the lead-up to a game against Memphis, he pointed to video of a pass-rushing rookie defensive end on the Showboats and said, "This Reggie White — he's terribly overrated."

"It didn't matter what city we were in — Gil would fuck it up," said Morris, the Gunslinger safety. "If we were in Tulsa he'd say, 'Now let's go beat the Predators!' If we were in Tampa, it'd be, 'OK, let's go beat Houston!' There were times he clearly thought he was still coaching Texas A&I."

"There was one game when Gil was trying to draw up a play, and he got frustrated because he struggled to make straight lines," said Rick Neuheisel, the San Antonio quarterback. "So he says, 'Aw, just go draw the play up in the dirt.' Gil was a lovely man. But when your coach watches from the stands with his wife, you're not going to out–X and O anyone."

Manges's organization was cheap, but not in the ordinary cheap-USFL-team sense. The Denver Gold, for example, were cheap in that they provided their players with jalopy rental cars and mediocre food. The Gunslingers were cheap in that their facilities were Grade F; their offices were inside a pair of trailers (one double-wide, one single-wide) encamped in the stadium parking lot; players needed to supply their own shoes at training camp; their front-office employees — ranging from general manager Roger Gill to president Bud Haun to Ruth Manges, the cheerleader coordinator — were friends and family members working for favors and a tiny bit of dough. "They had a PR director for a while, and the only thing he directed was the path from where he did nothing to the place where he ate his lunch," said Arland Thompson, an offensive lineman. "He sat around all day with his feet up, and he'd take a huge knife out and clean his fingernails with the blade. He was rumored to be Manges's helicopter pilot and general flunky. I just know he didn't do a thing."

Ruth Manges drove a brand-new baby-blue Cadillac, but the organization refused to spend money on team decals. Hence, she cut out the Gunslingers logo from a poster and Scotch-taped it to the door of the car. "Think about that," said Knaus. "The millionaire wife of the team's owner . . ."

"I'll give you an even better one," Thompson said, laughing. "We went back and forth from the hotel to the stadium during training camp on something called the Gunslinger Bus. It was an old converted San Antonio public school bus, which they repainted green and put the old Gunslinger logo on it. The driver of the bus was an old African American guy named Byron, and because the gas gauge on the bus didn't work, he'd have to get out periodically and stick a measuring stick into the gas tank to see what the level was. We ran out more than a few times."

"The front office spent a lot of time trying to figure out what color uniforms and helmets they needed to provide the visiting teams when they came to town," said Neuheisel. "I'll say that again, because I'm not making it up — *the front office spent a lot of time trying to figure out what color uniforms and helmets they needed to provide the visiting teams when they came to town."*

In 1984 the Gunslingers broke camp with the league's most eclectic gathering of souls. Steinke was old and confused ("But really nice," said Nick Mike-Mayer, the kicker). Tommy Roberts, the offensive backs coach, most recently sold insurance in the Rio Grande Valley. The head trainer, Bobby Oakley, came straight from Alice High School, and Aiken had been a microwave merchandising manager for something called the Magic Chef.* Larry Kuharich, the offensive coordinator, lasted a handful of games before being fired. Well, kind of fired. He was asked to remain with the team in the public relations department, but turned down the offer, collected his fully guaranteed $55,000 salary, and drove off in the company car. "He left his running shoes and a sack lunch on the desk," said Neuheisel. "He was gone."

Kuharich's replacement was none other than Roman Gabriel, the legendary NFL quarterback/sex symbol. "Roger [Gill] came into my office all excited one day, saying we'd hired Roman," said Greg Singleton, the director of media relations. "He wanted me to write a press release, but Gabriel hadn't

---

* On page 13 of the 1984 Gunslingers media guide, Aiken's bio begins thusly: "Tom Aiken joined the San Antonio Gunslingers as Marketing Director because he called it 'a deal I just couldn't turn down.'"

met face-to-face with Clinton Manges yet. I said we had to wait. Well, Gabriel comes to San Antonio, had a meeting with Manges — and we never saw him again. He never showed up, we didn't have a press conference. Nothing. No Roman Gabriel."

Neuheisel, the only quasi-name player, was an unspectacular passer who had recently led UCLA to a Rose Bowl victory over Illinois. For some reason his MVP showing (22-31, 298 yards, four touchdowns) convinced Manges the blue-eyed, blond-haired quarterback was worth $120,000 annually — even though his arm strength was suspect and he had been repeatedly benched with the Bruins. Several members of the coaching staff urged Manges to instead use the money on Dan Pastorini, the former Houston Oilers quarterback who was a free agent. Pastorini had a cannon and regional appeal. Manges, however, was smitten with the kid. "Rick didn't have pro tools," said Thompson. "He was smart and tough, which was wonderful. But Manges thought he was far better than he really was."*

Otherwise, the franchise's personnel philosophy was simple: "Grab anyone we can grab who will play for us," said Singleton, "and hope it works out." The Gunslingers brought two legendary former Houston Oilers, Earl Campbell and Kenny Stabler, for visits, reveled in the headlines — but never offered serious money. ("I used to hunt on Manges's ranch," said Campbell. "Was I going to play in the USFL? No.") Instead, the team depended on scrubs and has-beens, small-college standouts and large-college rejects. Tight end Richard Osborne was a five-year NFL journeyman whose last professional season came in 1979. "He was physically beat to shit," said Oakley. "A lot of the guys were." John Barefield, a linebacker and football veteran, spent offseasons working for a Denver beer distributorship and, for kicks, broke pecans between his biceps while stretching. Fullback Mike Hagen's side career was serving as a power lifter for Jesus. Juan Castillo, a former Texas A&I linebacker, vomited before every game. Tight end Joey Hackett said fewer than 10 words over 18 weeks (and gifted the clubhouse employees

---

* Neuheisel became the Man, locally, when he began dating Debra Daniels, a local news anchor and former Miss Arizona. He also drove a black Trans Am — which never hurt in the 1980s.

with homemade beer steins at season's end). Linebacker Jeff Leiding lived atop his wife's grandparents' funeral parlor ("You had to walk through the casket room to get to his door," one Gunslinger recalled). All four members of the starting defensive line chain-smoked cigarettes at halftime and, on occasion, along the sideline.

A major source of bodies was the San Antonio Bulls, a local team in the semipro American Football Association whose players made $50 a game. The Bulls' owner and general manager had been Gill, and when he came to the Gunslingers he maintained a minor-league approach to personnel. "Roger was one of the nicest people, and he could not say no to a football player," recalled Oakley, the trainer. "Our coaches would cut guys and Roger would re-sign them. We were always over the roster limit because Roger couldn't help himself. He believed in giving chances to the underdogs." That's how the Gunslingers found Peter Raeford, a former Division II standout at Northern Michigan who played briefly for the Bulls and, when the USFL came to town, was working as a baggage handler at the local airport. He began the 1984 season as a backup, and ended it as an All-USFL corner. In perfect Gunslinger fashion, Raeford was the son of a one-armed mechanic. "Some people just need the opportunity," Raeford said. "Most of the Gunslingers fit that description."

The oddest addition may well have been Karl Douglas, Neuheisel's backup and a 35-year-old who had last taken a snap *nine years earlier,* with Calgary in the Canadian League. "We were a bunch of kids mingling with guys at the ends of their careers trying to muster every ounce of football they had left," said Neuheisel. "It was bizarre."

During the preseason, Manges turned irate over stalled ticket sales. The Gunslingers were the newest show in town, yet less than 10,000 spectators attended the two exhibition contests at Alamo Stadium. "Our prices were too high," said Schaefer, the ticket manager. "For that market, seats were astronomical." Ruth Manges suggested the team feature a two-for-one offer for the season opener against the New Orleans Breakers. A crowd of 18,233 arrived, and the owner was giddy. "I told them that's the sort of thing you can only do once," said Schaefer. "But nobody listened to me, and we had two-for-one tickets every week. It lost its appeal and people stopped coming."

The *Sporting News* predicted San Antonio to go 2-16 in 1984, and early results were not promising. The team lost six of its first seven, failing to score more than 14 points in a single game. Spirits were low, the stands were empty, morale was nonexistent. One day during the span, Ruth Manges hurt her shoulder while bowling, and Haun asked Oakley to look her over in the training room. "I'm sorry Bud, but I need another 30 minutes to help some players. Then maybe I—"

"No," Haun barked. "You work for Mr. Manges. Do it now."

"That sort of thing," Oakley recalled, "isn't exactly a good way to bolster positive feelings."

On March 31, the Gunslingers flew to Detroit to play the defending-champion Panthers. Manges's team stood at 1-4, and the owner was frustrated. Four weeks earlier, during a nationally televised Monday-night game against Houston, the stadium lights went out and play was suspended for 48 minutes. What was presumed to be merely an electrical failure was, in fact, a deliberate sabotage. "Clinton Manges had all kinds of business dealings around town and some were aboveboard and some I guess were not," said Neuheisel. "And in some way, shape, or form he had crossed the guy in charge of the power in San Antonio, and so the guy, to get even, shut it down while they were on national television." Making matters worse, at halftime the Gunslingers held an automobile raffle-ticket giveaway. Tom Allison, the stadium's public address announcer, leaned into the microphone and said, "Tonight's winner of the 1984 Dodge Charger is . . . oh, my God! It's me!" Boos rained down, but the victory was legitimate. Allison had purchased a ticket.

Manges had yet to fully recover from the Monday-night humiliation. Now, midway through the flight to Michigan, he stood and asked for attention. Players presumed they were in store for a lame motivational pep talk. Incorrect. "OK, here's the deal," he said in a thick Texas twang. "Anybody who knocks [quarterback Bobby] Hebert out of the game gets $500. Anybody who knocks [wide receiver] Anthony Carter out gets $500, too."

The next afternoon, Carter caught a 22-yard pass in front of safety Larry James. "Barefield slams him out of bounds," said Larry O'Roark, a San Antonio wide receiver, "and in the process pulls Carter's left arm around his back. He had to leave the game . . . it was broken. John got paid."

The Panthers won 26–10, and the next week the in-desperate-need-of-a-spark Gunslingers traveled to Chicago to face the Blitz at Soldier Field. It was a warm day, with temperatures in the 40s, and midway through the third quarter San Antonio trailed, 7–3. Once again nothing was going right. The offense was stagnated. The play calling was boring. Then — magic. From somewhere among the 9,412 spectators, an empty whiskey bottle whizzed through the air and toward the San Antonio sideline. It soared closer and closer and closer and closer until — *Pop!* — the bottle nailed Oakley directly in the side of his skull. He dropped to the turf, in Knaus's words, "like a heavy sack of potatoes."

"He was out cold," said Jim Bates, the defensive coordinator. "Flat on his back."

Throughout the early part of the season, Oakley, an excitable 24-year-old, would respond to injuries by yelling, sans humor, "[FILL IN A PLAYER'S NAME] down! [FILL IN A PLAYER'S NAME] down!" Now, as Oakley lay unconscious, Greg Fields (the lineman who, two months earlier, punched his coach and threatened to kill him) stood over the body, grinned ear to ear, and hollered in his familiar baritone, "Oakley down! Oakley down!" Laughter erupted up and down the sideline.

"Holy cow, it was the funniest thing ever," said Neuheisel. "Are you supposed to laugh when your guy is hit in the head? No. But . . . man."

The Gunslingers lost in a 16–10 overtime nail-biter, but members of the team agree there was some unifying magic in "Oakley down!" They trounced Jacksonville, 20–0, the following week, then nearly beat the Stars before winning three of their next four. "We got on a roll," said Neuheisel. "Nobody thought we had much of a team. But we got hot, and we played hard."

The majority of USFL contracts stipulated that a player had to appear in three straight games to earn his full salary. It was deemed unimportant language until Manges came along. "They'd play us two games, then sit us," said O'Roark. "Unless you were marquee, you'd find yourself inactive every third game."

"The USFL teams had active and inactive rosters, and if you were on the inactive roster you got paid peanuts — but you couldn't be stashed there if you were injured," said Neuheisel. "Well, we had our quarterback meetings in the GM's office because our space was so small. And one day I opened the drawer looking for a pen, and I saw a list of who Roger was planning on putting on the inactive list. I told those guys, 'You *have* to get hurt today.' For the next seven weeks in a row I'd look to see who was about to be inactive, and then those guys came up with the goofiest injuries. Head bruises, twisted sternums. You name it. The team never caught on. God, they were *so* cheap."

One notable injury was correctly reported: though the man's name was never released to the media, the league office learned that a Gunslinger — listed as "out, groin" — was, in fact, missing a game because he (according to the team) "slammed his cock in a trunk." After much laughter, it was explained that the trunk was not attached to an automobile, but the trunk of a footlocker in the San Antonio clubhouse. Which was equally fantastical.

There was also the issue of a home field that beat and roasted the crap out of its victims. The Astroturf at Alamo Stadium rested atop a concrete slab, and San Antonio's players were always suffering leg injuries. "Your knees would be destroyed," said Neuheisel. "And the bottom of your shoes would melt — literally melt — from the 130-degree temperature."

This was bad. But not nearly as bad as the *paint* injuries.

Yes, paint injuries.

In an effort to save some money, Manges had the midfield Gunslingers' logo, as well as both end zones, designed with standard-grade industrial paint. "To be precise," said Ken Gillen, a defensive end, "it was industrial *crust*." Because the organization was too cheap to have the surface properly cleaned, multiple Gunslingers suffered pustules and skin infections related to their time at Alamo Stadium. "Everyone's spitting on it, hocking loogies on it — and you combine that shit with 100 degrees, it's nasty," said O'Roark. "I got the worst boil on the back of my knee from being exposed."

There was scabbing, too. The sorrowful paint, combined with the coarseness of the low-quality turf, led to some of the nastiest rashes anyone had ever seen. "They fucked up on that," said Morris. "It was this really cheap paint, and if you hit a stripe or the logo, it was like hitting cement. People joke about leaving body parts on a football field. Well, in San Antonio you left your epidermis."

"I've never played on a worse field than the one they had," said Jo-Jo

Townsell, the Express wide receiver. "It was nothing more than green concrete, with paint. Steve Young, bless his heart, he ended up getting the worst scab I've ever seen. He went for a scramble and he had to skid on that stuff. I swear, he had a scab from his ankle to his knee — three inches wide. It was the ugliest thing ever. I mean, it was serious ET-looking stuff. He kept trying to take a shower, but it was too much pain."

"You'd hear screams from the shower," said Neuheisel. "Hot water hitting skin burns."

The 1984 San Antonio Gunslingers failed to qualify for the playoffs. But the club's 7-11 record — a by-product of spunk and heart outweighing so-so talent and a seven-fingered senile coach — was a triumph. Only one player ranked in the USFL's top 10 in any statistical category, and it was Mike Ulmer, who placed second by averaging 11.2 yards per punt return. The defense led the league in interceptions and shutouts, earning the nickname "the Bounty Hunters."

"We didn't have shit," said Fields. "But we busted our asses like no other team I'd ever been on. The '84 Gunslingers ain't legendary. But they should be."

They also had fun. Manges had a thing for Taco Cabana, the city's best Mexican restaurant, and Gunslinger flights would feature large crates of icy beer and boxes upon boxes of tacos, burritos, and enchiladas via Taco Cabana. "The food would make its way down the aisle," said Knaus. "And we'd eat and drink like kings."

San Antonio was the perfect town to be a professional football player — big enough to support a team, small enough to be ubiquitous. Gunslingers players drank for free, ate for free, boasted their own cadre of groupies and an invisible ALL ACCESS pass to most anywhere. Multiple members of the Gunslingers snorted cocaine in the locker room before games — "Drugs were huge in the USFL," said Jeff Gaylord, a Gunslingers lineman. "I once went to my coke dealer, got a $1,000 bag, tooted it, stayed up all night, and played. That's just how it was in the 1980s in sports."

"We had one wide receiver who would kneel down and do blow right there," said Morris. "I'd never seen that before.

"A lot of times we'd go out, party until 4:00 a.m., then sleep in our cars in the stadium parking lot so we wouldn't miss practice. San Antonio was an amazing party town. Amazing. We'd practice until noon, be in the pool by 1, maybe lift later in the day, then every night we'd wind up at Fizz [a popular dance club]. We were the only football team in San Antonio, so wherever we went it was, 'Yo! It's the Gunslingers!'"

In 1984 Morris had an out-of-town girlfriend who visited on occasion from Kansas. He loved her, but also loved the massive amount of sex he was receiving from local women. "Some of us were fucking so many chicks, it was like we were running a brothel," he said. "I didn't want my girlfriend to know what was going on. Because, really, she was important to me. But I also didn't want to miss the calls from other women when she was at my apartment." Morris hatched a plan. He paid for a second phone line, then hired a maintenance man to drill a hole in his apartment and embed an answering machine within the wall. "So I had a regular phone line and I had a phone line for my bitches," he said. "We'd be laying in bed, me and my girl, and you could hear the answering machine in the wall, beeping. I'd pile clothes against the wall so she couldn't hear. The guys thought it was the funniest thing ever — 'Jim Bob has developed a serious system.' Man, we partied like we owned the Playboy Mansion."

The 1985 season kicked off with tremendous promise. Steinke retired, and was replaced by Jim Bates, the 38-year-old defensive coordinator and a sharp football tactician. The team used its first two draft picks on a pair of defensive backs — Issaic Holt of Alcorn State and Greg Greely of Nicholls State — who were lockdown defenders in college, then selected a terrific young halfback, Texas Tech's Robert Lewis. Alas, none signed.

"I, for one, was optimistic," said Peter Raeford, the defensive back. "I was honestly thinking playoffs."

As had been the case a season earlier, the Gunslingers opened with a streak of misery, losing four of five games before paltry home crowds. The insanity of 1984 had been replaced by even greater insanity. In an effort to drum up ticket sales, for example, the team held a Punt, Pass and Kick contest at Alamo Stadium, with the winner guaranteed a USFL contract. The

Gunslingers were not shocked that a 40-yard field goal sealed the win. They were not shocked that the kicker was a former college soccer star. They were shocked that the new Gunslinger's name was Julie Kellman, and that *he* was a *she*.

Regardless, Gill held a press conference to announce Kellman's signing, and even *Good Morning America* attended. "It was kind of historic, in that she was the first woman to have a professional football deal," said Aiken. "We let her practice for a day, but we had to have someone guard her. It was 60 men — some of them slightly crazy — and her."

Manges was never tempted to add Kellman to the roster. That didn't mean he failed to cross the line of reason when it came to his team's kicking game. San Antonio's punter was Ken Hartley, a 27-year-old product of Catawba College who, in 1981, averaged 29.6 yards over nine attempts with the New England Patriots. Manges rarely openly criticized Neuheisel, his golden boy quarterback. And he spent little time focused on the running backs and wide receivers. But whenever Hartley trotted onto the field, the owner fumed. Maybe it was because his nickname ("the Doctor of Love") and perfectly coiffed hair suggested a playboy image Manges aspired to. Or maybe it was because he was averaging 35 yards per punt. Whatever the reason, Manges wanted Hartley replaced.

He had just the man.

A week before the Gunslingers were scheduled to face the Breakers in a Monday-night game at Portland, Gill was ordered to write up a contract for James Edgar Roberts, the 39-year-old Manges ranch hand better known as Buddy. According to the team owner, Roberts boasted a cannon for a right leg, and would greatly improve on Hartley's mediocrity. To anyone within earshot, Roberts "was the best damn punter in the history of Freer High School!" and "a son of a bitch who can kick the ball a country mile!"

The coaching staff generally took Manges's boasts with a grain of salt. Now, though, Bates and Co. were being ordered to absorb a mystery punter onto the roster. Roberts arrived for his first practice, and was presented a No. 8 jersey. He was a smallish man, with stumpy legs and a pouch for a stomach. Two decades earlier he played football and baseball for New Mexico Highlands University, then did some high school coaching and managed a restaurant. His claim to fame came in once winning the Muy Grande Deer Contest in Freer, Texas, with a 15-pointer.

As he walked onto the field at Alamo Stadium for practice, Roberts was greeted by silence. He was introduced to Hartley and Nick Mike-Mayer, the kicker, and appeared genuinely ashamed. "I'm pretty sure he knew he didn't belong," said Jay Howard, the team's play-by-play announcer. "He was so embarrassed. But when Clinton Manges told an employee to do something, he had no choice. So the ranch hand was now a punter . . ."

All eyes focused on Roberts, and as he began to kick, the other Gunslingers could barely contain their laugher. His balls shanked left and right, traveling about 10 yards before plopping to the ground. Professional punters generally require two steps before booting the ball. Roberts needed four. "He couldn't punt for shit," said Bates. "Nice guy, but, dang, no punting skills."

Manges ignored the criticism ("Aw, he's just getting comfortable"), and made certain Roberts was with the club on its chartered flight to Oregon. On the afternoon of April 1, Roberts arrived at the Portland Civic Stadium, slipped into the visitors' locker room, and changed into his white jersey and silver pants. Though Roberts had no desire to appear in a real game ("It would have been humiliating," said Neuheisel), Manges told the coaches that he expected the newest Gunslinger to be involved.

Above the field in the press box, Greg Singleton was a wreck. It was hard enough handling the Gunslingers' public relations without a ranch-hand punter. But this . . . dear God. "We had a couple of local media guys on the trip, and I had to keep it covered up," Singleton said. "I left Buddy off the printed roster, and I told [special teams coach] Billy Schott to keep him in the background. Billy wouldn't let Buddy punt. He was against the whole thing. So throughout the game he made sure every now and then to make sure Buddy wasn't alone in the open."

At one point in the second half Jim Lefko, the *San Antonio Light* beat writer, turned to Singleton and said, "Hey, who's that No. 8?"

Shit.

"I'm not sure," he replied.

"Well, it's somebody," Lefko said.

The Gunslingers' roster was an ever-morphing collection of bodies. People came, people went. In 1984 the team had employed Alvin White, a quarterback nicknamed "the Grasshopper" for a wayward career that took him from Orange Coast College to Oregon State to the Southern California Sun (of the World Football League) to the Houston Oilers to the New Orleans Saints to the Denver Gold and, briefly, to San Antonio as Neuheisel's backup. White failed to last in 1985, but stuck around as the organization's assistant facilities manager.

"That's Alvin White!" Singleton said. "We needed a quick backup . . ."

"No," Lefko replied. "I'm pretty sure you're wrong. The guy on the field looks far too short."

"Nah," Singleton said. "It's him."

Somehow, the conversation faded, and Schott never inserted Roberts into the game (a rare San Antonio win). After enough protestations from Bates, Manges released him the following week. He returned to the Magic Kingdom, content in his anonymity, and died in 2012. "My dad was fine with the clothes on his back and a toothbrush," said Brandi Mutscher, Roberts's daughter. "He didn't need to be a football star to be happy."

Although many of the Gunslingers found Manges to be somewhat slimy and untrustworthy, they were always paid on time. Then, midway through the 1985 campaign, the financial cracks hit the foundation. Manges's oil business dried up, and while — like Bill Oldenburg in Los Angeles — he walked with swagger and spoke with bluster, there were rapidly decreasing finances behind the image. "We started as a train wreck," said Danny Buggs, a wide receiver. "But then we became a train running off the tracks trying to get on the interstate."

Around this time checks started to bounce. First, a trickle. A player or two would drive to the local bank, only to be told — *oops* — there was a problem. Then, more than a trickle. A waterfall. The funding was dry, and only the lucky few could receive their earnings. "I got paid in $20 bills one game," said Whit Taylor, the backup quarterback. "That was different." Before long, on payday players were jumping into their cars after practice to race to the Frost Bank in the nearby town of Freer. "It was like the Indy 500," said Bill

Bradley, the defensive backs coach. "Your only chance of getting any money was to beat the rush."

"I went for weeks on end without getting paid," said Schaefer, the ticket manager. "I still have checks — signed checks — that I couldn't cash. They'd say, 'Here's your check. But don't go to the bank just yet . . .'"

"It was ludicrous," said Dennis Farris, the assistant equipment manager. "Checks were bouncing all over the place. Somebody in the front office would actually tip off the equipment guys when the checks came in, so we'd rush off the field, run to the office, get the check, and haul ass to the bank and cash it."

Two days after an April 7 home upset of Birmingham, Manges called a team meeting. With the players kneeling before him in a circle, he apologized for the "issues" with the bank and guaranteed full payment ASAP. In the meantime, he invited the Gunslingers to visit Gill, who presented each man with a promissory note — "100 percent guaranteed, on my honor!" Manges said.

The following payday, the checks bounced again. Then, on April 11, the Gunslingers traveled to Jacksonville to battle the Bulls. The flight was uneventful, but as the players and coaches exited the plane, some were struck by an infuriating sight: Manges and his executive staffers, all of whom sat in the front of the jet, were greeted at the airport by a fleet of limousines. Rich D'Amico, a linebacker out of Penn State, couldn't believe it. "Look at this shit!" he screamed. "They're not paying us, and they have limos! What the . . ."

Schott, standing nearby, grinned sheepishly. "Don't sweat it," he said. "Clint's probably not paying them, either."

"Good point," said D'Amico. "Let's get a beer."

There was the game against Arizona, when Manges called the locker room at halftime and demanded Bates bench Neuheisel for Taylor, the newly acquired backup. "I spent the entire break arguing with him," said Bates. "That's a hard way to coach a football team." There was the game against Denver, after which Gillen, the defensive end, ripped the organization for failing to pay its players. "The next morning Roger Gill called me in and told me he had to trade me," Gillen recalled. "He asked if I had a preference, and I told him I'd love to go to Arizona. Not only did they send me to the

Outlaws, but Bud Haun wrote me a personal check for the amount due. And it was legitimate." There was Manges's decree that every member of the organization needed to attend one of MaLou Manges's shows at the Melodrama Theater. "We were the only ones there," said Singleton. "I don't even know how she got the gig, but we all had to pay to listen." There was the repainting of the locker room, minus the wisdom to hang WET PAINT signs. "We sat down for the day's meeting and had green paint on our $100 pants," said Thompson. "That sort of sucked." There was the presence of O. P. Carrillo, the Gunslingers' do-nothing vice president of personnel and the only Texas state judge to ever suffer impeachment (for abuse of power). He joined the team after being released from a federal penitentiary. Players referred to him as the club's "VP of Fajitas" — "because he loved fajitas," said Howard. "But, really, who doesn't?"

"O.P. was a funny guy," said Knaus. "He once said to me, 'Duvall County is the only county in Texas where we vote in alphabetical order.'"

Early on Saturday, May 11, two days before the Gunslingers were scheduled to play at Orlando, Aiken was sitting at his desk when Haun approached. "Tommy," he said, "we need to get the money for the flight to Florida."

Aiken excused himself, and returned later that afternoon with a $40,000 check, made out to Braniff Airways, the company that operated the Gunslingers' charter. It was written from the personal account of Ruth Manges. "She loved bowling," Aiken recalled, "and the check had images of a bowling ball and pins on it." The airline accepted the funds, and the squad was ferried to Orlando.

Two days later, roughly eight hours before kickoff against the Renegades, the team held its regular 10 o'clock breakfast meeting inside a conference room at the Orlando Marriott. A phone rang and Steinke, now working as the Gunslingers' vice president of operations, answered. Dianne Buckner, the team accountant, was on the other end of the line.

"What's that?" Steinke said. *"Whaaaaat?"*

His voice rose from a muted drawl to a shout. "You can't be serious!" he screamed. "You mean to tell me we might not have a plane back home?"

The room went quiet, until — *Crash!* Morris, the safety, lunged at linebacker Rick D'Amico, the team's player representative. "How is this shit happening?" Morris shouted. "You need to get us home! You need to handle this!"

"Food's flying everywhere, guys are trying to break it up," said Bradley. "The fighting spills over from where we are to the pool room. It was violent."

Tempers calmed, and at four o'clock the entire team was waiting in the hotel lobby for the bus to the Florida Citrus Bowl. Haun was paged over the intercom. He returned moments later, a blank gaze across his face. "We have no way home after the game," he whispered to Aiken. "The check bounced."

Huh?

"We're stranded."

Neither Haun nor Aiken dared say a word to the players, who went about the business of being a bad football team. The lowly Renegades won, 21–20, when Mike-Mayer missed a game-tying extra point with 46 seconds remaining in the third quarter. The locker room was understandably quiet afterward, and Bates spoke for a few moments about the setback before adding, sheepishly, "Now, there's one more thing. We can't fly home. The check bounced."

"I mean — what the heck?" said Neuheisel. "We couldn't fly home?"

That night, the Gunslingers slept in a budget airport motel, and the next morning Bill Hobby, the lieutenant governor of Texas and a Manges confidant, called Braniff to guarantee payment. "It was that kind of an existence," said Neuheisel. "Every day a carnival."

Bates was starting to lose it. Ever since walking on as an unrecruited linebacker at the University of Tennessee in the mid-1960s, the San Antonio coach believed in hard work, loyalty, dedication. Now he was asking his players to compete sans compensation. "Guys stuck in there and hung together," he said. "Well, one day someone came to me and said, 'Jim, the checks will be paid next week. Tell the players.' So I did. I got them together and said, 'I know it's been hard, but management says this week we'll get paid.' Then it goes Monday, Tuesday, Wednesday, Thursday, Friday — no pay. I told my coaches I could no longer deceive the players, and that I was going to resign." Bates held a meeting with his men. He held back tears and apologized for the difficult season. "You guys have fought, and I'm so proud," he said. "But I cannot deceive you any longer."

He stepped down on May 18. "I remember my eyes welled up," said Peter Raeford, the defensive back. "Not only was he a man of character, but he brought a lot of energy." That day, with Steinke back at the helm in an interim role, the Gunslingers fell to Oakland, 24–21. Afterward, the old-yet-new coach tried his best to motivate the team, pulling from the mothballs a stirring pep talk from his Texas A&I glory days. But this wasn't Texas A&I, and these weren't college football players. Leon Williams, a fiery defensive back who had spent the two previous seasons with Tampa Bay, interrupted. "You old motherfucker!" he yelled. "I don't wanna hear this shit! Just tell me when we're getting paid!" Bradley, Williams's position coach and a former All-Pro safety with the Philadelphia Eagles, hated the situation but loved Steinke. He screamed, "Shut your fucking mouth!" before ramming into Williams's torso and pulling him to the ground. "It was madness," said Singleton. "People flying all over the place, punches being thrown." D'Amico, a team leader, charged from the showers, naked and soaking wet. He was crying. "Stop this!" he yelled. "Stop this! This is terrible! We! Have! To! Stop!"

Meanwhile Howard, the play-by-play radio man, began to speak into his microphone. "OK," he said, "we're in the Gunslingers' locker room and . . ."

"No!" hollered Singleton. "Not now!"

". . . we're experiencing technical difficulties," Howard said. "We'll be back in a bit."

Silence.

"It caused a real divide," said Morris. "How do you play for no money? [Defensive coordinator] Tim Marcum was a friend. But he'd scream at us, like, 'You guys get in there and . . .' And I'm like, 'Tim, I haven't been paid in three weeks. Shut the fuck up.'"

Five games remained on the schedule, but for much of San Antonio's roster, the year was done. Another return flight, this one from Baltimore, was delayed for two hours because Manges needed to pay the airline for fuel. "A suitcase of money showed up," said Thompson. "Out of nowhere." Multiple player-only meetings were held, and the debate wasn't about *how* to play, but *whether* to play at all. "You love the game, but you play to eat," said Paul Hanna, a nose guard. "You can't eat without money." None of the paychecks were clearing, and hostilities were bubbling over. Votes were taken on multiple occasions — stay or leave? Some left. Most stayed, with the hope that dough would materialize. "It was very stressful," said Knaus. "I remember one meeting where a couple of the coaches walked in, and the players wanted us out. I said, 'Listen, don't think how you vote doesn't affect us, too.' Emotions got really heated. There were players who were best friends who became enemies."

Because Manges was now (wisely) keeping his distance, the target of player wrath turned toward Haun, a nice yet overmatched man whose previous managerial experience came as the owner of a feed store. Haun was one of the few team executives with a window office in the double-wide trailer, and as members of the Gunslingers arrived with increased frequency to collect their (nonexistent) checks, Haun — who could hear them approaching from outside his closed door — often climbed out the window, dashed to his car, and drove off. It was utterly absurd — the 71-year-old Korean War veteran, who came to work in a suit and tie, carefully lifting one leg, then the other, over the sill. Once, when he was unable to escape in time, Haun was confronted by Barefield. He held a baseball bat in his hand, and told Haun it was either the money or his head. "John got paid," said Raeford. "He was the one."

Actually he was one of two. Greg Fields, the man who locked himself inside a hotel room with the Atlanta Falcons, then slugged his coach with the Los Angeles Express, enjoyed his time in San Antonio, but grew increasingly unhinged with each missed pay cycle. Finally, as the season began to wane, and the losses mounted (the Gunslingers won but 2 of their last 11 games), he took action. Instead of chasing down Haun, or punching Steinke, Fields waited for the rare day Manges showed up at the facility. "I told myself, 'Man, I can't take this anymore,'" Fields recalled. "I needed to do something, so I followed him home."

According to Fields, he trailed Manges until they arrived at the Magic Kingdom, then parked his 1978 Buick LeSabre a couple of streets away. As the oil man walked toward his front door, Fields said he sprinted from behind and tapped him on the shoulder. Manges turned, and was shocked — and terrified — to see his six-foot-seven, 265-pound defensive end. Like Barefield, Fields was gripping a baseball bat.

"Hey, Clinton Manges, don't tell me you're broke man," he said. "You're not gonna have me believe that stuff."

Manges stood silently.

"Listen, pay me," Fields said. "I'm gonna be a big headache for you. Pay me off so you don't have no big headaches."

There was a long pause — the short owner with the graying hair; the young pass rusher with the pulsating muscles and the baseball bat. Manges was well aware of Fields's propensity for violence. He had signed off on bringing the guy to San Antonio.

"I'll be right back," he said, and walked into his house.

Fields waited 10 minutes. When Manges returned, he was holding a thick wad of $100 bills. "Do you want to sign something?" he asked — and Fields shook his head no.

"Don't worry," he said. "I'm not coming back."

Greg Fields retreated to his car, $17,000 in hand. He returned to his apartment, packed his belongings, and drove cross-country to his home in San Francisco.

"Craziest shit ever," Fields said. "Craziest. Shit. Ever."

Not quite. A couple of days after Fields's departure, Manges attempted to make things right by hosting a soiree at his ranch home for hundreds of friends, family members, and employees. All of the Gunslingers were invited, and while they were angry and ornery, a big party was a big party. "We were hungry," said Bradley.

En route to the event, Bradley and a couple of other coaches stopped at a package store and purchased four cases of Cold Duck, the inexpensive sparkling wine found in hundreds of American fraternity houses. "We walk in with this cheap shit, and they're serving bottles of Dom Perignon, wines from all different years, different vineyards," said Bradley. "We found the basement wine cellar, and a bunch of us went down there and started drinking these $6,000 bottles from France, $4,000 bottles from somewhere else. Then, when they were empty, we'd fill them with the Cold Duck and bring the bottles up to the party." One of Bradley's greatest moments with the Gunslingers came when Manges's well-heeled guests raved about the libations — never realizing it was the oenophile's equivalent of cow urine. "We laughed our asses off," said Bradley. "It was our revenge."

San Antonio's last game was played on June 23, and a surprisingly robust crowd of 19,603 packed into Alamo Stadium to see the Gunslingers beat the inept Breakers, 21–13. They wrapped the season with a 5-13 record.

Because the facility held only 24,000 spectators, before every home game the building's caretaker, a large Hispanic man known as Big Chief, unfolded 6,000 blue metal chairs, one by one by one, and placed them on the outskirts of both end zones. The seating arrangement — while uncouth and clunky — fulfilled USFL capacity requirements.

A couple of hours after the stadium emptied and the lights were dimmed for the final time, Bradley returned with his nephew, Ray, a friend nicknamed Bear Cat, and a truck rented from Hertz. Bradley was owed $27,000 by Manges, and he knew the money would never materialize. So he slipped Big Chief some bills and backed the vehicle onto the field. Then, over the course of three back-and-forth trips, the men loaded every single chair onto the truck. "I drove them down to Fort Worth and sold them for $8 each," Bradley said. "The Gunslingers never knew what happened to their chairs.

"Served 'em right."

15

HOOKERS HAVE BUSINESS CARDS?

When I say I coached in the USFL, a lot of people will look at me like, "What is that?" I always thought we never got credit for what the USFL was — a pretty darn good league.

— Jim Mora, coach, Philadelphia/Baltimore Stars

BECAUSE THEY PLAYED in San Antonio, with a largely anonymous cast of ordinary football players, the Gunslingers didn't generate many national headlines throughout the 1985 season.

Elsewhere, however, people seemed to be paying attention. Or at least — entertainment-wise — they *should* have been . . .

• In New Jersey, as Doug Flutie was showing himself to be a merely ordinary professional quarterback, Herschel Walker *finally* set the league ablaze. Interestingly, it was Donald Trump — football moron — who deserved much of the credit. Following the Generals' March 30 loss at Arizona, during which Walker carried just nine times for 53 yards, Trump stormed into his team's locker room and demanded an audience with Walt Michaels, the head coach. "You've got the greatest runner in the history of football, and he ran nine times!" Trump screamed. "This guy has to have 25 to 30 carries a game."

The very next week, Walker used his 34 carries to gain a USFL-record 233 yards against the Gamblers. When asked whether Trump made the difference, the gruff Michaels snarled, "No comment." It was the first of 11 straight weeks with at least 100 rushing yards for Walker, who found himself on the cover of the May 27 *Sports Illustrated* alongside the headline RUNNING WILD. "In one game I decided when the Generals gave Herschel the ball I was going to take him on," said Sel Drain, an Outlaws safety. "He made all that money, had all that hype. I'd show him. Well, he came around on a sweep, I hit him high and he didn't even slow down. It was like tackling a bull, and he ran right over me. From that point on, I only went for his ankles."

On June 10, the Generals hosted the Jacksonville Bulls in front of 36,465 fans at the Meadowlands. With a victory, New Jersey would seal a playoff berth. But the focus was Walker. He entered the evening's contest 138 yards shy of Eric Dickerson's professional football single-season rushing record of 2,105. With 6:38 left in the third quarter, the Generals led 24–10, and faced first and 10 from their own 45. Walker — standing deep in the I-formation — took the handoff, drifted right, pivoted left past a lunging errant tackle, charged straight through the arms of defensive end Joe Costello, sprinted five steps and headfirst into the torso of linebacker Vaughan Johnson, who fell to the turf. He then spun, escaped strong safety Chester Gee, and sprinted down the field away from free safety Van Jakes, who ran a 4.4 40 but looked as if he were stuck in molasses. As Walker crossed into the end zone, Charley Steiner, the Generals' radio voice, yelled out, "Look out record! Walker touchdown!"

Never one to boast or celebrate, Walker handed the football to Vince Stroth, the burly offensive guard, who spiked it to the turf. The Generals beat the Bulls en route to an 11-7 mark, and Walker's 2,411 yards remain an all-time professional mark. "It was exciting," said Steiner. "But it was manufactured exciting. He played 18 games against lesser competition. There were moments when Herschel would outrun people or run people over. Would he have dominated the NFL that way? Of course not. Still, it was fun to be a part of."

The week after the game, Walker presented every New Jersey offensive lineman and tight end with satin jackets and chains featuring gold medallions inscribed with "2,411." "I've never forgotten that," said Sam Bowers, a tight end. "Decency isn't easily forgotten."

• In Birmingham, the Stallions' biggest standout wasn't quarterback Cliff Stoudt or halfback Joe Cribbs, but a flighty free safety named Chuck Clanton. Before arriving in the USFL, Clanton spent three years at Auburn University, interchanging magnificent defensive plays as a cornerback with dispiriting displays of poor judgment. He missed classes, curfews, meetings, appointments with guidance counselors. The Tigers' second game of the 1983 season came against Texas, and Clanton played well in a 20–7 loss. Afterward, he drove to Atlanta to party with friends, and failed to return in time for curfew. Pat Dye, Auburn's coach, enforced his standard punishment for such transgressions — a player was forced to run up and down the steps of Jordan-Hare Stadium, two cement blocks in hand. If the athlete was 10 minutes late, he had to run 10 times. Clanton was 200 minutes late . . .

"I ran three, then left," Clanton said. "I quit the team. The biggest hit was financial. Players would get tickets to the games, then we'd sell them for $400 a pair. I'd miss that. But would I miss Auburn football and all the bullshit that went with it? No way."

Clanton spent the remainder of his junior year as Auburn's biggest intramural flag-football star, then decided to take his parents' car from his home in Pensacola, Florida, to Birmingham to audition for the Stallions. "I'll never forget the date — January 12, 1984," Clanton said. "I was driving, and I met a truck driver along the way. I told him what I was doing, and he said he'd attach my car to the back of his truck if I'd give him four tickets to a future game should I make it. I sat in my car behind the truck for hundreds of miles, pounding Budweiser, listening to the radio. I passed out, and when we arrived in Birmingham I woke up drunk, slammed on the brakes — then realized it was OK, I wasn't driving."

Though only 5 foot 11 and 192 pounds, Clanton was a gifted athlete with a sprinter's speed and a spider's instincts. His biggest flaw was slippery hands, which he corrected, strangely, by playing in scuba gloves. He made the Stallions, started opening day, and emerged as the USFL's elite ball hawk. He had 10 interceptions through the first 13 games, and was a leading candidate for the league's Defensive Player of the Year. Then, in early June, Clanton arrived at the Stallions' facility with his left hand wrapped in tape and gauze. When asked by the press, Clanton blamed "this huge butcher knife . . . I was cutting watermelon on the counter."

Which was a complete lie.

Here, some 30 years later, is Clanton's telling: "My girlfriend was pregnant with my daughter, and we were not getting along. She was always yelling at me. I was trying to focus on football. And she would go to the clubs at night. We'd both go. But she'd stay until midnight, I'd leave at 10. Well, she called me at midnight one time and said, 'Come back to the club and bring me something to eat.' I was in bed, but I got up, drove to Burger King, picked her up food, brought it to the club. She was a bitch about it. She ordered me to get her different food. 'No — I'm not doing that.' She threw the food in a trashcan. So I put her out. I said, 'You've gotta go. Go home to your mom. Just go away.' She wouldn't leave, so I called the police and they told her she had to evacuate. OK, wonderful. That night's sleep was the best I ever had.

"The next morning I went to practice, and when I came home she was there. She'd broken in. I said, 'You have to go right now.' She wouldn't listen. She said, 'I'm not going anywhere.' Then she picked up a kitchen knife. If she hadn't screamed before charging me, I'd be a dead man. But I rose my left hand as she tried stabbing, and the knife went through my left hand. Through the bone, through the cartilage. Straight through. I pulled my hand back, and you could see both sides of the knife."

According to Clanton, his girlfriend ran out the door and never returned. He wrapped his hand in a bouquet of paper towels and drove off for the practice facility. He walked into the training room, held up his bloodied palm, and said, "Fellas, can someone here pull this out?"

"Your girlfriend do that?" a teammate asked.

"Yup," said Clanton.

"We told you she was psycho," the teammate said.

The knife was removed, and Clanton rushed to the hospital. He wound up missing two games, but still intercepted 10 passes for a league-best 249 yards and three touchdowns.

The 1985 season was even better. With his hand scarred but healed and his girlfriend gone, Clanton was the heart of the USFL's best defense. He picked off 16 passes (which broke the professional record of 14, set by Night Train Lane of the 1952 Los Angeles Rams), and led the Stallions to a 13-5 mark and first-place finish in the Eastern Conference. Much like the Gunslingers, the Stallions were unhinged. Unlike the Gunslingers, the Stallions were unhinged and talented. Rollie Dotsch, the head coach, refused to impose a home curfew, and the team responded as one would expect. "We went out *a lot*," said Clanton. "Name the drug, name the drink — it was with us. Shit, we had guys bringing cocaine and guns on the planes. Liquor, coke, heroin — whatever you could think of, our team did."

Clanton vividly recalled the lead-up to the Stallions' April 7, 1985, game at San Antonio, which figured to be a piece-of-cake triumph over a flawed franchise. On the flight to Texas, Clanton said teammates were spread out throughout the cabin, drinking and sniffing cocaine. "Someone said to me, 'C'mon, have a snort,'" he recalled. "But my drug wasn't coke. It was marijuana. I had it delivered in bundles by UPS to my house." Birmingham jumped out to a 7–0 lead when Clanton returned an interception 49 yards for a touchdown. The Stallions then sat back and did nothing. "The Gunslingers ran a no-huddle 15 straight plays and we had to use two timeouts on the first drive," Clanton said. "I could just see everyone breathing hard, sniffing hard in their noses. I was like, 'Y'all fucked up. Y'all fucked up.'" The final score was 15–14, San Antonio, and the return trip was unpleasant. The Stallions' regularly scheduled charter airplane was docked in Denver, so the team was placed in the rear of a commercial 747. A furious Dotsch stood before his players and announced, "There will be a lot of changes on Monday."

Clanton was sitting in the last row alongside defensive tackle Doug Smith, his former Auburn teammate. The two shared a bottle of Jack Daniel's, and as he imbibed, Smith's laughter turned increasingly loud.

Jimmy Smith, the veteran wide receiver, was a former Pittsburgh Steeler who didn't believe in chuckles and giggles after defeats. "Y'all need to keep it down back there!" he yelled. "This isn't a time for talking. We just got our asses kicked."

Doug Smith, all six foot four inches and 294 pounds of him, rose. "Who said that?" he barked.

"You guys need to shut up!" said Jimmy Smith. "Just shut up!"

"That motherfucker needs a parachute," Doug Smith said to Clanton. "He needs to be off this plane." He marched toward Jimmy Smith, reached back his arm, and smacked the wide receiver across the mouth. "Then he grabs Jim by the throat," said Clanton. "I'm yelling, 'Someone grab Doug! Someone grab Doug! He's gonna kill Jimmy!'" Doug Smith now had his hands wrapped around Jimmy Smith's throat. "I will kill this motherfucker!" he screamed. "I will kill this motherfucker!"

It took six or seven players to declaw Doug Smith. "His shirt was ripped, he reeked of alcohol," said Clanton. "I can only imagine what everyone on the plane thought of us."

That should go down as the season's most disturbing moment. It does not. A few weeks earlier, an *Orlando Sentinel* columnist named Joan Ryan entered the Stallions' locker room after a game against the Renegades, seeking an interview with Cribbs. She wore a blouse and a knee-high skirt, and as Ryan waited for the running back, she felt something running up her leg. It was the handle of a razor, held by one of the Stallions players. "What are you doing?" Ryan yelled, then stormed out of the room. As she exited, she passed Jerry Sklar, the team president, who — according to Ryan — stood grinning. "I never had an experience like that," said Ryan. "Players were laughing, making comments. I felt genuinely frightened for what might happen."

Ryan later complained to the league. Sklar denied wrongdoing. No, he didn't think women belonged in the locker room. But he insisted he did nothing untoward. Then, he made the mistake of explaining his feelings to a United Press International reporter. "If she thinks she feels humiliated, how do you think those naked players feel to have a strange woman in there?" he asked. "It's a trumped-up deal by [a woman] looking for a story."

• In Philadelphia . . . no, in Baltimore . . . no, in College Park — the geographically challenged Baltimore Stars struggled early on, winning just one of their first five contests before small crowds and a conflicted fan base. It

was hard to blame the players, who practiced at Philadelphia's Veterans Stadium and lived in (or near) the city, but played their home games 133 miles
away on the University of Maryland campus. "We were gypsies," said Dave
Opfar, a Stars defensive lineman. "On the schedule we had nine home games
and nine road games. But in reality, it was 18 games on the road."

"Every Tuesday [head coach] Jim Mora and I would take the 7:00 p.m.
train out of Philly to Baltimore," said Michael Kaine, the Stars' public relations director. "We'd meet with the Maryland reporters the next morning for
a press conference. The Baltimore media was not receptive to us — understandably. We didn't practice there, we didn't live there. We didn't even play
there. We had a sales office in Baltimore, and the city was in our name. But
that was it. Our players literally never went to Baltimore. Still, Jim was a pro's
pro. He gave it everything he had."

The wayward existence turned even more complicated on June 3, 1985,
when the city of Philadelphia served the Stars with eviction papers from
the Vet, and sued the team for $5.8 million, claiming fraud and a breach of
lease (the Stars signed a 20-year lease with Veterans Stadium, but argued
the switch to fall rendered the document null and void). The papers arrived
on Myles Tanenbaum's desk on a Monday, and by Tuesday morning the organization was relocating to temporary headquarters on the campus of the
University of Pennsylvania. "They were moving our stuff as we were preparing for a game," said Mora. "Our coaches were on their knees in the offices
without any furniture, trying to game-plan." Veterans Stadium provided the
Stars with all the comforts a professional football team would expect. Penn
did not. The team operated inside an abandoned ROTC structure. Position
meetings were held in different corners of the same room. Shoulder pads and
equipment dangled from nails hammered crudely into the plywood wall. "It
was like working out of a factory," said Antonio Gibson, the safety. "And it
smelled really bad."

The eviction was the final straw for many of the Stars players. They were
tired of being here, then there, then here, then there. The stadium at College
Park was awful. The fans didn't care. They had recently received their 1984
championship rings, and the diamonds — large and flashy — were made of
cubic zirconia. "We started flying on this shit airplane," said Pete Kugler, the

nose tackle who had won a Super Bowl with the 49ers. "It was an old DC8. If we flew cross-country we had to stop three times to refuel. The seats were tiny, and on one trip I had to sit next to [300-pound offensive lineman] Irv Eatman. We couldn't fit. I turned on the AC one time and this black goo started dripping out. My neighbor was an American Airlines pilot, and I asked him to identify the goo. He told me it was the nicotine oozing from the vents." What sort of way was this to treat the defending-league champions? The NFL certainly didn't operate in such a manner. "There was a lot of bitching," said Carl Peterson, the Stars' president and general manager. "It's hard enough playing football without all the distractions."

The Stars fell to the Birmingham Stallions 14–7 on June 8, and afterward Mora refused to mince words. Sitting inside Legion Field's cramped visitors' locker room, he explained, "I know it's hard, I know it's challenging. But we've spent most of this season bitching and moaning, and it's time to stop. You are the defending champions of this league. Now you need to play like it."

The Baltimore Stars won their last two games in convincing fashion, and entered the playoffs with a 10-7-1 mark.

• In Portland, the Breakers — new to yet another location that, water-wise, kept the nickname appropriate — arrived from New Orleans to be greeted by prostitutes.

"That's 100 percent true," said Steve Weaver, the team's director of marketing. "We had training camp in California, played our first game on the road, then came to Portland for our first home practice. These two buses stop at the facility — a converted elementary school — and a bunch of hookers get off. They wanted to leave their cards with the players. Honestly, I didn't even know hookers had business cards."

Coming off a better-than-it-sounds season, the Breakers seemed to have promise. Then, in the opening-day loss at Arizona, halfback Marcus Dupree suffered a devastating knee injury that ended his year. Though the news was upsetting, the team's staff had tired of the star's sense of entitlement. "He was a big baby," said Russ Blunck, the assistant director of public relations. "One day ESPN was in town to interview him, and he wouldn't do it without first getting a blackberry shake. He was a nice kid. But immature and exhausting." Now, he was gone, and the team lost the centerpiece of its entire marketing strategy. After Dupree, the second-most-famous Breakers were Tim Mazzetti, the kicker, and Dan Ross, the tight end who had starred with the Cincinnati Bengals. Neither man was a household name. The organization tried signing Dieter Brock, the big-armed Canadian Football League quarterback. He said no. They sent two first-round draft picks to Tampa Bay for quarterback Wayne Peace, then watched as he quit after one day in town.

Portland turned out to be a lousy place for a lousy team. The stadium was a dump, the weather was cold and wet, the attendance dipped by the week. "We stored our supplies beneath the bleachers," said Katherine Ossey, the director of merchandise, "and it was so dank and damp that you'd be surprised if there was no mold."

Despite the obstacles, no franchise has ever employed a more across-the-board joyful and enthusiastic staff. The director of public relations was John Brunelle, a 23-year-old who had recently spent four years as the quarterback at Pacific University in Forest Grove, Oregon. His assistant was Blunck, his top wide receiver with the Boxers. "We were kids, and if we weren't being asked to judge cheerleader auditions, we were on the field playing catch with Matt Robinson and Marcus Dupree," said Brunelle. "We halfway thought we were members of the team. I always had this hope in the back of the mind that they just might need an emergency third quarterback." Dick Coury, the head coach, hosted barbecues at his house and continued his tradition of having a different fan design a play for each home game. The cheerleaders, who were required to sell four season tickets and earned a whopping $15 per game, made regular public appearances. "For that year we were local celebrities," said Stephanie Daley, a member of the Heartbreakers. "It was magnificent."

Weaver was a Bill Veeck disciple who never met a promotion he wouldn't try. During one home game he distributed 1,000 cardboard signs to fans located in a specific section. Printed on one side, in red, was the word RUN. Printed on the other side, in green, was the word PASS. Before certain designated plays, Pete Kettela, the offensive coordinator, looked toward the section and followed the collective wisdom. "Pokey Allen was the defensive coordinator," said Weaver. "He told me afterward, 'Boy, those fans have no idea what they're doing.'"

Sadly, with poor attendance and meager merchandising efforts, the Portland Breakers hemorrhaged money. Ed Trachtenbarg, the team's controller, would gaze over the weekly spreadsheets with dread. As was the case in San Antonio, checks bounced. "My job became sitting at my desk and having enormous men come in and say, 'Am I getting paid today?'" Trachtenbarg recalled. "Halfway through the season the answer was always yes. Then it became, 'Well, I don't really know.'"

"They started giving us checks every other week," said David Bayle, a tight end. "It was fishy."

With four games remaining, Breakers players stopped receiving money. As did hotels and airlines and caterers and the $15-a-pop Heartbreakers. "I don't think I lied too much," Trachtenbarg said. "But I'm not sure."

The Breakers' wrapped the season with a 21–13 defeat at San Antonio. The team finished 6-12, and upon returning, players and staff cleared out their goods, knowing it was over. "Pokey Allen and I snuck into the hallway right by our offices," said Weaver. "There was a big Portland Breakers sign, probably two and a half feet high and three feet wide. We unscrewed it, put a sheet over it, and put it in my car.

"It's still in my garage."

● In Los Angeles . . . where to begin? With Bill Oldenburg and Jay Roulier both out of the picture, the team's owner was (egad) the USFL, which was charged with paying the remaining $7 million in salaries of a franchise nobody much cared about. Wrote Jim Murray of the *Los Angeles Times*: "Well, the L.A. Express really isn't owned. It's kind of a public trust or the embarrassing ward of a prominent family — the kind they would lock up in the attic of the family castle in the old days, or tie to the bed, or pack off to India with the stipulation that he never tell anybody his real name . . . The Express has had more owners than a 1949 DeSoto." A month before kickoff, the USFL's other franchises begrudgingly agreed to foot the bill for the team, and installed Richard Stevens, the commissioner of the modern pentathlon at the 1984 Olympic Games, as chairman. It would be a Jack's 99-Cent Store approach to football management. Were a player injured, no replacement was signed. If fans failed to show, the advertising budget rose from $0.00 to $0.00. Stadium repairs? Equipment upgrades? Um, no. Three players were even sold off to (gasp!) the NFL. (The Cleveland Browns, for example, paid the Express $200,000 for running back Kevin Mack.) Wrote Matthew Barr in *Los Angeles Magazine:* "What killed them was the injured-player provision. Without it, the team could not build, could not grow. It was a death sentence."

Training camp, held in Long Beach, was fine. Kind of. The lawn went unmowed and the clogged toilets went unplunged and the cafeteria stopped serving beef and fresh fruit. Mike Ruether and Gary Zimmerman, two offensive linemen, walked out for a day and a half over not receiving payments. Rumors swirled that the team would relocate to Hawaii, and season-ticket sales stalled at 6,000. One afternoon John Hadl excused himself from a meeting to take a phone call. He returned with a horrified expression across his face. "Fellas," he said, "they've kicked us out of the hotel because we haven't paid our bill."

The head coach walked up to the blackboard — normally used to design plays — and listed members of the team who could spare a bed or two for teammates. He looked at one wide receiver whose mother lived nearby. "Your mom's place, how many can it fit?"

"Well," the player said. "Maybe four."

"Is there any possible way she could squeeze in five?" Hadl replied.

"No," he said. "I'm pretty sure four is the max."

"That's what we came to," said Steve Young. "Trying to figure out how many football players someone's mother could fit in her home."

On another morning, a painter charged into the Express offices and asked to speak with Mona Williams, the team's bookkeeper. "If I don't get my money, I'm going to kill you!" he screamed. "If that's what it takes to get my money, I will hurt you!" Williams excused herself and returned moments later with a check. Shortly thereafter a self-employed carpenter came to collect the $1,000 he was owed. He was out of work with a broken leg, and in dire need of the funds. "It started to affect me personally," Williams later said. "It got to the point where I couldn't take it very well."

Zimmerman has never forgotten the day he and his teammates were practicing when the moving vans arrived. "They took the office furniture, they took the supplies . . . anything of value," he said. "It was the greatest thing, because we had no film and we had no game plan. So we'd come at ten, shrug, and leave by one."

The Express opened at home against the Houston Gamblers on February 24, 1985, and any hopes of Los Angeles finally embracing the Oldenburg-liberated Express were immediately dashed. The team's cheerleading squad, once a robust 28 women, was pared down to a pathetic 18, and the only reason their squeals could be heard was because the crowd, listed at 18,828, was significantly exaggerated. (The first 10,000 fans received a free Steve Young poster. There were leftovers.) Although Gamblers-Express was originally scheduled to air on ABC, it was bumped in favor of Generals-Stallions, and therefore could be witnessed either in person or not at all.

What those select few in attendance viewed, however, was something out of a futuristic football-video bonanza. In what would come to be known as "the greatest game no one saw," Young and Houston quarterback Jim Kelly combined to throw for 829 yards (a pro football record 574 for Kelly alone). "It was Muhammad Ali–Joe Frazier," said Scott Boucher, Houston's offensive guard. "Only with no one there." The Express held a 23–13 advantage entering the fourth quarter, and when the score reached 33–13 D. J. Mackovets, the Gamblers' media relations director, led reporters from the press box and toward the field to prepare for the postgame interviews. On the Houston sideline, chaos reigned. Jerry Argovitz was grousing aloud about his team's flaccid defense, yelling insults toward Jack Pardee. Hosea Taylor, a six-foot-four, 249-pound defensive tackle, overheard his obnoxious boss and stomped toward him, fists up. "Hosea had to be held back," said Ray Alborn, the defensive line coach. "He had murder in his eyes." Then — *Bam!* Within a span of nine minutes, Kelly led the Gamblers on drives of 74, 43, and 84 yards.* It was one deep laser after another, and the fleet of Houston wide receivers made mincemeat of the secondary. "We started running the no-huddle offense with 10 minutes left in the game," said John Jenkins, the offensive coordinator. "Nobody had ever done that before. It was like watching an Olympic track meet."

When, with seconds remaining, Young's pass to Mel Gray was intercepted, Houston sealed a 34–33 win for the ages. "That was the most amazing single battle I ever engaged in," Kelly said years later. "Empty stadium, exciting teams, nonstop scoring. Steve and I still talk about that one."

Theoretically, the shootout should have served as a neon marketing campaign for the Express. The team was thrilling, engaging, fun, dynamic. Instead, it was little more than a yawn. Nobody cared. Three days later the franchise dismantled its entire cheerleading unit, whose members made $35 a game. Stevens insisted the firings were about taste, not money. "I don't know what we're going to end up with," he said. "We may end up with dancing dogs. Who knows . . ."

Cathleen Lally, the cheerleader chief, threatened to sue for wrongful termination. Only how does one sue an entity without money? A few weeks later the team rehired eight cheerleaders and found less revealing uniforms for the women to wear. The team brought the outfits to Moore Dance Wear for alterations, but never received them back. The Express, it turned out, owed the company money. "We've still got the uniforms," a Moore employee told the *Los Angeles Times* months later. "I'd rather not talk about it."

The Express dropped their next two games on the road, then returned home to play before a silent crowd of 10,410 fans at the Coliseum. That they beat the Gunslingers, 38–7, was far from the main story. Talk all week was that the club might fold at any moment. Instead, players were told the games would continue. To celebrate, Wymon Henderson, a cornerback, distributed cigars throughout the locker room.

---

* The word "drive" presumes a long stretch of plays. Such was not the case. Kelly hit Richard Johnson for a 52-yard touchdown pass on the first "drive," Vince Coutville on a 40-yard touchdown pass on the second "drive," and Ricky Sanders on a 39-yard touchdown pass on the third "drive."

It was the last true joyful moment of the year.

If the Gunslingers were football's most discombobulated operation, the Express were a close second. The same organization that canned the cheerleaders to save a little dough spent $4,000 on a *daytime* fireworks display. "All we saw was sparks and smoke," recalled Tom Vasich, a season-ticket holder. Meanwhile, the water at the office was turned off for failure to account for a $139.84 bill. Young, the highest-salaried player, paid money from his own pocket for someone to finally cut the grass at the practice facility. Jo-Jo Townsell was in charge of bringing the footballs. The chartered flights were on an airline, Key Air, that no one had ever heard of. "It once took us eight hours to get back from New York," said Zimmerman. "The pilot got lost." Midway through the season some 20 players walked off the team after not receiving paychecks. Several of the coaches, also without their money, ceased working, and Young and his backup quarterback, Frank Seurer, drew up the game plans and practice schedules. "We'd bring our own pens and pads and scribble it all down," said Young. "There'd be these big fights about the strategy, because I'd put in all these passes, and the running backs would be screaming, 'Where are the runs? Where are the runs?'"

An all-time Express-low 4,912 came to watch the Bandits destroy the home team, 24–14. Four weeks later, the attendance futility mark was smashed, and 3,059 people (or, according to writers covering the game, 1,500-ish people) saw the visiting Denver Gold (led by the immortal Bob Gagliano at quarterback) win, 27–20. Young had skipped his college graduation ceremony at BYU to play in the game, and surely regretted the decision when Seurer, his backup, was chosen by Hadl to start.

The final three weeks were brutal. The Express traveled to San Antonio for a June 9 game against the Gunslingers, and 4,963 attendees watched two awful teams play awful football in awful heat. The on-field temperature hit 110 degrees, and by the fourth quarter 18 of the 33 active Express players were hooked to IV drips along the sideline. "A few of us had stayed out extra late the night before, and might have had a couple of frosty brews," said Danny Rich, the linebacker. "One of the guys was [offensive lineman] Derek Kennard. The humidity had to be 80 percent, and I looked at Derek and thought he was gonna die." That the Gunslingers won, 31–27, was an afterthought to what happened at the airport two hours later. The Express players and coaches were sitting on the plane, waiting for takeoff, when the truck used to load baggage slammed into one of the engines. "It sounded like a bomb going off," said Young. "We had no idea what happened."

"They keep loading the stuff on the plane," said Rich. "I mean, we're staring out the window, seeing the hole. And they weren't going to say anything."

Klosterman, the general manager, had served as an Air Force pilot. He assessed the damage, then stood and announced, "Guys, that truck put a small hole in the engine. But I flew in World War II, and I can tell you this plane is safe enough to fly."

Every member of the team filed out the door, exited the airport, and retreated to the nearest bar. They flew back to California the following morning.

On June 15, the Arizona Outlaws were slated to visit Los Angeles for the Express's final home contest. Only, instead of holding the game inside an empty Coliseum, the Express opted to bring professional football to Pierce College's John Shepard Stadium, a community college facility with 5,000 seats. Located in the Los Angeles suburb of Woodland Hills, Pierce's high-school-level operation would have been a preposterous site for the game, were the Express not so utterly preposterous.

In the lead-up to Saturday evening, members of the Express visited local shopping malls and chamber of commerce breakfasts, talking up ticket sales. It was simultaneously noble and pathetic, and the USFL paid for a slew of bleachers to be placed around the field, boosting capacity to an optimistic 15,000. The ambition of the USFL was to impress enough to find an owner for the franchise. "We hope if we make a good showing, someone will buy the team," said Herb Benson, a spokesman. "We don't have a lot of selling points right now." The goal of Pierce College was to convince the Express it could serve as a future landing spot for the organization. "I was hoping that some pro team would come in and build a facility and we could play in it and they could play in it," said Jim Fenwick, the school's head football coach. "You let yourself dream."

Kickoff was scheduled for 5:00 p.m., and the Express barely made it. En route from the team's Manhattan Beach facility to the field, the driver pulled the team bus to the side of the road, folded his arms across his chest, and announced, "This bus isn't moving another inch until I get paid!"

Players chanted, "Don't pay him! Don't pay him! Don't pay him! Don't . . ." Hadl, plopped down in the front seat, removed his checkbook, but the driver insisted upon cash. Young reached for a baseball cap, passed it around, and asked for contributions. Nobody offered so much as a dime. Finally Jim Hammond, the head trainer, raised his hand. He had just been paid, and was willing to cash his check. The driver approved the plan, and received his $600. "We made it to the field," Young recalled. "But barely."

The facility was a site to behold. With the temporary metal bleachers it looked, according to Tom Hoffarth, a writer covering the game for the *Daily Breeze*, "as if it was a stadium under construction for a movie shoot." The field was a resting place for rocks and potholes, and large brown stretches were camouflaged with green spray paint. The locker rooms were the size of coffins. Some necessary equipment was provided. Some (medical tape) was not.

Before kickoff, sportswriters covering the game were asked to participate in a field-goal-kicking contest. A handful of scribes climbed down from the press box via a wobbly wood ladder, then took the field. It was a low moment in the history of sports journalism — both the unwise blending of participant/observer and a fleet of out-of-shape, ink-stained media members booting footballs 3 yards through the air. "Who won?" Hoffarth later asked. "Does it matter? The winner was probably offered a one-game contract with the Express. Who wanted that?"

A solid 5,000 fans attended the game. Or 6,000. One person said 8,200. The stadium lacked turnstiles, so there was no official count. The press box, rickety and small, seated 15 people. The game was not televised, but Randy Rosenbloom and Ron Glazer provided commentary for KWNK-AM radio. Ten people listened.* Because the scoreboard was positioned at a unique

---

* This is an estimated guess. It could have been 11.

54.7-degree angle, the setting sun reflected in such a way that rendered it unreadable and useless. Hence, Herb Vincent, the assistant public relations director, sent Steve Vanderpool, an (unpaid, of course) intern in the public relations department, to squat at its base and radio the information to the press box.

Those in attendance were enthusiastic. Those employed by the Express felt as if they resided at the bottom of an ashtray. This was the end of the road. The team only dressed 37, and had 13 available offensive players. "Our listed depth chart had one guy backing up all five offensive linemen," said Vincent. "We were, um, thin." There was a single healthy running back, Tony Boddie, and therefore the Express employed the one-back offense. Young, already suffering from a pinched nerve in his neck, was sacked twice in the final minute of the 21–10 defeat, and wobbled off the field as if he were mimicking a three-legged mutt.

The star of the game was Arizona quarterback Doug Williams, who threw for 250 yards and a touchdown. But he felt less like a professional football player, more like a junior-high-schooler waiting for Mom to pick him up after practice. He and Young met afterward, both wondering how their careers had reached this point. "What have we gotten ourselves into?" Williams asked with a sad grin.

Young could only shrug.

"I thought maybe the cheerleaders would decorate the team bus like they used to do in high school," he said later. "I feel like I've come full circle." A couple of days after the game, Vanderpool ran into Klosterman at the team offices. "I'm here for my expenses check," the intern said with a smile, knowing he wouldn't be paid.

"Take this," Klosterman replied, and handed the kid a silver helmet.

One week later, the season — and the Los Angeles Express — came to a merciful end. Unbeknownst to many at the time, the Kansas City Chiefs wanted to see Seurer in action, and asked Hadl if he would allow Young's backup to play in the Week 18 matchup at Orlando. With his team 3-14 and decimated by injuries, the coach could think of no logical reason to deny the request. Seurer entered the game to start the third quarter. Moments later, Boddie limped off the field with a leg injury, and Hadl found himself gazing up and down the sidelines for a replacement. Mel Gray, the only other active halfback, was also hurt. "I was the first person he looked at," recalled Young, who jogged into the huddle with a wide grin, patted Seurer on the rear, and said, jokingly, "I'm the new halfback. Now give me the pill."

In USFL lore, Young ran for 150 yards and five touchdowns. In reality, he did a little blocking and never touched the ball. "I'd come back to the huddle and say, 'Frank, c'mon, I'm wide open!'" Young said. "The truth is, it was a good experience for me to know what it feels like. In the future, when Jerry Rice yelled at me, I understood."

The Express, naturally, lost 17–10, and afterward Lee Corso, Orlando's coach, praised Los Angeles for performing valiantly with a depleted roster. The sentiment was relayed to Hadl as the Express headed for the airport. It had been a rough year, but the coach could at least take pride in trying his best and holding things together.

Alas, he wouldn't be let off the hook so easily.

The pilot hired to fly the Express home refused to allow the players to board.

He had yet to be paid.

16

CANCER

Our league was hijacked by Donald Trump.

— Mike Hope, Los Angeles Express
executive vice president of marketing

THE HEADACHES BEGAN in November of 1984.

John Bassett, owner of the Tampa Bay Bandits, believed them to be Donald Trump–induced. He was, after all, fed up with the New Jersey Generals and their carnival barker front man. From the beginning, Bassett had been a believer in the USFL's mission — spring football, sensible salaries, regional appeal. When he heard and saw Trump, well, he hated it. Hated *him*. The arrogance. The bravado. The volume. The scam. The stupid antitrust lawsuit, which was moving at a snail's pace and wouldn't be heard until 1986. So, again, he presumed the headaches were caused by all the drama Trump had brought to the league. That's why, when he went for a physical in his hometown of Toronto, he wasn't particularly concerned. "They gave me the works," he said. "No problem, said I could play for my team. Then the doc said he'd also give me a CAT scan."

That came a few weeks later, and the results were returned on February 23, 1985, the same day Tampa Bay opened its season against the new Orlando Renegades. Because Bassett was in Florida with the team, his wife, Susan, took the call from the physician. The news was devastating: eight years after battling skin cancer, Basset now had two spots on his brain.

He started radiation treatments, but had little desire to share the news with the Bandits players and coaches. In fact, he took the opposite approach. One evening Bassett ferried every offensive and defensive lineman, via limousine, to a fancy restaurant and treated the men to a decadent five-course meal. "He told us we never get credit in the press, and we deserved some attention," said Fred Nordgren, a nose tackle. "It was so classy." Football, Bassett stressed, was their job, and it needed to be the only concern. That's why he waited until late March to inform the team, and did so only because his worsening physical condition made it impossible to avoid. "John told us in a meeting," said Zenon Andrusyshyn, the Bandits punter. "It was devastating. I was the head of the team chapel program with [quarterback] John Reaves, and John would come to chapel, sit in the back, and just cry. John Bassett wasn't just an owner. He was a part of it all. He was one of us."

"John was at practices, at team functions," said Jim Russ, the trainer. "So players struggled with his decline."

The Bandits dedicated the remainder of their season to Bassett, and Bassett dedicated the remainder of the season to both enjoying his team and either ruining or derailing Trump's plans for the USFL. At first, he fought to keep the league in the spring, urging his fellow owners to stay the course, to think long-term, to ignore the ranting monster and defy his irrational nonsense. Then he suggested the league play eight spring games and eight fall games in an oddball football experiment. The concept never gained traction. "John Bassett had Trump 100 percent figured out," said Charley Steiner, the Generals' announcer. "He saw through everything. But his brain cancer was a fundamental game changer in the life and, sadly, death of the league. When he got sick, Donald stomped all over him."

Bassett's pleas fell on deaf ears. It was simply too late to fight Trump, and the damage of the spring-to-fall drama (and ultimate decision) made it impossible for the USFL to stop its course change. "John being sick, there was nobody that would, or could, stand up to [Trump]," Chuck Pitcock, the Tampa offensive lineman, recalled. "[John] was the one person in the USFL who could stand up and be accountable. He didn't give a rat's ass about Donald Trump."

"[Donald] was abusive to anybody who didn't agree with him, particularly

John Bassett," said Tad Taube, the Oakland owner, "who at that point became kind of a sad figure because he had a stage-3 or -4 brain tumor and he was not going to make it."

By the middle-to-end of the season, the nation's top USFL fan bases largely lost interest in their teams. Though it went unstated, the message screamed, *What's the point? You're leaving.* In Denver, the Gold went from averaging 33,953 in 1984 to 14,446 in 1985. The Philadelphia Stars drew 28,668 to the Vet in 1984, but only 14,275 headed out to College Park, Maryland, in 1985. The most striking plummet concerned the Houston Gamblers, an exciting operation that found itself in a world of trouble. The Gamblers brought an average of 28,152 fans into the Astrodome for their debut season, but now drew a scant 19,120 as the ownership group shed money at a precipitous rate. It wasn't that the product worsened from the first year to the second. No, it's that fans knew the Gamblers were likely to go extinct. "We lost 15,000 season ticket holders by our announcement to move to the fall," Jerry Argovitz, the team's owner, recalled. "All to appease Donald, who was promising that if we follow him we might somehow benefit when we win a lawsuit against the mighty NFL."

Realizing his peers had given up on staying in spring, Bassett switched gears. In a May 5, 1985, *New York Times* piece, Bassett said, come season's end, he would remove the Bandits from the USFL and begin a new spring league. "After three years," he said, "we're the only team with the same owner, the same town, the same coach that hasn't been moved, sold or gone out of business, so why should I go to fall?" He insisted he would find a new network deal, new owners, new stadium leases. "I think the current frustration of this thing with the fall has caused the cancer to flare up," he told the *Times.* "I've decided I'm going to do only the right thing, and I'm staying in the spring because it's right. I've got the spring league done. I'm just filling in the blanks. I had a master plan to save the USFL. They wouldn't listen to me."

The Bandits averaged a league-best 45,220 fans. But there was scant light. For example, Bassett supplied filmmaker Mike Tollin with day-to-day full access to the team for a behind-the-scenes documentary. Coach Steve Spurrier, guarded and increasingly disliked by players, did his all to stop the project, and Tollin resorted to hiding a wireless mic inside the pads of Ron Sally, the third-string quarterback, to record meetings. "Everyone was terrified of Coach," said Sally. "I wasn't."

Bassett's health worsened, and he distanced himself from the club — and the league — toward season's end. He attended a USFL meeting, and in *The $1 League* Jim Byrne described the scene: "Once tall, handsome and athletic, his illness and the treatment had caused a dramatic change to his appearance. His face was puffed and he had lost most of his hair. His physical bearing was that of a man twenty years his senior. The owners and league officials had to hide their alarm." Tampa Bay lost five of its last six regular-season contests, and talk turned to a new owner and potential merger with the Orlando Renegades. Andrusyshyn, a three-year member of the organization, paid a visit to Bassett in Toronto. "I went to see him in the hospital," he said. "He had giant bandages around his head. I asked him how he was doing, and he shook his head. Then I told him we were all praying for him, and I'll never forget his response . . ."

"So what?" Bassett said.

"What?" replied Andrusyshyn.

"I said, 'So what?'" Bassett said. "I don't believe that stuff anymore."

"It was so dark and so depressing," Andrusyshyn recalled. "Everything about it."

Bassett ultimately relented and sold the team to Lee Scarfone, a local architect who agreed to move the Bandits to fall. It was one of the truly painful moments of Bassett's life. He loved and lived for the Bandits. "He would have done anything to make it work," said Spurrier. "It was his baby." Bassett died on May 15, 1986, and no media member seemed to understand the man and his USFL devotion quite like Dick Schneider, sports editor of the *Fort Myers News-Press*. "I believe that Bassett never wanted the USFL to become another NFL," Schneider wrote. "He wanted this league to be special. To be more fun. To be a place where you could sing stupid fight songs, watch new employees enter the business in a chariot, eat clam chowder. A summer alternative. The rest of the USFL owners should have listened to him, listened when he said that the only reason other teams felt they needed big-name players is because they lacked the skill of marketing."

With Bassett out of the picture (first via illness, then death), his ideas faded to black. The Bandits would not leave the USFL or become the center-piece of a new spring league.

Instead, on June 30, 1985, with their owner 2,258 miles away in Toronto, Steve Spurrier's Bandits fell to the Invaders in the first round of the playoffs. The game drew 19,346 to the Oakland Coliseum, and — for Tampa Bay die-hards — felt like a wake. Though nothing was official, most players and fans seemed to believe the franchise was dead.

They were correct.

With the looming move to fall, the entire USFL playoff schedule felt aca-demic. Save for Memphis's 48–7 home win over Denver before 34,528 spec-tators, the first-round attendance figures were dreadful, and the TV ratings even worse. The semifinal games were held on the weekend of July 6–7, and when Oakland and Baltimore emerged as the two teams to meet for the third USFL crown, the sports world yawned. Literally, the sound of yawning could be heard from millions of disinterested households. The promise and excite-ment of everything that occurred but two years earlier were gone, replaced by confused indifference. Would the USFL live? Meh. Would the USFL die? Meh. *Pass the cheese spread and let's see what movies are playing.*

The championship game — still known, blandly, as "the championship game" — was scheduled for Sunday, July 14, at the Meadowlands in East Rutherford, New Jersey. The league did its all to stamp a happy face on the event, with sunny talk of the fall 1986 season and an entity ready to directly challenge the NFL. It was bluster. Because there would be an enormous gap between the end of the current season and the start of the next (alleged) campaign, and because it would be both unfair (to the employees) and finan-cially unrealistic (to the teams) to maintain the status quo, the majority of franchises commenced with the process of letting go of players and coaches. Teams were required to pare down to 35 players by August 1, then pay each man 30 percent of his salary from March to the following June. According to an Associated Press report, "Many want to dump their highest-paid players in order to save money." Bobby Hebert, Oakland's quarterback, already paid the Invaders $50,000 to allow him to walk immediately after the title clash, and Anthony Carter, his top receiver, was reportedly negotiating with the Miami Dolphins. Even Trump's Generals shed their coaching staff. A league source told *Sports Illustrated*'s Paul Zimmerman that the USFL might lose 20 to 25 players to the NFL, and the estimation was preposterously low. This would not be a trickle, but a flood. "The current position is that if they want to go, we won't hold them," the source said. "We'll save the money on their contracts and use it to invest in future draft choices."

Making matters worse, the week of the title game Steve Young and his agent, the feisty Leigh Steinberg, told commissioner Harry Usher that they either expected the Express to find a new owner ASAP, or have the quarterback liberated to join the NFL. It was embarrassing; an effort by Steinberg to put the heat on a league his client no longer wanted to call home. Helping his cause was the mess known as Los Angeles football. In the days since the final loss, the debt-strewn organization had been selling off all of its equipment and gear, from helmets to socks to filing cabinets. "I loved the USFL — I really did," said Young. "But you could see the writing on the wall. I was young, but I was smart enough to see going up against the NFL was suicide. I wanted out."

The USFL's official championship game program featured a front-of-the-book "Letter from the Commissioner" that could have been renamed "Ten Paragraphs of Bullshit." If Usher learned one thing in his time on the job, it was that the owners were a mess, the finances were a mess, the future was a mess. In a book proposal he later submitted to publishing houses, Usher referred to the USFL as "a magnificently bizarre land" and bemoaned "the idiosyncrasies and fringe lunacy I encountered." Yet the letter in the game program was one of sunny optimism, celebrating "the first time in more than two decades that a professional football championship has been decided in the greater New York area" and calling the league "a tribute to our players" (many of whom were about to be unemployed). "After tonight, the league will turn its attention to the 1986 season when we will begin play in the fall," he wrote. "That is the traditional season for football and there is no doubt in my mind that we belong there."

Usher's note ended with, "See you next season" — a declaration of either dread disguised in hope, or hope disguised in dread. Would there be a "next season"? Who the hell knew? Certainly not the commissioner.

Still, there was a contest to be played, and while Invaders-Stars hardly evoked on-paper comparisons to Steelers-Cowboys or Packers-Raiders, it was a fantastic matchup. First, while the Stars' relocation to Maryland was an exercise in frustration and geographical indifference, this marked the franchise's third-straight appearance in the title game. The Stars were the USFL's dynasty, and in halfback Kelvin Bryant and middle linebacker Sam Mills it possessed, arguably, the league's two all-time best players. "Just making that third game was a tribute to our survival instincts," said Herbert Harris, a Stars wide receiver. "If you look at everything we went through — the hurdles, the obstacles, the struggles, the difficulties, the drama. There was no way we were expected to be there."

"We were more prepared for every down than any other team, USFL or NFL," said Mike Lush, the Stars safety. "We also were a family. That matters in football. If you love the guys you play with, you'll do anything for them."

For many casual football fans, the Invaders were merely the Invaders — Oakland's team, funky uniforms, name ripped off from "Raiders." Yet upon closer inspection, this was a quasi rematch of the first title game, when Michigan downed Philadelphia at Mile High Stadium. Though the Panthers and Invaders had merged, it was hardly a traditional absorption by Oakland. Of the 50 players on the roster, a quarter were former Panthers, and the league-best 13 wins could largely be attributed to the play of Hebert and wide receivers Carter and Derek Holloway, all Detroit refugees.

When the merger first came to fruition, nobody knew what to expect. "I figured they'd be coming to us," said Carter. "I mean, I signed with the USFL specifically to play in Michigan. We were one of the USFL's great teams. What were they?" Carter's vision wasn't to be. The players arrived at Mesa (Arizona) Community College, the Invaders' property, for training camp, and tensions were initially palpable. The Panthers thought their guys were better. The Invaders thought their guys were better. "It was a matter of feeling everything out," said Ray Bentley, the linebacker. "Knowing where you stood."

A particularly heated battle took place between the two top-shelf starting quarterbacks, Hebert and the Invaders' Fred Besana. Throughout the preseason Besana was the far superior player, and many in camp presumed he would be named starter. Then, a couple of days before the opener against

Denver, Charlie Sumner, the head coach, summoned for both quarterbacks. "I'm sure Bobby saw it differently, but to me I won the job and it wasn't even close," said Besana. "I was just happy Coach was finally going to let me know." Sumner spoke to each man privately — first Hebert, who exited the office with a suppressed grin. Then Besana. "Freddie, I'm going with Bobby as the starter," Sumner said. "It's just . . ."

Besana interrupted. "You have to be kidding me," he said. "That competition wasn't even close."

Sumner closed the door and tried explaining to Besana that he was still important, that as a veteran he needed to maintain professionalism. "I was hurt and I was angry," Besana said. "But I wasn't going to ruin team chemistry over something that was clearly out of my control."

Before long, the Invaders were rolling through the league, drawing few fans (an average of 17,509 to a stadium that seated 54,615) but playing inspired football en route to a 13-4-1 mark. "We were an NFL-caliber team, without question," said Stan Talley, the punter. "I don't know if anyone realized it outside of our locker room. But we knew."

Approximately one month before the title game, the USFL expressed expectations that more than 55,000 fans would attend. When that figure was suggested, however, it seemed as if Trump's Generals might play in their own stadium. It also was stated without the knowledge that, come kickoff, the New York–New Jersey metropolitan area would be overtaken by torrential rainstorms. The weather, combined with two teams supported by limited fan bases, resulted in a half-empty stadium, an attendance lie (the 49,263 people who watched apparently included 15,000 ghosts), and limited media coverage. Whereas the first USFL championship game drew newspapers from near and far, this time most outlets relied on the reports of wire services. In fact, while the Associated Press covered the actual game, its primary story from that night concerned a private pregame owners' meeting in one of the stadium's suites. Usher, according to the report, wanted 12 teams for 1986. Trump insisted upon 8 — and wasn't backing down. It was supposed to be information shared only among members of the league, yet Trump made himself available for reporters, thereby embarrassing the commissioner with such off-the-record tidbits as the Denver Gold possibly moving to Hawaii and the Gamblers relocating to . . . *Queens, New York?* Norm Frauenheim, a columnist for the *Arizona Republic,* wrote a next-day piece headlined USFL ON STRAIGHT PATH TO BECOMING TRIVIAL PURSUIT.

By the time the national anthem was being sung, several of the owners in attendance had bolted from the stadium. The league planned a magnificent pre-kickoff balloon release, and the ensuing imagery remains vivid. "It was raining hard, and all the balloons went up and came right down," said Gary Plummer, the Oakland linebacker. "I remember thinking how poetic that was."

Despite the precipitation, for the second time in three seasons the USFL put forth a dandy. The Stars and Invaders were evenly matched, and Baltimore's 21–17 third-quarter advantage was erased with 46 seconds remaining in the period, when Hebert hit Carter for a 7-yard touchdown pass. The 24–21 lead held into the fourth quarter, but the Stars marched down the field, and on first and goal from the Oakland 8, Bryant — standing atop the I-formation — took the pitch from quarterback Chuck Fusina, sliced right, then glided into the corner of the end zone. It was his third touchdown of the evening, and the type of scamper Bryant specialized in — smooth, casual, almost effortless in appearance.* As the huge crowd roared (well, the small-ish crowd shouted a bit), Bryant handed the ball to center Bart Oates, who spiked it into the artificial turf. Bryant then lined up across from safety Antonio Gibson, fullback David Riley, and wide receiver Victor Harrison and performed their trademark dance number for, perhaps, the final time as teammates.

The game wasn't over. The Invaders botched the kickoff return, and took over on their own 4-yard line with 7:47 left. Mora turned to one of his assistants and said, "If they can drive the length of the field on us and score, they

---

* It also brought extra joy to the Stars, as the player responsible for containing Bryant — Invaders safety David Greenwood — was the same man who'd referred to him using the N-word two years earlier. On this particular play, Greenwood was pulverized on a block from fullback David Riley.

deserve the game." Led by Hebert, Oakland methodically marched through the Baltimore defense, mixing runs from halfback John Williams with quick passes. At one point Lynn Swann, working the game for ABC, said, "Have to remind you all that this is as close to a rematch of the first championship game played in the USFL as you could get. And that ballgame, with very little time left to play, it was Bobby Hebert to Anthony Carter for a touchdown that brought them back and gave them the championship." Shortly after those words were stated, Hebert rifled a dart to Carter as he crossed the middle of the field. The wide receiver turned his body to catch the ball, pivoted to dodge a tackle, spun to avoid another, then dashed to the Stars' 13, where he was pummeled by cornerback Jonathan Sutton.

Moments later, following a Holloway reception at the 6, Oakland faced third down and 2. There was 2:50 remaining, and the Invaders lined up in a tight formation, with Williams and fullback Tom Newton positioned side by side to Hebert's rear. The quarterback received the snap, turned to his left, and handed to a streaking Williams, who rumbled for just 1 yard behind Newton's block of Sutton. It was now fourth and 1, with the championship game on the line and . . .

Wait.

A yellow flag soared into the air before fluttering toward the ground. Newton had shoved, shoved, shoved — then *punched* Sutton in the head. The infraction was unnecessary roughness, and the penalty pushed the Invaders back 15 yards.

"Ref had to throw the flag," said Gordon Banks, the Invaders receiver. "It's one thing to be physical. It's another thing to punch a guy. I don't want to throw Tom under a bus, but that cost us something very big."

"Tom drove that guy out of bounds about 10 yards," recalled Plummer. "I don't see how there couldn't be a penalty."

Newton argued, but to no avail. It was a call made of circumstance, but also reputation. Before joining the USFL, Newton had spent six years as a backup with the New York Jets. His self-anointed nickname was Wild Man, and it encapsulated the merging of intense physicality and repeated boneheadedness. In an interview with the *Poughkeepsie Journal,* Newton once likened himself to Clubber Lang, the crazed boxer played by Mr. T in *Rocky III*. He professed to often being lazy and "not having my mind together," and too often let emotions overtake results.

"That penalty killed us," said Talley. "No doubt about it."

Facing third and 17 from the 20, Hebert called Oakland's final time-out. Earlier in the game, TV cameras caught Sumner smoking a cigarette along the sideline. Now, as he and Hebert conferred, Newton stood inches away, helmet on the ground, eyes averting his teammates' glare. He was banished. The following two plays were desperate heaves to Banks, both of which were incomplete. As the final ball bounced off the turf, Keith Jackson, the legendary broadcaster, lowered his voice to a funeral-appropriate measure. "The Stars have played in all three of the USFL championship games," he said. "They go to the fall in 1986, so this is the end of spring and summer football. And the only thing I have to say is good luck. I wish them well."

Afterward, as the Stars rushed into their locker room and celebrated with exploding champagne, a large number of Invaders — perhaps fueled solely by the loss, perhaps fueled by a terrifying future of driving a rig or serving hamburgers — entered their clubhouse and berated Newton, who stood silently by his stall and accepted the blame. "It was ugly," said Banks. "Tom messed up and some guys wanted to tell him how they felt."

"There was definitely yelling," said Ron Lynn, the defensive coordinator. "You have to understand, it was a hard way to lose."

Slowly, the players on both teams undressed, showered, and exited the stadium. Many hugged. Some cried. A few paused to look around, to take one last glance at the splendor that was spring football. Hebert stood silently for 20 minutes beneath the hot water from the showerhead. "I think we all knew as soon as that game was over we'd be flying home and never coming back," said Jeff Wiska, an Invaders guard. "It was the end for most of us."

"I got back to my hotel room, sat on my bed, and sobbed," said Geno Effler, the Invaders PR director. "My fiancée was with me, and she said, 'Are you sad because you lost?'"

"No," Effler replied, "I'm sad because these guys will never play with each other ever again, and I'll never be with them again."

For the victorious Stars, there would be no parade in Baltimore, no commemorative newspapers or magazines. Carl Peterson, the Philadelphia general manager, had some connections with the White House, and 25 Stars visited with George H. W. Bush, the vice president. There was a small ceremony. Finger food was served. President Ronald Reagan was nowhere to be found.

Like a punctured football, the league deflated by the minute.

"You didn't want the joy to stop," said Hebert. "But it was dead."

17

LEGAL EAGLES

Courage is being scared to death, but saddling up anyway. I was scared shitless that Donald J. Trump was now in complete control.

— Jerry Argovitz, owner, Houston Gamblers

NOW WHAT?

The question sat there, and nobody was 100 percent certain how to answer.

With the conclusion of the USFL's third and final spring season, there were big things to do. Huge things to do. Enormously important things to do. Like ... um ... ah ... eh ... um ...

Yup.

The antitrust lawsuit filed against the NFL would finally be heard inside a New York federal courtroom in the spring of 1986, but in the meantime there was a chaotic tidal wave of confusion. What should happen to the players? The coaches? The teams? The uniforms? The ... everything?

For all his bluster about the USFL switching to a fall schedule, Donald Trump sure wasn't big on details and specifics. Neither, for that matter, was Harry Usher, a lovely man with a thick résumé and glorious hair who appeared increasingly incapable of handling the task placed before him. Other than assuring journalists that — win or lose its lawsuit — the USFL would go on in the fall of 1986, Usher was a sprinkler lacking water. When he spoke, nobody seemed to listen. When he didn't speak, people wondered if he ever opened his mouth. In July, he declared the USFL's next season would likely include 12 teams, not 14. Which was a departure from Trump's statement that the USFL should have 8 teams, not 12. There was talk of the Denver Gold moving to Honolulu, as well as the Denver Gold merging with the Portland Breakers. The Los Angeles Express were said to be headed to the San Fernando Valley — or the graveyard. The Tampa Bay Bandits and San Antonio Gunslingers were prepared to become one and relocate to . . . *Chicago?* Then, on July 23, the Gunslingers waived all 46 of their players because the owner, Clinton Manges, was broke. "I regret that the players are unpaid," Usher said meekly. "I hope that they are paid in the near future." They never were.

Less than two weeks later, the penniless Breakers released their 39 players, though John Ralston, the team president, insisted the invisible, three-towns-in-three-years franchise was rock-solid. "We're in the habit of changing lemons into lemonades," he said, using a completely nonsensical metaphor for the situation. "We're confident we'll field a much stronger team in 1986 than we did in 1985." Shortly thereafter, in an interview with the Associated Press, Joe Canizaro, the team's owner, blamed the players themselves for not earning their (contractually guaranteed) salaries. "Maybe if we'd had a winning season in Portland," he said, "we'd gotten those investors."

The Bandits, forever the rock of the USFL, turned their attention 90 minutes east, and aspired to combine with the Orlando Renegades and morph into the Orlando Bandits. There were plans for a new Boston team, not to be confused with the old Boston team, which was also the old New Orleans team and the old Portland team. In Birmingham, Marvin Warner, the Stallions' principal owner, pulled out over personal legal entanglements, and the organization — deep in debt — laid off 13 administrative workers in the month after the season ended. Joe Cribbs, the star halfback, asked out of his contract, noting that "playing before 18,000 and 19,000 [fans], that's not what I call professional football." Still, the Stallions could survive. Or not survive. Nobody was sure.

Truth be told, as the lawsuit lingered, the USFL's future was pure rumor and speculation. One day the nonhuge announcement was made of a fall TV deal — with something called RCA American Communications, an outlet that specialized in satellite syndication. (Wrote the *Philadelphia Daily News:* "Now all the struggling league has to do is find a network or group of stations willing to air the product.") On another afternoon, the Gold and the Bulls decided to join forces and play out of Jacksonville, with Doug Flutie said to be the team's next quarterback (the Flutie part never materialized). A few days later, the Gold hired the Radcliffe Auction Company to rid itself of all possessions. Several college and high school coaches swooped in to bid on the football equipment and medical supplies.

On August 1, 1985, less than a month after the title game, Usher formally declared that Trump's Generals and Jerry Argovitz's Gamblers had merged into the *new* New Jersey Generals. The roster would mix 22 Houston players with 15 New Jersey players, and bring together the otherworldly talents of quarterback Jim Kelly and halfback Herschel Walker. "It's probably the best team in football," Trump raved to the *New York Times.* And while that sentiment was preposterous, Kelly bought in. "What I'd really like to do," he said, "is play for the New Jersey Generals and Donald Trump and merge with the NFL and take the run-and-shoot with Herschel Walker in the backfield and just kick ass."

"A lot of NFL scouts told me we would have gone through defenses like Ex-Lax," said John Jenkins, the Gamblers-Generals offensive coordinator. "We were trained and ready to go."

Where, exactly, would the Generals play? Nobody knew.

Did Herschel Walker intend to spend 14 months on a sideline? Nobody knew.

Wrote John Delery of the *Morristown (N.J.) Daily Record:* "The Generals' future in the USFL remains as murky as the Hackensack Swamp."

Trump brought in Argovitz to run the franchise on a day-to-day basis, and the former Gamblers owner talked a big game ("If we played the Chicago Bears, and they used their Super Bowl defense against us, we'd score 40 points!"), but was immediately miserable. "Was Donald a good man to work with?" Argovitz said. "No, terrible. He was all about himself, about his team, about his success. He didn't care about the USFL, so much as he cared about Donald Trump.

"We had a discussion one day. He said, 'How do you think I'll be remembered in this league?' We had a very friendly relationship, and I said, 'Donald, you're going to be remembered as the man who destroyed the USFL.' He didn't like that. He said, 'But I hired Doug Flutie! I hired Doug Flutie!' And I said, 'Yeah, and you sent us all a letter demanding we pay for him.'"

By February, the USFL officially slimmed itself down to eight teams — the Liberty Division would house the New Jersey Generals, Baltimore Stars, Birmingham Stallions, and Memphis Showboats; while the Independence Division was the terrain of the Arizona Outlaws, Tampa Bay Bandits, Orlando Renegades, and Jacksonville Bulls. Usher stated, declaratively and confidently, that a championship game was to be played on February 1, 1987, in Jacksonville, and that the Generals — who shared a stadium with the NFL's New York Jets and New York Giants — would, um, somehow figure the whole thing out. "That's a problem," he acknowledged.

"We were preparing for 1986 in Baltimore," said John McGuire, the Stars' video director and jack-of-all-trades. "We held a press conference at the mayor's office to announce a lease agreement with Memorial Stadium. We brought in old Colts to work in community outreach. The Colts agreed to give all the memorabilia from Baltimore back to the city, and there was another big ceremony at city hall. I was optimistic."

One and a half years earlier, on June 5, 1984, the NFL had held a relatively low-key supplemental draft of USFL (and CFL, to avoid the appearance of poaching one league) players. It was in preparation for what Rozelle and Co. considered to be the fledgling outfit's inevitable death and, truly, a way of normalizing what would otherwise turn into an unruly free-agent free-for-all. As USFL officials seethed from afar, 84 players were selected over three rounds, beginning with Steve Young to the Tampa Bay Buccaneers, Mike Rozier to the Houston Oilers, Gary Zimmerman to the New York Giants, and Reggie White to the Philadelphia Eagles.

Now that draft came into sharp focus. Less than 24 hours after the Bulls wrapped their season, Rozier agreed to a four-year deal with the Oilers. Jacksonville also let loose the contract of Keith Millard, its standout defensive end, and he wound up with the Minnesota Vikings. "The whole time I was in the USFL," Millard scoffed at the press conference, "I felt I should've been in the NFL." The Vikings also obtained Anthony Carter, Oakland's splendid wide receiver, by paying him $2 million for five seasons.

Just three years earlier, quarterback Bobby Hebert was an afterthought in what would go down as the Year of the Quarterback. Now, he was the subject of a bidding war among the Seahawks, Raiders, and Saints. The native Louisianan agreed with New Orleans to five one-year contracts that paid approximately $800,000 annually, and at the press conference joked, "I'm happy to be in a place where people can pronounce Hebert (A-bear). I don't have to worry about them saying Herbert or Hee-bear."

By the first week of September, 78 former USFL players were on NFL rosters, leading *Sports Illustrated*'s Rick Telander to write, "They arrived in the NFL by means of buy-outs, wait-outs, wheel-and-deal-outs and — in the particular case of players from the USFL's moribund San Antonio and Portland franchises — virtual throw-outs." For much of the previous three years, the NFL publicly acted as if the USFL didn't exist, and issued derisive statements about its limited pool. Yet now, with the sudden availability of prime rib, that take changed. The San Diego Chargers, for example, gobbled five refugees, causing Ron Nay, the director of scouting, to admit, "We always knew there was talent over there."

As the chaos swirled, the remaining USFL teams attempted to present a business-as-usual exterior. There was a college draft (Mike Haight, the Iowa offensive tackle, went first to Orlando) that meant nothing. The Generals held a mini-camp at Giants Stadium, and Kelly and Flutie — meeting for the first time — took turns throwing to receivers. It was awkward. "We shook hands," Flutie confirmed. "We didn't slug each other." The Stars maintained the office in Baltimore. The Jacksonville Bulls, now coached by Mouse Davis, flew in quarterback Ed Luther to learn the Run and Shoot offense. "It was a lot of fun," Luther recalled. "I was looking forward to it." In Tampa, the Bandits, with former Invader Fred Besana at quarterback, were unrecognizable yet functioning. Then Bret Clark, a defensive back who never received his $150,000 signing bonus, sued the organization. A lien was placed on the team. "So one day everything is confiscated," said Fred Nordgren, a defensive tackle. "The weights, the equipment — all taken away in trucks. I knew we were done." The Renegades, 84 miles up Interstate 4, *were* shockingly unchanged — mainly because the NFL had little to no interest in its players, and those who remained were all familiar faces. "Around that time I got a call from [Orlando owner] Don Dizney," Lee Corso, the Renegades' coach, recalled. "He said 'I have good news and I have bad news. The good news is the NFL is starting to buy a whole bunch of the USFL's players. The bad news is they don't want any of ours.'"

What hung over everything was the lawsuit.

The lawsuit.

The lawsuit.

The lawsuit.

The lawsuit.

Donald Trump made promises, and they sounded amazing. "He is an incredible self-promoter," Philip Hess, counsel to the chairman of City Planning, New York, once said. "He talks, talks, talks. He makes you feel like he's making you part of the greatest deal mankind has ever had the privilege to develop and if you don't jump in with him, there's a line waiting to." The lawsuit, Trump guaranteed, would bring down the NFL and prop up the USFL. In the absolute worst-case scenario, the NFL would pay out a ton of money. In the more likely scenario, the NFL would pay out a ton of money *and* absorb many of the USFL's remaining franchises. It was win-win-win-win, and all the owners had to do was come along for the ride and trust in Donald J. Trump.

So they did.

"People were fools to follow him," said Ted Diethrich, former Wranglers' owner. "Donald was looking out for one man — Donald."

Back when the legal action was initially introduced, Trump stood side by side with Roy Cohn, the infamous McCarthy-era attorney. Yet in the 19 months that had passed, Cohn was diagnosed with the HIV virus, and Trump dumped him. His replacement, as selected by Trump without any input from the league's other owners, was Harvey Myerson, a 47-year-old attorney who went by the nickname "Heavy Hitter Harvey." Myerson, according to a *New York Times* profile, "has come to personify the two sides of modern corporate law practice: its glamour, visibility and profitability as well as its instability, riskiness and potential for treachery." He was described in *Time* magazine as "short, chunky and menacingly combative," but that passage left out rude, harsh, arrogant, dismissive, and unlikable. His rate was $400 per hour, and his salary from (warning, take a deep breath) Finley, Kumble, Wagner, Heine, Underberg, Manley, Myerson and Casey — a 700-lawyer Manhattan-based firm — was $1.4 million. Myerson owned a Ferrari, a Rolls-Royce, five homes, a vast collection of cigars (he always reeked of smoke) and 20th-century art. Though married, he lavished his wealth upon a string of mistresses — giving one an $86,000 Cartier ring and a $24,000 full-length mink coat. "To Myerson," a former partner once said, "there is just no distinction between persuading a jury and persuading his wife or clients or partners of something."

In short, he was the personification of slime.

Trump embraced Myerson. The other owners did not. Argovitz, now working for the Generals, had devoted a good amount of time to convincing Trump to (a) hold the trial anywhere but New York, and (b) employ a more agreeable, likable attorney. "First, the trial should have been in Houston," Argovitz said. "People loved the Gamblers in Houston, and we had the best antitrust lawyer in the country [Joe Jamail, later known for winning $10.53 billion for Pennzoil in the company's 1985 lawsuit against Texaco]. He was a personal friend of mine, and he told me, 'If we get this lawsuit here, we'll kick ass and take names.' We could have filed it anywhere. But Donald put himself in charge of the litigation, and he said, 'New York is my town. We'll win in New York!'"

"Anywhere but New York," said Steve Ehrhart. "It was the one place *not* to have the trial."

From Argovitz's viewpoint, a hopeful scenario turned increasingly dire thanks to Trump's bombast-fueled missteps. "First he wanted to have Roy Cohn, which flabbergasted me," Argovitz said. "The man was in the early stages of AIDS, and Donald said — and I'm quoting directly — 'If we hire Roy, they won't even hire a lawyer. They'll settle.' Dear God, it was like taking a knife to a gunfight. Roy Cohn? Ridiculous. Then he finally hired Harvey, and that was an even bigger mistake, if that's possible. Harvey was just a pounder — he pounded on people. He yelled, he screamed, he hollered, he pointed fingers. And I didn't think he was very good at making legal points."

In the days before the trial, Argovitz said he offered a few suggestions, to which Myerson snapped, "I thought you were a dentist. When did you get your law degree?" The words stung, and spoke to the arrogance of an attorney who, like Trump, believed he owned the world. "Myerson was the worst," said Ehrhart. "He was absolutely beholden to Trump. He was thinking Donald was going to be his ticket. It was a disastrous decision to hire that man."

Maybe so — but the NFL was secretly terrified. Facts were facts, and the NFL *did* have contracts with all three major television networks. The NFL *did* put pressure on the networks to ignore the new league and deny it any help in the fall relocation. The NFL *did* commission a paper, titled "Spending the USFL Dollar," that suggested a strategy for driving up USFL salaries. And, potentially most damaging, in 1984 the NFL *did* hire a Harvard Business School professor named Michael Porter to research, write, and present a paper at a league meeting titled "USFL vs. NFL." The 46-page document, which the USFL got hold of, was presented to some 65 NFL executives, mostly second-level-management people. It was a manifesto on how to snuff out an opponent that opened with a quotation from Sun Tzu's military-strategic epic, *The Art of War*:

> To win every battle by actual fighting before a war is won, it is not the most desirable. To conquer the enemy without resorting to war is the most desirable. The highest form of generalship is to conquer the enemy by alliance. The still next highest form of generalship is to conquer the enemy by battles. The worst form of generalship is to conquer the enemy by besieging walled cities.

"It was literal — they were trying to kill us," recalled Ehrhart. "It didn't take a lawyer or a genius to see that."

Among Porter's most explosive passages was a section, OFFENSIVE STRATEGIES, that advised the NFL to:

- Actively encourage sending undesirable players to the USFL.
- Actively influence problem players and overvalued players to jump leagues.
- Help encourage strong unionization of the USFL.
- Sign away the USFL's biggest stars.
- Move the college draft up earlier in the year to compete with the USFL's draft.
- Drive up player costs with fringe benefits and unified bargaining power.
- Attempt to co-opt the most powerful and influential owners with promises of NFL franchises.
- Attempt to bankrupt the weakest teams to reduce USFL size and credibility.
- Provide ABC with a mediocre schedule of *Monday Night Football* games to punish the network for televising the USFL.

A couple of weeks before the start of the trial, the NFL hired a company named Legal Creative Services to produce a true-to-life mock version of what would, in all likelihood, transpire. There was a judge, a jury, a courtroom, as well as a sobering result: were Legal Creative Services to be believed, the USFL would win a unanimous decision. Robert Fiske, the NFL's lead attorney, met with the owners at the 1986 league meetings. He was asked by Norman Braman, owner of the Eagles, "What would you do — fight this or settle it?"

"I'd settle it," Fiske replied.

A glum silence filled the room, punctured only by the voice of Paul Tagliabue, the NFL's chief outside counsel and the future commissioner. "I have a different opinion on that question," he said. "And I would like the chance to give you my answer."

Standing before some of the most powerful men in organized sports, Tagliabue declared the USFL's case to be "a sham." He thought it clear that not only was the new league its own worst enemy, but Trump was the NFL's greatest legal asset.* Tagliabue also insisted that settling set an awful precedent. What would prevent other rival leagues in the future from suing, too? If the NFL caved so easily, there would be nothing for anyone to fear. "If you felt you didn't do anything wrong," he said, "you should be prepared to stand in defense."

The NFL would fight.

Beginning on May 12, 1986, the USFL trial was, in the words of author Michael MacCambridge, "a melodramatic tour de force, a draining eleven-week ordeal." It was held in Courtroom 318 at the federal courthouse in lower Manhattan, and featured a jury of one man and five women (on the second day of the trial, Wendell James, a Mount Vernon, New York, postal clerk and the jury's lone football fan, asked to be excused, and he was replaced by a sports-ignorant British-born woman, Margaret Lilienfeld). Before the jury were nine charges against the NFL — three violations of common law and six of the Sherman Antitrust Act — as well as a damage-award request of $1.69 billion, plus additional punitive damages. It was the trial of the year, and the *New York Times*'s Michael Janofsky did not overstate its importance. "A victory by the NFL or a USFL victory with a minimal damage award would probably leave the USFL to continue along a parallel course with the NFL for several years. But beyond that, as the history of competing sports leagues shows, the weaker league fades into extinction if a merger does not occur."

Myerson and Trump felt emboldened by a strategy that, some 30 years later, still made absolutely no sense to anyone (USFL or NFL) involved. From Argovitz and Myles Tanenbaum to Alfred Taubman and Tad Taube, the league was overflowing with the sagas of successful, wealthy businessman-owners done in by the evil NFL. There was the story of Reggie White, the Memphis Showboat defensive end who was, according to Ehrhart, being illegally wooed by the Philadelphia Eagles. "They were messing with his

---

* Years later, Tad Taube, owner of the Oakland Invaders, expressed agreement with Tagliabue's take. "The owners collectively committed suicide," he said. "Donald certainly had influence, but they were committing suicide on their own by making so many awful decisions."

contracting, trying to bribe his agent," said Ehrhart. "It was very wrong." There was the story of Herschel Walker, who many in the USFL believed was pursued by the Dallas Cowboys while under contract with the New Jersey Generals. There were dozens of tales to tell from upstart men in an upstart league just trying to excel under the umbrella of the American dream. "I spent a full month in a New York City hotel room, waiting to get deposed," said Jim Kelly. "Waiting and waiting and waiting."

Over 42 days in court Myerson never called any of them (save Walker) to the stand. In fact, he instructed every non–Donald Trump USFL owner to stay as far away from Manhattan as possible. The stated reasoning: he was worried that the NFL might depose them. The real reason, as suspected by several owners: Trump craved the spotlight and thought his peers to be obstacles toward dominance. "Myerson turned it into Trump versus the NFL," said Ehrhart. "The NFL deposed me for two or three days before the trial, and then Myerson never called me to testify. In fact, later one of their lawyers said, 'We were still waiting for you . . . when were you going to testify? You had all this information about important stuff.' Myerson didn't care. He was in Trump's pocket. Whatever Donald said, went. And Donald wanted to be the star of the trial."

Among the witnesses who *did* testify were Al Davis, the renegade Los Angeles Raiders owner who spoke on behalf of the USFL, and legendary announcer Howard Cosell, who also stood behind the new league. (Larry Felser of the *Sporting News* rightly described Cosell's appearance as "merely pathetic, a sour has-been striking out in all directions." He failed to mention that Cosell arrived intoxicated.) There was Chet Simmons, the ex-USFL commissioner, and Harry Usher, the current commissioner, whose boneheaded testimony suggested, in the words of Judge Peter K. Leisure, "that the reason the USFL decided to move . . . was to position the league for a merger."

The NFL's lead attorney, Frank Rothman, utilized an approach that was the 180-degree opposite of Myerson's. He didn't beat people down. He didn't scream, rant, snarl. A distinguished 59-year-old with broad shoulders and gray hair, Rothman was the former CEO of MGM/UA Entertainment, and he exuded a natural dignity. He sat back, let Myerson do his dance (as the entity that filed the suit, the USFL was first to call witnesses), then meticulously went about making the NFL's case that the USFL, by moving to fall, dug its own grave. "They had everything their way at the beginning," Rothman said. "They had the jury they wanted. They hammered away at the Harvard [presentation]. Myerson was pitching the little guys versus the big guys. I would go back and tell the NFL people, 'Listen, when we get our turn we can start turning this thing around. We have to be patient.' But, actually, it didn't take that long."

Beginning with the trial's opening day, Rothman asked himself a single question: *Who is my bad guy?* He sought someone the jury would find difficult to believe and even harder to like. He sought someone with false bravado, with arrogance, with indifference. He didn't want the jury to think about a sad little league going up against a powerful machine. No, he wanted the jury to see that the USFL, sympathy be damned, was its own Frankenstein. "The more I developed the strategy," he said, "the more I wanted Donald Trump as my fall guy. I would call it Donald versus Goliath. I would make their scheme Donald's plan, which it was. I would show that Donald Trump is not a little lightweight; he is one of the richest men in America . . . He was such a lousy witness for them, and a great one for us."

"The way Myerson set it up was perfect for [Rothman]," said Ehrhart. "It was big, bad Donald Trump trying to screw the poor little NFL people, who had worked so hard to build their league up."

Early in the proceedings, the USFL called Pete Rozelle, the NFL's commissioner, to testify as the trial's first witness. Over the course of five interminable days, Myerson hammered Rozelle, pounded Rozelle, grilled Rozelle. In particular, he focused on Trump's claim that the NFL commissioner promised him a franchise should he abandon/damage the USFL. There was, both sides agreed, a meeting held between Trump and Rozelle at the Pierre Hotel in March 1984. What happened, however, was of dispute.

"Didn't you tell Mr. Trump you wish he had been able to buy the Baltimore Colts and hadn't gone into the USFL?" Myerson asked.

"No," Rozelle replied.

"Did you tell him that if he hadn't gone to the USFL, the USFL would have died?" Myerson asked.

"No," Rozelle said. "Never."

Trump's testimony was decidedly different. He said the hotel rendezvous was Rozelle's idea, and recalled the commissioner saying, "You will have a good chance of an NFL franchise and, in fact, you will have an NFL franchise." The tradeoff, according to Trump, was that the USFL remain in the spring and "not bringing a lawsuit."

Trump insisted he and Rozelle were friends. Rozelle insisted he and Trump were certainly not friends. Trump insisted Rozelle wanted him in the NFL. Rozelle insisted he would rather have maggot-infected fungus overtaking his cranial lobe.* "Rozelle told me I should be in the NFL, not the USFL," Trump said. "At some point, he said, I would be in the NFL. Then he would reiterate that the USFL was not going to make it."

Rozelle couldn't believe what he was hearing. He made clear that it was Trump who *reserved and paid for* the Pierre suite. He told Rothman: "[Trump] said, 'I want an NFL expansion team in New York.' And he said, and I'm quoting him exactly, 'I would get some stiff to buy the New York Generals, my team in the USFL.'" Unlike Trump, Rozelle was a meticulous note-taker, and he presented his documented recollections from the meeting.

Rothman's cross-examination was a breathtaking ode to knowing your subject, and taking him apart, piece by piece. Wrote Richard Hoffer of the *Los Angeles Times:* "Rothman characterized Trump as the worst kind of snake who was selling his colleagues down the river so he could effect a merger of a few rich teams." It wasn't Trump's words, so much as his swagger and irritability. The USFL was the little league trying to be big, but Trump didn't seem little. Or sympathetic. Or, for that matter, believable.

"He did not do the USFL well," recalled Patricia Sibilia, a juror. "Donald Trump and I actually got into a staring match. I would watch the people on the stand, trying to read them. So he and I started looking at each other, and he tried to stare me down. It was an obvious try at intimidation. And what's funny, in hindsight, is that this so-called business genius ruined it for them. He was not believable in anything he said. He came off as arrogant and unlikeable."

---

* Not those exact words.

Rozelle's cool, controlled testimony was Kryptonite to Trump's apparent unhinged allergies to truth. Rothman asked, repeatedly, what motivated Trump's actions, then showed the jury multiple documents — signed or written by the Generals' owner — that alluded to a "merger" and "merger strategy." Trump denied his motive was to have the USFL and NFL become one, but lacked credibility. "It was so obvious that's what this was all about," said Sibilia. "No question."

When Rothman suggested Trump's ultimate goal was to wind up with a valuable NFL organization, the reply was staggering. "I could have gotten into the NFL a lot easier than going through this exercise," he said. "I could have spent the extra money and bought the Colts on many occasions."

A historic level of eye rolling filled the courtroom. Trump was lying. He was never a serious candidate to purchase the Colts. Never. "Who do you believe?" wrote Dave Goldberg of the Associated Press, "Donald Trump or Pete Rozelle?"

It wasn't a tough one.

"It was a hard thing to watch unfold," said Argovitz. "Donald didn't love the USFL. To him, it was small potatoes. Which is terrible, because we had a great league and a great idea. But then everyone let Donald Trump take over. It was our death."

Though often immune to criticism, Trump seemed aware that the trial was not going as planned. He was being used by the NFL, and it stung. "All the reporters would rush to the nearby payphones at breaks to call in information," said Bob Ley, covering the trial for ESPN. "One day I walked into one phone booth, Donald walked into an adjacent one. And he's absolutely motherfucking someone on the other end of the line."

On July 29, 1986, at exactly 3:55 p.m., Patrick Bowes, the court clerk, announced that a verdict was at hand. The jury members — tired, battered, and emotionally drained following five days (31 total hours) of deliberation — stepped forward into Courtroom 318. Behind the scenes, the six jurors had engaged in several heated battles. There was yelling, there was barking, there was crying. "It was very high pressure," said Sibilia. "The court puts pressure on you because they want it all to come to an end. The two sides put pressure on you because they think they're right. We all had our own notes

that highlighted different things. It was never hostile, but it was challenging."
Three jurors favored the NFL, three favored the USFL. Patricia McCabe, a
reference clerk for AT&T, suffered from heart murmurs. Miriam Sanchez, a
high school English teacher, had excruciating headaches, as well as heart pal-
pitations. "We were not getting anyplace," Sanchez said. "We were screaming
at each other, calling each other names. I was called frivolous. It's the worst
name I've ever been called."

"At one point, Mrs. Sanchez and [juror Bernez R. Stephens, a West In-
dies–born nurse's aide from the Bronx] were sitting on the sofa in the jury
room, while the rest of us were sitting around the table," Lilienfeld recalled.
"We were having a strong debate about a particular point. I said, 'Why don't
you two join us up here, you seem to be a minority.' When I said that, one
of them jumped up and said, 'Yes, I *am* a minority!' She had misconstrued
my words. I meant nothing more than she was in the voting minority. It had
nothing to do with race or ethnic background. But that was an example of
the tension in that room."

Although Rothman believed the case went well, he had been involved in
past lawsuits that he also thought went well, only to suffer shocking defeat.
"It wasn't like we all knew what was about to be announced," said Gary Myers, who covered the trial for the *Dallas Morning News*. "It was real suspense."

So now, with the room packed and quiet, Leisure asked McCabe if the
jury was ready.

"Yes, your honor," she said. "We are."

She handed a piece of paper to the judge, who stared downward and
cleared his throat. Everyone in the courtroom was standing. Myerson and
his colleagues were positioned at a table near the front of the room. Rothman
and Fiske, his co-counsel, were directly behind them at another table. "The
first question," Leisure said, "is, 'Do you find that the NFL monopolized the
business of professional football, yes or no?'

"The answer is, 'Yes.'"

Myerson grinned like a child receiving a bag of M&Ms. This would be
outstanding . . .

A series of 27 questions ensued, asking whether the different NFL clubs
(excluding Davis's Raiders) were beholden to the monopoly.

"Yes" was stated 27 times.

Myerson could barely contain his euphoria. *Holy crap! Holy crap! Holy crap!* Several reporters on hand darted from the room to call the news into their offices. This was earth-shaking. The National Football League was found guilty of violating an antitrust law. It had, according to the jury, monopolized professional football and willfully acquired its monopoly power. Yes, the jury would clear the NFL on the eight other charges. But for a brief spell, radio and television outlets reported that the United States Football League had . . . won!

What followed was . . .

"Shocking," said Argovitz.

"Unfathomable," said Myerson.

"Emotional," said Rozelle.

"Confusing," said Usher.

"Very, very strange," said Larry Csonka, senior vice president of the Jacksonville Bulls. "I mean, *really* strange."

"I [was in] graduate school at LaSalle," said Chuck Fusina, Stars quarterback. "I was in a finance class and a guy comes up to me and said, 'Hey, Chuck, you guys won the trial.' I said, 'Hey, that's great.' I ran up to the teacher and said, 'Let me use the phone.' I wanted to call my wife and tell her the good news. I called my wife and she was kind of down. I asked her, 'What's the matter?' She said, 'Did you hear the settlement?'"

"My wife and I were traveling to Notre Dame, and we heard the news on the radio," said Jim Russ, the Tampa Bay Bandits' trainer. "I looked at my wife, my wife looked at me. I said, 'Holy cow, we won! We won!' Then the radio faded out."

"We were in a conference room when the verdict came down," said Ginger Lacey, a public relations assistant with the Orlando Renegades. "Bugsy Engelberg was our general manager, and he got the call that we won. He just starts yelling and screaming. Bugsy was a big guy, and I thought he was about to have a heart attack right there."

"I had a lot of friends in the NFL, and when they heard the first part of the verdict they were terrified," said Carl Peterson, the Stars' general manager. "It was clear to everyone listening that the NFL was in big trouble."

Rozelle hadn't made it to the courtroom in time for the verdict. He was stuck in traffic, listening to the trial on WCBS Radio, which was broadcasting live. His car was on 23rd Street, and the announcer said, "The National Football League has been found guilty." A furious Rozelle ordered the driver to turn the car around and take him back to league offices.

And yet, in less than five minutes, USFL joy was replaced by USFL horror, and NFL horror was replaced by NFL joy. After confirming that, yes, the NFL had violated the law, the jury awarded damages of . . . *$1*.

Yes, one dollar.

"Actually, $3," said David Cataneo, who covered the trial for the *Boston Herald*. "Damages in antitrust laws are tripled."

Rozelle had the car turn around again and speed to the courthouse. Trump, already there, was sitting alongside John Mara, the son of New York Giants' owner Wellington Mara. When the words "one dollar" emerged from Leisure's lips, the younger Mara pulled out a $1 bill from his wallet and handed it to the Generals' owner. Trump's sunken expression was worth the price.

Thanks to Myerson and Trump and a strategy that made little to no sense, the USFL walked out of the courtroom with $3 to its name. "I covered that trial, and you had to hate Trump," said Chris "Mad Dog" Russo, who hosted the Renegades' postgame show. "I just never saw how anyone liked him." Sibilia could not get past two things: (1) that the USFL's dysfunction was the greatest culprit in the league's failings, and (2) Trump was awful. "He was extremely arrogant and I thought that he was obviously trying to play the game. He wanted an NFL franchise . . . the USFL was a cheap way in."

Later on, Sanchez, also one who empathized with the USFL, suggested the preposterous $1 figure was actually a signal to Leisure that the jurors could not agree on a total, and they wanted that decision left to the judge. She said she believed the USFL was entitled to $300 million — trebled to nearly $1 billion. "It was definitely a compromise," Sanchez said outside the courtroom. "One dollar was all we could do, or we'd have a hung jury."

When the sentiment was expressed to Myerson, he exploded. "She said what?" he yelled. "That's not the procedure! I don't understand this!" The combative attorney promised the USFL would live to fight another day, and that the appeal would be an enormous success.

It never happened.

Even though the NFL was eventually forced to pay the USFL $5.5 million in attorney fees, the money was far too little to keep the young entity afloat.

Twenty-three years after David Dixon placed his first call about starting a spring football outfit, the United States Football League was officially dead.

18

SUSPENDED

In the end, the USFL was like a really fun, good-looking college girl-friend. You dig her at the time, don't cry when it's over, and forever look back fondly.

— Tom Vasich, Los Angeles Express season-ticket holder

IN THE DAYS that followed the court decision, the United States Football League announced it would be "suspending operations." This, almost everyone knew, was a euphemism for a funeral.

Or, as Joe Cribbs, former Birmingham Stallions halfback, rightly noted, "We're dead."

On August 7, 1986, the vast majority of the 530 players under contract with the USFL were released from their deals and declared free to sign with franchises in the National Football League and Canadian Football League. A couple of days later Mike Brown, assistant general manager of the Cincinnati Bengals, callously shrugged and told *Sports Illustrated* that he was uninspired. "We're not looking for much help from over there," he said. "There are a half dozen USFL players who will be stars. And a dozen or maybe a score who will be backup players. That's about what the impact will be. Not much."

"We might get somebody who could be our 44th or 45th player," added Bill Tobin, the Chicago Bears' personnel director. "There aren't many players who are going to help the world champions."

As if on cue, a mad scramble commenced to secure talent. The first expatriate to jump, post-lawsuit, was Baltimore safety Antonio Gibson, who agreed to a two-year deal with the New Orleans Saints that featured a $30,000 signing bonus and salaries of $130,000 and $160,000. Gibson was beyond giddy, noting that his USFL experience made him a highly valuable chip. "I knew my market value," he said. "This was my year to get out."

In a blink, newspaper transaction sections were overwhelmed by NFL teams acquiring USFL players. The big names jumped with great fanfare. Jim Kelly was in Buffalo with the Bills. Kelvin Bryant, the Stars halfback, went to the Washington Redskins along with Outlaws quarterback Doug Williams and a foursome of wide receivers (Houston's Ricky Sanders and Clarence Verdin, Oakland's Derek Holloway and Jacksonville's Gary Clark) who, coming out of college, were thousands of miles off the NFL radar. "I told [star Redskins receiver] Art Monk, 'Man, you wouldn't have even made our team in Houston,'" said Verdin. "We played three years together, and if he said five words to me after that it was a lot."

Maurice Carthon, the Generals' marvelous fullback, became a New York Giant. "Donald Trump came to me and said, 'Fans don't come to see you play. They come to see Herschel and Flutie,'" Carthon recalled. "I said, 'I'm glad you're telling me this, because I just signed a contract with the Giants.' He changed his tune real quick; told me to stick around for fall; that he'd call my agent and get it done. No chance — I was gone." Defensive end Reggie White traveled from the Memphis Showboats to the Philadelphia Eagles; Gary Anderson, the Bandits' otherworldly halfback, morphed into a San Diego Charger; and Jo-Jo Townsell, Steve Young's go-to wide receiver with the Express, crossed coasts to wear the green-and-white Jets duds. Young found himself a centerpiece of the lowly Tampa Bay Buccaneers. "It was a huge talent drop-off from the Express," he said. "Huge." All told, the NFL featured 158 ex-USFL players in 1986, and their arrivals brought forth unprecedented levels of both excitement and awkwardness. "Oh, the animosity was real," said Verdin. "Guys who'd been in the NFL for a decade were being let go for USFL players. They thought we were a semipro league."

That juxtaposition was on clear display in Dallas, where Herschel Walker signed a five-year, $5 million deal with the Cowboys and was greeted with the hype of a Hollywood premiere. This did not sit well with four-time Pro

Bowler Tony Dorsett, the team's all-time leading rusher and a man being paid $450,000 annually. Dorsett ripped the Cowboys for pulling a "publicity stunt," threatened to walk out, and demanded a trade. "Tony Dorsett," he said, "is second to no one.

"If this team does not pay me like they are paying their other back, I would suggest strongly that the team try to trade me or pay me because . . . I can be a very disruptive force."

Dorsett proceeded to have the worst two seasons of his career. By 1988 he was gone.

The most profound USFL-to-NFL experiment took place in New Orleans. The Saints were now coached by Jim Mora, straight off of three straight title games with the Stars, and he (more than anyone) understood the riches the defunct league provided. Along with Gibson and quarterback Bobby Hebert, New Orleans's roster would include 10 more USFL refugees. Five of his players were ex-Stars, including Sam Mills, the middle linebacker who, coming out of Division III Montclair State, had been cut by multiple NFL and CFL franchises. When the five-foot-nine Mills first entered the Saints' locker room, he was greeted by guffaws and raised eyebrows. "I imagine some guys looked at him and wondered what he was doing in camp," said Mora. "But Sam played for me for three years, and he deserved a chance." The Saints improved from 5-11 in 1985 to 7-9 in 1986, and Mills wound up a five-time Pro Bowler and one of the better middle linebackers in NFL history. In 1987 New Orleans made the playoffs for the first time in its 21-year existence. "Sam, me, [linebacker] Vaughan Johnson—all of us USFL guys had a real connection in that locker room," said Hebert. "They called us a farm league. We were no farm league. We were talent."

The New York Giants won Super Bowl XXI in 1987, and the team's starting lineup featured Carthon at fullback, ex-Panther Chris Godfrey at guard, ex-Star Bart Oates at center, and Sean Landeta, also a former Star, as punter. It has been said that nothing in professional sports compares to winning a Super Bowl. Landeta, however, insisted his two USFL championships equaled his NFL crown. Actually, exceeded it. "How about this?" he said. "We win in Philadelphia, and in the locker room there's champagne everywhere, pouring the stuff all over each other, rubbing it in our hair. Huge party, followed by this amazing parade down Broad Street. Then I win the Super Bowl. There was no champagne allowed in the locker room, and [New York City mayor] Ed Koch told the Giants they couldn't have a parade in the city. So we had this celebration at the stadium in the freezing cold."

"Obviously, the prestige of the Super Bowl was much bigger," said Oates. "But the personal satisfaction? Exactly the same."

During a 2009 interview with Mike Tollin for a documentary about the old league, Donald Trump dismissively referred to the USFL as "small potatoes"—a tag that outraged too many to count. Small potatoes? *Small potatoes?* "That infuriated me," said Steve Young. "There was nothing small potato about the USFL. He just didn't have the vision to see it."

"I was in a room with a bunch of USFL coaches when the $1 was announced," said Bill Bradley, working at the time as the Memphis Showboats' defensive coordinator. "They all started to cry. Small potatoes, my ass. It wasn't small potatoes to all of us who lived and bled for that league."

The USFL produced 60 Pro Bowlers and two Super Bowl MVPs, as well as four Hall of Famers (Young, White, Kelly, and Gary Zimmerman, the Express offensive lineman). Dozens of NFL head and assistant coaches got their starts in the USFL, and Steve Spurrier went from guiding the Tampa Bay Bandits to becoming one of the great coaches in college-football history. The Buffalo Bills reached four straight Super Bowls in the 1990s behind the personnel genius of general manager Bill Polian (Chicago Blitz), the coaching of Marv Levy (Chicago Blitz), the quarterbacking of Kelly (Houston Gamblers), the blocking of center Kent Hull (New Jersey Generals), and the tackling of linebacker Ray Bentley (Michigan Panthers and Oakland Invaders). Halfback Emmitt Smith of the three-time-champion Dallas Cowboys regularly found himself finding daylight behind left guard Nate Newton (Tampa Bay Bandits). Though Doug Flutie's NFL career was somewhat spotty, he is a three-time Grey Cup champion and widely regarded as one of the finest players in Canadian Football League history. "You can't argue about what we brought to the NFL," said Gary Plummer, the linebacker who jumped from the Invaders to the Chargers. "And I'll tell you what—leaping leagues was easy. In the USFL *everyone* was hungry, so they'd bite you, kick you, scratch you, claw you, try to rip your nutsack off. It was professional wrestling, because you were fighting for football survival. By comparison the NFL was cake."

"[Former Washington Federals halfback] Craig James and I were playing for the New England Patriots in a preseason game against the Redskins," recalled Tom Ramsey, the onetime Express quarterback. "I threw Craig a swing pass and he took it 50 yards for a score. I sprint to the end zone and we're jumping up and down and I'm screaming, 'USFL guys! USFL guys!' It meant something."

On January 17, 1988, Harry Usher wrote an open letter to Pete Rozelle, congratulating him on the NFL's success while noting "the residual sense of pride that the owners, coaches, the players and the fans feel for the success of the USFL alumni. It was a very credible league, Peter, that deserved better treatment." The NFL adopted both the two-point conversion and the coach's challenge from the USFL, and when the league expanded into Jacksonville and Tennessee (first Memphis, then Nashville), USFL veterans took enormous pride. "We opened up the markets," said Steve Ehrhart. "Whether the NFL acknowledges it or not, the USFL brought football to a lot of new places."

What often gets lost in the aftermath of success stories, however, is the $1 smothering of dreams. For every Jim Kelly and Herschel Walker, there are hundreds of professional football players and coaches (and administrators and cheerleaders and popcorn vendors) whose careers ended the moment the USFL ceased to exist. What ever became of Nat Hudson and Ronnie Estay? How about Todd Dillon and Johnnie Dirden? Ken Bungarda, anyone? Sylvester Moy? Sel Drain? "It was a sickening feeling," said Marcus Bonner, a Gunslingers halfback. "Like someone was punching you in the stomach and stealing your joy."

"I have a team photo that I've looked at every day for thirty years," said Bruce Miller, a Breakers defensive back who never reached the NFL. "It still hurts. There were so many of us who moved their families, who enrolled their kids in school — and then it died. I've never seen more grown men cry."

"It was the worst," said Mark Raugh, the Maulers and Showboats tight end. "Most of us were not rock stars, were not superstars. We gave everything we had on every down, just so we could play the following week. When it ended — it was a kick to the head."

Kevin Nelson, the Express halfback, had a couple of workouts with NFL teams, but failed to gain a roster spot. He devoted the next 24 years to his job as a Riverside County probation officer. "I sometimes wonder what could have been," he said. "But it's sort of depressing." San Antonio's Peter Raeford was one of the USFL's elite defensive backs, finishing his two seasons with seven interceptions. "No one from the NFL called me," he said. "No one. I started working jobs trying to support myself. To go from being All-USFL to picking up cigarette butts in a parking lot . . . it's humbling." Marcus Marek, the greatest linebacker in Boston–New Orleans–Portland Breakers history, attended camp with the Chicago Bears, whose roster was overflowing with Pro Bowl–caliber players at the position. "I tried to learn the 46 Defense in a week and a half," Marek said. "It was impossible." His last three decades have been largely spent working in a Massachusetts fish market. Ken Dunek, the Stars tight end, presumed he'd find an NFL gig. "No," he said. "Nobody cared." He went into magazine publishing. Whit Taylor, a backup quarterback in Michigan and San Antonio, transitioned from playing to working as an assistant coach at Vanderbilt University. One night, when he was out to dinner, someone broke into his home. "My championship ring was stolen," he said. "That was a crusher."

A few years ago Tom Davis, a Denver Gold offensive lineman, received a random call from a reporter who informed him he was the only person in team history to appear in all 54 regular-season games. Did he want to share his story? "Seriously, man, who the fuck gives a shit?" Davis responded. "It's forgotten history." Davis's bitterness is understandable. When the Gold folded, so did his football life.

At least five former USFL players went into professional wrestling, including Tampa Bay's Ron Simmons (half of the WWF's "Doom" tag team) and Larry Pfohl ("Lex Luger") and the Gunslingers' Jeff Gaylord, who was known as the Hood and the New Spoiler and credits the league for giving him his start. "I tore up my knee when the Express center dove into it, and I couldn't play," he recalled. "So I started going to this gym in my free time in San Antonio. One of the local guys trained me and he said, 'You ever think about wrestling?'"

Though Mike Lush, the Stars free safety, caught on with the Vikings and Colts in 1986, the USFL's death was the demolition of a dream. "I was a nobody and the Stars made me a somebody," he said. "That league gave me life." Not long after his football career ended in 1987, Lush began noticing changes in mental and physical capacities. He underwent multiple back and shoulder operations, and started to have issues with his memory. "But because the USFL folded, I lost my severance pay and my workman's compensation," he said. "I went from loving football to wishing I'd never played a down. Nothing is worse than losing your mind. I'd choose cancer over the way I feel."

Few took the league's demise worse than Bugsy Engelberg, who began his USFL existence as the Bandits' director of football operations before becoming general manager of the Orlando Renegades. The portly Engelberg was a character. He walked with a waddle. He cursed 100,000 times per minute and had, Larry Guest of the *Orlando Sentinel* once wrote, a "pumpkin silhouette." Engelberg referred to everyone as "Bubba," and his go-to saying was, "You know who loves you, Bubba."

"He'd dog cuss me one minute, then the next minute I'm his best friend," said Alex Narushka, the Renegades' team controller. "Bugsy was unforgettable."

He loved the USFL and, in particular, the hunt for players other teams somehow missed. Few football executives had Engelberg's nose for talent, instincts for recognizing a seemingly flawed athlete's hidden skills. "I don't know how he did it," said Lee Corso, the Renegades' coach, "but he found us some really good players nobody else had ever heard of."

When the USFL folded, Engelberg, just 41, was crushed. While he also owned and operated a commercial collection agency, football fulfilled him. "I think the USFL outcome hit him awfully hard," said Mike Reid, a former NFL player and Engelberg's close friend. "He believed in that league." Though widely respected, Engelberg failed to find an NFL or CFL job and missed the buzz of the game. He was married, with two young sons. Family was not enough to outweigh depression.

On July 29, 1987, a little more than two years after the Renegades' final game, Engelberg drove to the Oviedo, Florida, house of his longtime friend, University of Central Florida head coach Gene McDowell. He pulled his Dodge Diplomat into the garage, closed the door, rolled down all the car windows, and kept his engine running. He was found by McDowell at 9:15 that evening.

The vehicle, having run out of gas, was silent.

Wait.

For the USFL, there was one final spark.

In 1987 the NFL's players went on a 24-day strike, and while the league's Week 3 contests were canceled, for the next three games all 28 teams were required to find replacements to step in, don unfamiliar uniforms, and play for an average of $5,000 per Sunday.

Reenter: the USFL.

More than 100 of the league's former players wound up on NFL replacement clubs. John Reaves, Tampa Bay Bandits quarterback, was now throwing passes for the Buccaneers, and served as one of 19 ex-USFL players on the team's 53-man roster. "They called us the Scabaneers," recalled Fred Nordgren, the former Bandits defensive tackle who joined the Bucs. "We didn't care. We were back together and getting paid to play football again." Kansas City's halfback was Ken Lacy, the Michigan Panthers' all-time leading rusher, and the Houston Oilers' starting wide receivers were Joey Walters (98 catches for the 1984 Federals) and Leonard Harris (101 catches for the 1985 Gold). Showboats quarterback Mike Kelley was temporarily the man in San Diego, until he was replaced by Rick Neuheisel, ex-Gunslinger. The Bears' quarterback was Mike Hohensee, the all-time Washington Federals passing leader. In need of a center, the Cleveland Browns turned to Mike Katolin, ex-Outlaw, to anchor their offensive line. "[Katolin] had the look of a fighter about him," Ed Meyer wrote in the *Akron Beacon Journal*. "The kind of look that tells you he would slug anyone who had the guts to call him a scab."

This was pure nonsense. People were free to call Katolin and the other USFL-seasoned replacement players scabs, traitors, slime-balls, scumbags — they could not care less. Clare Farnsworth of the *Seattle Post-Intelligencer* later came up with nicknames for several of the squads (Los Angeles Shams, Chicago Spare Bears, San Francisco Phony-Niners), and the participants laughed and laughed. Hell, what difference did it make? The uniforms were real, as were the paychecks. And even if the stadiums were 70 percent empty, people were once against paying to watch them perform. "It was about extending the dream," said Dan Manucci, the ex-Wranglers quarterback now replacing (interestingly) Jim Kelly with the Bills. "There was nothing malicious. We just wanted to keep playing."

On September 22, the day the NFL's players first went on strike, approximately 20 wealthy onetime football team owners — spread across the country from coast to coast — were surely experiencing some sort of collective heartbreak and/or heart attack. When, back in 1984, the USFL commissioned McKinsey & Co. to piece together its study on the demographics of switching to a fall schedule, the main piece of advice was . . . *wait*. Wait a couple of years. Wait until the league feels a bit stronger. Wait until some of the owners are weeded out. Wait until the NFL's players inevitably strike — *then pounce*. "We knew there would be labor unrest in the NFL," said Jerry Argovitz, owner of the Houston Gamblers. "It was coming. We saw it coming. Now imagine if we were still around at that time. It's a game changer, because so many of those angry NFL players would wash their hands of that league and come play in the USFL. It would have been an enormous shift in the football landscape."

"If the USFL had just stayed in spring until 1987, everything ends differently," said Charley Steiner, the Generals announcer. "First, there would have been this mass exodus of players to the USFL, where they would have been able to work. And second, the NFL would have been more inclined to bring in some new teams and circumvent the possibility of the strike. That's where they screwed up. The USFL could have had a definite future."

Alas, it was not to be. Donald Trump had a plan, and it involved neither virtue nor wisdom. Either the USFL would join the NFL, the New Jersey Generals would join the NFL, or the whole thing would die in a blaze of glory. Patience? Trump had no familiarity with the word.

• • •

Thirty-one years after the trial, as the USFL's most explosive owner was being sworn in as the 45th president of the United States of America, the league's former players and coaches found themselves befuddled, bewildered, perplexed, surprised, nostalgic, and even slightly amused. Many voted for Trump, often because he was not Hillary Clinton. ("I couldn't support that woman," said Ray Bentley, the Panthers linebacker. "I'm sorry.") Several sat out the election. Almost all of the USFL veterans interviewed for this book considered the Donald Trump of the mid-1980s and the Donald Trump of 2017 to be eerily familiar. Thirty-three years after insisting his fellow owners would pay for Doug Flutie, he was insisting Mexico would pay for a border wall. Thirty-three years after being accused of cozying up to Pete Rozelle, he was being accused of cozying up to Vladimir Putin. Thirty-three years after Roy Cohn and Harvey Myerson, his chief advisers were the equally controversial Steve Bannon and Stephen Miller. Thirty-three years after insisting the USFL needed to move to fall ASAP (then lacking a concrete plan for implementation), he was insisting America needed a ban on immigration ASAP (then lacking a concrete plan for implementation). "He's just like he was back then," said Danny Buggs, a Bandits wide receiver. "Big talker, lots of bluster, not much behind the words. It's like going back in time."

"Trump killed us," said Zenon Andrusyshyn, the Tampa punter. "We could have grown and thrived as a spring league. Even as a farm league for the NFL. But he didn't care. It was a power grab."

Although with the $1 verdict the USFL vanished and never returned, it lives on in some idiosyncratic ways. Shortly before the death became official, Dom Camera, the league's director of marketing, had to call the myriad USFL employees to tell them their services were no longer needed. One of the first people he reached was Gary Cohen, who along with Mike Tollin produced the weekly highlight show *This Is the USFL*. Camera opened by informing Cohen that he had both bad news and good news. "The bad news is we're done," Camera said. "The good news is you get to keep the library."

By "the library," Camera was referring to the entirety of USFL video footage. Cohen wasn't feeling it. "I didn't want the library," he said. "What was I going to do with tapes of football games played by guys no one knew, wearing uniforms people didn't care about? But Dom just said, 'Trust me. You'll thank me later.'"

He wasn't lying. Every year millions of American men buy televisions in order to watch football. The various companies that produce TVs are aware of this, and try to run advertisements for their contraptions that feature games. Unfortunately, the NFL only sells footage to its official television company. That means if, say, Zenith is the NFL's TV of choice, Panasonic, Sony, and myriad other entities can't use league action. "So every year — every single year — I get calls from the companies, wanting to purchase USFL stock footage," Cohen said. "I averaged about $100,000 a year for a long time. Dom was right."

Don't blink, or you might miss ubiquitous snippets of USFL game footage. That game Julie Taylor was watching in the student lounge on *Friday Night Lights*? Blitz-Bandits at Tampa Stadium. The "Bubble Bowl" game in the *SpongeBob SquarePants* episode "Band Geeks"? Bandits-Showboats at the Liberty Bowl. A Scientology advertisement stars Anthony Carter scoring a touchdown for the Panthers; Russ Feingold, a United States senator running for reelection in 2010, ran a spot with Gamblers receivers Clarence Verdin and Gerald McNeil dancing in the end zone; *Two for the Money*, the 2005 Al Pacino and Matthew McConaughey film about sports gambling, features repeated USFL action.

More important, the USFL continues to exist via the memories of its players and owners and, perhaps most profoundly, its fans. Yes, greed and arrogance destroyed a beautiful journey. But those vices couldn't fully extinguish the joy.

Which allows me to end with a story I was told in the early stages of this project . . .

Back in the spring of 1983, an 11-year-old Philadelphia Eagles fan named Jason Sean Garber learned of a new football team, the Stars, that was coming to the region. Jason lived and died for the sport, so when he heard Carl Peterson, the former Eagle, was in charge of the upstart enterprise, he sat down at his desk and handwrote a letter with a list of players the team should sign. One of the names was John Bunting, an 11-year NFL linebacker who had recently been released by the Eagles.

After dropping it in the mailbox, Jason thought little of the letter. Then, a few weeks later, a package arrived in the mail. The boy opened it, and found a slew of Stars stickers and pins alongside a note from Peterson.

Dear Jason:

Received your letter. I'm going to take your advice and sign John Bunting.

Thank you for the suggestion.

Carl Peterson

On March 7, 1983, John Bunting became a Philadelphia Star.

ACT FOUR

Here's what I don't understand. There are players and coaches from the USFL in the Pro Football Hall of Fame, but their plaques don't mention the league. Were they not being paid? Did they not play in a professional league? For me, it's a lack of respect for a really great place to be a pro football player.

— Dave Lapham, offensive lineman,
New Jersey Generals, 1984–85

19

REUNION

THE MEN ARRIVE with nervousness and trepidation and uncertainty. This is not altogether surprising, when one considers the passing of three and a half decades and the simultaneous accumulation of fat and wrinkles with the subtraction of hair follicles and muscle mass. Yet along with the outward appearances of yesteryear, what seems to have vanished most is the athlete's natural confidence.

Back in the early 1980s, they arrived in Philadelphia as aspiring football gods, glistening forearms, sweat-glazed torsos, decked out in the red-and-gold uniforms of the brand-new Philadelphia Stars of the United States Football League. The helmets were shiny and unscratched; the pads were straight from the box; the jerseys and pants never before worn. Within a short span, it became clear that the City of Brotherly Love would embrace its new Stars as, well, stars, and the glorious run that followed only sanctified a mutual love affair. From Chuck Fusina and Sam Mills to Kelvin Bryant and Irv Eatman, the Stars were the league's finest collection of talent; a team that would surely stand the test of time.

Yet in sports, there is no *surely*. Time marches on. Stars morph into forgotten stars, and those same heroic gladiators who drew autograph seekers and groupies and endorsement deals now exist as mere mortals. Fusina, once mustached and Montana-like, is gray and mortal. He is a partner in a Pittsburgh-based baseball- and softball-manufacturing business. Eatman, the biggest of the big, lumbers around on bad knees, anxious for the nearest chair. Bryant, the flashiest member of the team, stays out of sight in his rural North Carolina home. His gold earring is gone. So are his cutbacks. "It's not sad, getting older," says Scott Woerner, a Stars safety. He pauses. "Well, maybe it is sad. I don't really know how to explain it . . ."

As he speaks, Woerner is standing inside Chickie's and Pete's, a South Philadelphia bar that offers authentic pretzels and two genres of cheesesteaks. He is here, along with 40 or so former teammates and coaches, for the June 10, 2017, Philadelphia Stars reunion — only the second to be held since the league's 1986 demise. That the numbers mark no special rationale (it's been 34 years since the USFL was formed, 31 years since it died) matters not to the participants. They file in one by one, grab a requisite white HELLO MY NAME IS name tag from the front table, and seek out faces deemed unrecognizable by the passage of days into weeks, and weeks into years, and years into decades. It is not altogether unlike a 30-year high school reunion. You want to see your old girlfriend — kind of. Sort of. Maybe. "It's weird," says Dwayne Wilson, a wide receiver from Rutgers who caught nary a pass in his one professional season. "Wonderful. But weird." In his right hand, Wilson holds a rolled-up photograph of the 1983 Stars. He says he wants everyone to sign it. But the image also serves as a handy gap filler between introduction and facial recognition. As an old teammate looks over the picture, Wilson looks over the teammate, desperate for a spark of nostalgia. Then, inevitably, "*Heeeey* — Big Brad!"

This goes on for a while. The room is dark. The music is loud. The old Stars logo appears on various television screens, and memorabilia — pictures, media guides, hats — are spread throughout. There are handshakes, nibbles on pretzels, the ordering of beers. It is a nice event. But merely that. Nice.

Then, almost on cue, something beautiful takes hold. Carl Peterson, the team's general manager, enters through the front door with the regal gait of a king. He is 74, with slicked-back silver hair, attentive brown eyes, and a perfectly fitted blue blazer and slacks. He is *someone* — that much would be clear to those unfamiliar with the game of football — and as the men spot him their expressions change; their tones leap; their smiles widen.

"*Carl!*"

"*Carrrrrrllll Peterson!*"

"*Miiiisssttterr Peterson!*"

One by one, they approach. Extended hands are rejected; only hugs are allowed. Peterson flashes the two Stars championship rings on his right hand, as well as the 1984 USFL runner-up watch wrapped around his wrist. Of all the people in the room, he has enjoyed the most post-USFL glory, spending two decades as the highly successful president and general manager of the NFL's Kansas City Chiefs. It would have been easy, one imagines, for Peterson to have left the Stars in his long-ago wake, never to be thought of again. Instead, he refers to his USFL days as "the greatest experience of my life." The men surrounding him — they're "his" guys. The two Stars titles — "his" bliss.

And now the reunion has transitioned into a full-fledged party. There is no magic formula embedded within the pretzels — no one turns young or goes back in time. But the discomfort departs. When Jim Mora, the Stars' coach, reminds a circle of men about the play that won a nail-biter against New Jersey, their smiles are all teeth and gums. A couple of monitors show Philadelphia's 1984 demolition of Arizona in the second championship game, and Bryant's dash into the end zone is accompanied by *ooh*s and *aah*s. Many here have attended past football-related reunions. High school, college, a small handful of NFL. This, though, is different. The USFL is a shared experience unlike any other shared experience. "A leap of faith," says Brad Oates, the offensive lineman turned attorney, "to try something new and different." There was no blueprint to starting a spring football league, just as there was no blueprint to playing atop the diseased Astroturf of San Antonio; to Steve Young's $40 million contract; to the torrential rains of Jacksonville and the 120-degree heat of Arizona and the madness of Donald Trump. So while they look back at the football, they look back at far more than the football. "It's not about the game," says Mora. "It's about the ride."

After a couple of hours of banter and recollection, of jokes and nicknames, of laughs and cries, Peterson climbs some steps, stands at a balcony, and clears his throat. From below, the old Stars grow quiet. Peterson taps the mic, dabs at his eyes. Earlier in the day, his horse Tapwrit won the Belmont Stakes — a remarkable moment that he was forced to miss to attend the reunion. "It was a life highlight, that race," he says. "But I wouldn't want to be any other place than right here, right now."

His words are drowned out by applause. Then more silence. "I've been in this business for 40 years," Peterson says. "I know how to measure players.

Height, weight, speed. But the one thing you can't measure is the size of a man's heart. And I will tell you, there are no bigger hearts in the United States than the ones in this room."

Among those listening, Woerner stands out. He is the only attendee to arrive in his old Stars jersey, and it goes over well. Now, he looks up toward Peterson, smiles widely, and hoists his drink.

"God bless you, Carl!" he yells.

A pause.

"God bless the United States Football League!"

FINDING GREG
(ACKNOWLEDGMENTS)

The morning of August 11, 2016, was a hot one, and as my son, Emmett, and I walked the streets of San Francisco, we wondered aloud whether this was all a big waste of time.

"Do you think we'll find him?" Emmett, age nine, asked.

"Maybe," I said.

"Do you think we'll find him?" he asked again.

"Dunno," I said.

"Do you think we'll find him?"

We strolled over some broken bottles shattered atop a sidewalk, and past an antiquated psychiatric hospital, and through a field that smelled of raisins and urine. We took a break for ice cream and another for sandwiches.

Finally, four hours later, we reached our destination — the childhood home of Greg Fields, former Los Angeles Express defensive end/coach puncher.

"Are we here?" Emmett said.

I looked down at my sheet of paper. Through a year of USFL research, I failed to track down Fields's phone number. All I had were two iffy addresses — one of which (314 Holladay Avenue) we were standing directly in front of.

"Yup," I replied. "We're here."

The boy smiled.

Throughout the research for this book, Emmett was my nonstop companion. A kid who was born more than 20 years after the USFL's final game turned into a league buff. He knew the Blitz and Wranglers were traded for each other; knew Donald Trump owned the Generals and Doug Williams quarterbacked the Outlaws and Reggie White collected sacks for the Showboats. On the day I discovered that former Jacksonville Bulls quarterback Ed Luther coincidentally lived in our Southern California hometown, Emmett anxiously joined me for the two-mile drive to his house. We even filmed a mini-documentary, *Finding Luther* (it's available on YouTube). On a different occasion, I learned (and this is a *really* weird one) that former Philadelphia Stars offensive lineman Irv Eatman is related. Well, sorta related. He is (hold your breath) the husband of my brother-in-law's stepmother's daughter. Or, as Emmett likes to call him, "Uncle Irv!" We were *deep* into the USFL.

So now, standing before Fields's alleged house, the son and I felt accomplished. We had driven seven hours north to San Francisco, followed by the meandering walk. The home is charcoal-gray, with 12 wooden steps rising toward the entrance. We approached the front door, giddy with nervous excitement. Would Fields invite us in for popcorn? Would he tell us to get lost? Did he have all his old USFL gear? Was it an experience he wished to forget?

Pfft.

We knocked.

And knocked.

And knocked.

The house was dark and empty. I looked down at Emmett's glum expression. He seemed both heartbroken and relieved. "With the USFL," I told him, "nothing comes easily."

Most things with the USFL do, however, ultimately come. The second Greg Fields address was located in San Francisco's projects. I went later that night, this time leaving Emmett with a friend. I parked near a sidewalk, climbed three flights of steps, and rang a doorbell outside a hallway leading to a cluster of apartments. Moments later a woman approached. She was in her 50s, African American, somewhat tired-looking. Our faces were separated by a thin pane of glass.

"Yes?" she said.

"This is weird," I said, "but are you related to a former football player named Greg Fields?"

"That's my brother," she said.

"Wait!" I said. "What?"

"Can I help you?" she said.

"Yes!" I replied. "So my name is Jeff Pearlman, and I'm a writer, and I'm working on a USFL book, and I really want to speak with Greg because he played for the Los Angeles Express and was really good and punched a coach and . . . and . . ."

"I'll get him your number," she said, seemingly unmoved. "Maybe he'll call."

Less than 24 hours later, Emmett, Greg Fields, and I found ourselves sitting around a table inside the food court of a shopping mall in Sacramento, where the former defensive end is now retired. It was a weird meeting. Though he walks with a noticeable limp and looks at least a decade older than his 62 years, Greg stared down a solid 10 women in their 20s, often noting how he'd love to "tap that." At one point, after I asked whether he'd like a bite to eat, Greg motioned toward Emmett, then the nearby Cold Stone Creamery. "Get me one of those cherry things," he said. "On a cone."

I handed my son $10, and he returned with an enormous ice cream — which Greg Fields proceeded to spit all over me during our interview.

It mattered not. The search to find Greg Fields, while rocky and unruly, was worthwhile.

The same can be said for this book project.

Football for a Buck is my eighth title, and it goes down — without question — as the most fun of the bunch. As I am prone to include in the acknowledgments of every offering, writing a book is a nightmare. In this case, however, the nightmare was lessened by a subject that brought tremendous joy and spark to my life. I loved the United States Football League as a child, and I love it just as much as an adult.

Also, for the first time, I was able to share a full reporting experience with my children. I am convinced Casey Pearlman is the only ninth-grader in Southern California to boast the holy trinity of a Chicago Blitz T-shirt, a Tampa Bay Bandits sweatshirt, *and* an Oakland Invaders keychain attached to her backpack. (Also, weirdly, her math teacher is the son of Stan Talley, Oakland's punter). And Emmett, well, Emmett morphed into my USFL co-pilot, not only traveling with me to find Greg Fields and Ed Luther, but requesting daily interview updates. There is one child in the history of the United States of America who has ever owned his own stitched Los Angeles Express Greg Fields No. 99 jersey. That boy's name is Emmett.

Books like this spawn from a team effort, and my team remains as strong as oak. My editor, Susan Canavan, took a shot on the USFL when no one ("Nobody wants a fucking USFL book" was literally stated to me) would listen. I can't thank her enough for that. David Black is the Pat Ryan of agents, and Michael J. Lewis is not only the Al Toon of family blogging, but the best proofreader in the entire eastern portion of Rhode Island. Casey Angle, veteran fact-checker, has survived the inevitable IQ drop that accompanies reading all of my unedited works. Jack Cassidy continues to put the Thor in "thorough," Stanley Herz's head always bobbles, Jenny Xu is a two-letter wonder, Melissa Dobson is the fantastic purveyor of *Au,* and Joan Pearlman remains convinced this book is about the United States Floor Lamps. Mike Tollin, director of the excellent USFL *30 for 30, Small Potatoes: Who Killed the USFL?,* could not have been more gracious, and the packages of papers and notes provided by Kathy MacLachlan, Tamara Tanzillo, and Mike McCarthy were gifts from the Chicago Blitz gods. Steve Ehrhart is Mr. USFL, which made him Mr. Awesome, and Robert Ricker and Adam Lazarus were perfect guides to the USFL past. Major thanks to Laura Cole, Norma Shapiro, David Pearlman, Greg Singleton, Jordan Williams, Isaiah Williams, Richard Guggenheimer, Kai Williams, Bear Webb, Sarah Leonard, Christy Gisser, Frank Zaccheo, Ross Heise, Zach Mentz, Kevin Bartner, Joseph Merkel, Taylor Kennedy, Steve Young, Leah Guggenheimer, Leigh Steinberg, Doug Plank, Travis Puterbaugh, Patrick Shuck, Sandy Glaus, Deek Pollard, Jerry Argovitz, Matthew Gourlie, Robyn Furman, Michael Moodian, David Anderson Jr., Tom Keiser, John Brunelle, Glenn Geffner, Diego Cordovez, Michael Chapman, Gregg David, Kevin Mulvoy, Chris Berman, Jessica Berman, R. J. Ochoa, Noah Fried, Joe Southern, Gary Cohen, Todd Arky, and Jenny Freilach. Oh, and John Teerlinck — holy shit. One can walk this globe 1,000 times and never uncover a more unique and riveting man.

My wife, the spectacular Catherine (Earl) Pearlman, recently published her first book, and while *Ignore It!* and *Football for a Buck* share few commonalities, there's bliss in being able to talk writing highs and lows with the five-foot love of my life. Plus, after 15 years of one-sided literary anguish, I've started receiving the much appreciated, "I feel you, bro" sighs in return. As always, Catherine's patience and compassion throughout this whirlwind experience have been gold nuggets.

It's one thing to write. It's another to write while being loved.

Lastly, a nod to Rodney Cole, who passed during the writing of this book. Rodney wasn't merely my father-in-law, but a man who still marveled at the splendid beauty of a bird in flight and the healing power of a single piece of Hoffman's chocolate.

Worry not, Rodney.

My hats never touch the bed.

NOTES

1. Origins

page

1   *"It dawned on me that the NFL never in its history"*: Mitch Lawrence, "Traditional Art Collector 'Radical When It Comes to Football,'" *Rochester Democrat and Chronicle*, Feb. 26, 1984.

2   *"I did reason correctly that football"*: Dave Dixon, *The Saints, the Superdome, and the Scandal*, pp. 185–86.

3   *The headline was splashed across newspapers*: "New Football League Announced," Associated Press, June 25, 1966.
    "I always liked the name United States Steel Company": Paul Reeths, *The United States Football League*, p. 8.

5   *"My God, this will work!"*: Lawrence, "Traditional Art Collector 'Radical.'"
    Wrote Jim Byrne in his book about the USFL: Jim Byrne, *The $1 League*, p. 8.

6   *"We just started virtually calling people on the Forbes Top 400"*: Reeths, *The United States Football League, 1982–1986*, p. 11.
    "We are going to play football, the most wildly popular": Byrne, *The $1 League*, p. 8.
    a franchise in Southern California, for example: Max J. Rosenthal, "The Trump Files: His Football Team Treated Its Cheerleaders 'Like Hookers,'" *Mother Jones*, July 5, 2016.
    Every team, wrote Byrne: Byrne, *The $1 League*, p. 10.

10 *With two dozen reporters in attendance*: "U.S. Football League Becomes a Reality," *Santa Cruz Sentinel*, May 11, 1982.
    In a small article on the third page of the sports section: Michael Strauss, "Football League Planned," *New York Times*, May 12, 1982.

11 *"I was at ESPN at the time"*: Reeths, *The United States Football League*, p. 27.

12 *He was a unanimous choice by the owners*: Byrne, *The $1 League*, p. 18.

13 *"Doesn't the U.S.F.L. season"*: "Unseasonable; No Time Out," *New York Times*, Jan. 9, 1983.

15 *In a follow-up memorandum*: Byrne, *The $1 League*, p. 19.
    "We had a gentleman's agreement": Reeths, *The United States Football League*, p. 72.
    The Philadelphia franchise: Dan Herbst, "Name That Team: USFL Club Christenings," *USFL Kickoff*, vol. 6, 1983.

2. Albert C. Lynch

21 *The Blitz even traveled:* "Super-Tough Prisoner Convinces Blitz's Allen," *Decatur Herald and Review,* Aug. 21, 1982.

23 *"I thought I'd let them believe what they want":* "Garo or Not, Can He Kick?" *Detroit Free Press,* Nov. 10, 1982.
    The headline appeared: Bart Barnes, "Allen's USFL Team Signs Tight End Drafted by Bears," *Washington Post,* Aug. 7, 1982.

26 *"The opportunity with a new league is very exciting":* "USFL Conducts Draft, Dan Marino No. 1 Pick," United Press International, Jan. 5, 1983.

28 *"Everybody laughed":* Bryan Lennon, "Pioneer Spirit Leads James to USFL Pact," *Anniston Star,* Jan. 13, 1983.

31 *First, there was Jeff Gaylord, a nose tackle:* "New League Won't Take Back Seat for Zaniness," Gannett News Service, March 6, 1983.

33 *"Frankly, I had an inside source":* Reeths, *The United States Football League,* p. 42.

34 *It was the first time he had heard of Sam Mills:* Zander Hollander, *The Complete Handbook of Pro Football,* p. 118.

35 *and owned by Ron Blanding:* William Oscar Johnson, "It Was Up, Up and No Way," *Sports Illustrated,* May 14, 1984.
    "It was the most violent football ever played by mankind": Reeths, *The United States Football League,* p. 69.

37 *"He couldn't throw the ball 15 yards":* Ibid., p. 88.
    "They seem well set in the pass catching": Paul Domowitch, "Caution: Builders at Work," *Sporting News 1983 USFL Guide.*

3. Herschel

40 *"[Walker is] a 220-pound running back with Jim Brown's strength":* Dave Anderson, "USFL Fouls in Walker Bid," *Pittsburgh Post-Gazette,* Feb. 22, 1983.
    He aspired to join the United States Marine Corps: Curtis Walker, *Fallen Generals,* p. 11.

41 *The team even sent Walker a USFL contract:* Tom Callahan, "Frank Merriwell Turns Pro," *Time,* March 7, 1983.
    "Allen," wrote the New York Times, "has never been": Anderson, "USFL Fouls in Walker Bid."

44 *On the evening of February 8, he met in Orlando:* Byrne, *The $1 League,* p. 38.
    "The Dallas people had an intermediary": Reeths, *The United States Football League,* p. 61.
    "Isn't he the guy on the Monopoly game?": Herschel Walker with Brozek and Maxfield, *Breaking Free,* p. 168.

45 *"I told Herschel that if he came to me by 7 a.m.":* "Herschel Approved Contract Personally," *Tyrone (PA) Daily Herald,* Feb. 26, 1983.
    When asked about USFL rumors, Walker made stuff up: "Walker Denies Signing Contract with Generals," *New York Times,* Feb. 20, 1983.

46 *"Herschel played with fire and got burned":* Callahan, "Frank Merriwell Turns Pro," *Time,* March 7, 1983.
    He later curtly told Walter to take his millions: Douglas S. Looney, "A Runner on a Roll," *Sports Illustrated,* Jan. 27, 1985.

48 *"The money is great, but I love the enjoyment"*: William N. Wallace, "Walker Reports; Seeks Challenge," *New York Times,* Feb. 27, 1983.

4. Seasonal

55 *Reggie Collier of Southern Miss*: Robert Weintraub, "The Original Dual-Threat," *SB Nation,* April 24, 2013.

56 *He was, in fact, a former University of California backup*: "Besana Glad to Get One More Chance," *New York Times,* June 6, 1983.

57 *A well-liked and respected man, Bernhard thought*: Paul Attner and David DuPree, "A Federals Reprise: 'What Went Wrong? What Didn't?'" *Washington Post,* May 27, 1984.

58 *The Federals held part of their training camp in Jacksonville*: David Remnick, "Federals: They Beat the Crowd," *Washington Post,* Aug. 10, 1986.
    The organization's first-ever cut: Dave McKenna, "Spring Backward," *Washington City Paper,* May 2, 2008.
    On more than one occasion: Darren Rovell, "After 63 Years, Globetrotters Drop Rival Generals as Primary Opponent," ESPN.com, Aug. 14, 2015.

59 *In the days leading up to the season opener*: Reeths, *The United States Football League,* p. 78.

61 *Groundskeepers dyed the grass green so it would look*: Ibid., p. 78.

63 *In an effort to generate some interest*: Johnson, "It Was Up, Up and No Way."

65 *There were even a handful of celebrities*: Matthew Barr, "The Life & Death of the L.A. Express," *Los Angeles Magazine,* November 1985.

66 *The coach defended himself afterward*: Barry Stanton, "General Walker a Forgotten Man," *Journal News,* March 7, 1983.
    "I don't want to question Coach Fairbanks": Paul Zimmerman, "Teaching an Old Dog New Tricks," *Sports Illustrated,* March 14, 1983.

67 *"I ran the ball a little better than I expected"*: Stanton, "General Walker a Forgotten Man."
    Now, thanks to the new league: Barry Stanton, "Boddie Grabs the Spotlight," *Journal News,* March 7, 1983.

5. Weathering a Storm

70 *The surprising 2-1 Arizona Wranglers came to Legion Field*: Byrne, *The $1 League,* p. 53.
    "I would've liked to not play": Walt Jayroe, "Poor Conditions Help Stallions Break Away from Wranglers," *Arizona Republic,* March 27, 1983.

71 *When asked about the catastrophe*: Byrne, *The $1 League,* p. 54.

72 *Hence, Dick Coury, the head coach, came up with an idea*: Judson Hand Jr., "Coury Has Good Sense of Humor and He Needs It," *Asbury Park Press,* April 14, 1985.
    One, in particular, involved a spread formation: Katie Castator, "Taking Rams Offense to the Air," *San Bernardino County Sun,* August 5, 1986.

73 *"I gave serious thought about giving it all up"*: Ron Hobson, "Coury Coaches a New Wave of Winners," *USFL Kickoff,* vol. 1, 1983.

75 *"You talk about putting it together with a marketing blueprint"*: Travis Puterbaugh, "Catching Up with Jim McVay," tampasportshistory.blogspot.com, June 7, 2010.

76   *"That's all part of the scene"*: Nick Moschella, "Bassett Calling the Signals Perfectly in Selling 'Bandit Ball,'" *Fort Myers News-Press*, June 14, 1984.
  The colors — silver, red, and black — rivaled the Los Angeles Lakers': Ibid.
  When Jim Brown, the NFL's all-time leading rusher: Nick Moschella, "Bassett Is the USFL's Pied Piper," *Fort Myers News-Press*, June 14, 1984.
  "The scent hung around the press box for weeks": Dick Schneider, "To Bassett, It Was All Fun and Games," *Fort Myers News-Press*, May 15, 1986.
  "It was a skin and sex spectacular": Moschella, "Bassett Calling the Signals Perfectly."
77   *The most memorable promotional ploy*: Tim Povtak, "Tampa Bay Scoring Surge Breaks Denver's 3-Game Winning Streak," *Orlando Sentinel*, April 16, 1985.
78   *From afar, it made no sense*: Howard Balzer, "Signings in the USFL: A Chain Reaction," *Sporting News*, Jan. 24, 1983.
80   *Hebert was sacked 15 times*: Dan Donovan, "Former Steelers Having a Ball in USFL," *Pittsburgh Press*, July 16, 1983.
  "From junior high, through college and now": "Ex-Steelers Lead Panthers," *Port Huron Times Herald*, July 8, 1983.

6. Title Dreams

83   *He described his father as "hard-nosed and mean"*: John Husar, "Wisconsin Looking Up at Its David Greenwood," *Chicago Tribune*, Sept. 1, 1980.
84   *"[Corker] is a lean, mean sacking machine"*: Joe Santoro, "Miami's Corker Hopes Talent Will Overshadow Reputation," *Fort Myers News-Press*, Aug. 1, 1986.
86   *"Walker set a record of sorts by telling the same"*: Tom Callahan, "The Odd Season," *Time*, July 25, 1983.
88   *the older league did nothing to help Junior*: William N. Wallace, "A Difference, Too, for the USFL Party," *New York Times*, July 16, 1983.
  "The National Football League looked down": Callahan, "The Odd Season."
89   *"So instead of taking 35,000 at almost $12"*: Harry Usher, with Peter H. King, book proposal for *End Over End: An Inside Account of the Dizzying Demise of a Professional Football League*.
90   *Ed Sabol, the founder of NFL Films*: Reeths, *The United States Football League*, p. 111.
  nicknamed "the Game with No Name": Ibid., p. 113.
  That's why he arranged for beat writers to travel on team planes: Joe Lapointe, "Lots of Low-Lights in a Week with the USFL in Mile High City," *Detroit Free Press*, July 19, 1983.
91   *"The first huddle at the Denver Broncos"*: "Elway Makes Biggest Splash as Camps Open," *Lubbock Evening Journal*, July 13, 1983.
  The city of Denver hardly helped: Michael Knisley, "A Few Title-Game Snags," *Sporting News*, July 18, 1983.
92   *"I started wondering if I could play anymore"*: Alan Goldstein, "Star QB," *Pittsburgh Post-Gazette*, Feb. 9, 1985.
93   *He was born and raised 32 miles south*: Rick Telander, "Big Bucks for a Bayou Boy," *Sports Illustrated*, July 1, 1985.
94   *When the Panthers selected Hebert with their third-round pick*: Curt Sylvester, "This Time, Contract Dispute Involves Hebert," *Detroit Free Press*, Jan. 5, 1984.
  At the end of his first day in camp: "Obituary: Michael Christopher 'Mike' Hagen Jr.," *Missoulian*, Oct. 14, 2015.

95 *His outfit at press conferences always included a Mickey Mouse T-shirt*: Greg Bishop, "The End of the USFL," Sportsillustrated.com, 2015.

   Two years earlier, in the lead-up to Super Bowl XV: Ray Didinger, "Bunting: No Super Strain," *Philadelphia Daily News*, July 16, 1983.

96 *"The first Super Bowl didn't sell out, either"*: Lapointe, "Lots of Low-Lights in a Week with the USFL."

97 *The play call was split right A44*: Ralph Wiley, "The Panthers Are No. 1, Thanks to No. 1," *Sports Illustrated*, July 25, 1983.

98 *"I am No. 1 and the team is No. 1"*: William N. Wallace, "Panthers Guess Right," *New York Times*, July 19, 1983.

99 *One fan was removed on a stretcher*: Howard Balzer, "Denver Police vs. Spectators — It Was a Nasty Confrontation," *Sporting News*, July 25, 1983.

   "If it had been my decision, I would have let them": William N. Wallace, "Fans' Action Mars Title Game," *New York Times*, July 19, 1983.

   Like many papers across the nation: Steve Serby, "Rocky Horror Show," *New York Post*, July 19, 1983.

7. Craziness

104 *"It's a confusing trade"*: James Litke, "And Now, a Trade to End All Trades," *Journal News*, Oct. 1, 1983.

105 *"I've rebuilt four franchises"*: Myra Gelband, "Dose of Their Own Medicine," *Sports Illustrated*, Oct. 10, 1983.

106 *Five days later Memphis released 20 of the 21*: Reeths, *The United States Football League*, p. 178.

108 *"He's a guy that I thought could really make this league special"*: Ibid., p. 153.

109 *"Had that secretary not walked into the room"*: Chris Brown, "Bills All-Time Draft Memories: Jim Kelly," buffalobills.com, April 6, 2010.

   The Gamblers traded four future draft picks: "Millions Attract Kelly to USFL," *Tennessean*, June 10, 1983.

   "We considered three different offers": "Never Got Chance to Sign QB Kelly, Bills Maintain," Associated Press, June 12, 1983.

110 *"If someone wanted a perfect redneck"*: Doug Williams with Hunter, *Quarterblack*, p. 108.

   "I wasn't going to be their slave anymore": Ibid., pp. 108–9.

111 *"It's the highest paying [contract] in the history"*: "Big Deal for Fouts, Chargers," *Philadelphia Daily News*, June 30, 1983.

112 *"It was storming outside, and she started screaming again"*: Williams, *Quarterblack*, p. 109.

113 *A couple of days later, the* Pittsburgh Press *sports section*: Ron Cook, "Maulers Get First Pick in Draft," *Pittsburgh Press*, Jan. 3, 1984.

   Although it was technically against NCAA bylaws: Gerald Eskenazi, "Rozier's Agent Is Dismissed," *New York Times*, Jan. 17, 1984.

114 *Rozier, on the other hand, was a product*: Bob Sansevere, "Mike Rozier's Hometown," *Lead Daily Call*, Jan. 3, 1984.

116 *"The marriage had been consummated"*: Bruce Keidan, "Rozier: Pro Before His Time?" *Pittsburgh Post-Gazette*, Jan. 10, 1984.

The story that never ended refused to end: "Rozier Fires Agent, Says Attorney," *Green Bay Press-Gazette,* Jan. 16, 1984.

Stoudt told the Associated Press that it was "a great career move for me": "Stoudt Signs with USFL's Stallions," *Akron Beacon Journal,* Jan. 13, 1984.

A product of Auburn University: Jerry Argovitz and J. David Miller, *Super Agent,* pp. 236–37.

118   *Collinsworth made the announcement at the halftime:* "Florida Beckons," *New York Times,* June 27, 1983.

He was carted onto the Tampa Stadium field: Dick Schneider, "To Bassett, It Was All Fun and Games," *Fort Myers News-Press,* May 15, 1986.

"Our inclination is that we have no choice": Jeff Miller, "Hiring Infuriates Bengals' Paul Brown," *Jacksonville Sun Times,* July 13, 1983.

The coach, blessed with a sharp sense of humor: Reeths, *The United States Football League,* p. 168.

119   *A high school marching band:* Terry Galgano, "Lewis Is a $1 Million Showboat," *Anniston Star,* Feb. 6, 1984.

8. The Saviors

121   *"I found that, for me, owning a professional":* Sam Goldaper, "Generals Are Sold to Trump," *New York Times,* Sept. 23, 1983.

They took their $2 million profit and bolted: Chris Dufresne, "Derailing of the Express," *Los Angeles Times,* July 14, 1985.

"It is not practical to be an absentee owner in this day": Chris Dufresne, "Oldenburg Takes Over as Owner of Express, and Hires Klosterman," *Los Angeles Times,* Dec. 23, 1983.

122   *This is Manhattan through a golden eye:* Kranish and Fisher, *Trump Revealed,* p. 177.

When the office first opened in November: "The Unraveling of a 'Billionaire,'" *New York Times,* June 3, 1984.

123   *Meanwhile, the Generals and Express wasted little time:* Dave Raffo, "Last Year They Were the One-Star Generals," United Press International, Feb. 20, 1984.

as was Dallas Cowboy defensive tackle Randy White: Walker, *Fallen Generals,* p. 102.

124   *Shula had a hostile relationship with Joe Robbie:* Dave George, "Thrown for a Loss," *Palm Beach Post,* Sept. 1, 2005.

"Trump was sell, sell, sell": Ibid.

125   *After buying the Express:* Byrne, *The $1 League,* p. 104.

"Klosterman's directive from Oldenburg": Barr, "The Life & Death of the L.A. Express."

"[Oldenburg] wanted me to design a car": David J. Miller, "USFL Preview: Before the Fall," *Sport Magazine,* March 1985.

126   *"My nickname was 'Limo'":* Bill Murphy, "'Wild Man' Newt Gives Jets Depth," *Poughkeepsie Journal,* Oct. 2, 1979.

When it became clear the Express needed a strong special team: Chris Dufresne, "Express Was on Fast Track," *Los Angeles Times,* Aug. 12, 1992.

To know Bill Oldenburg was to dislike Bill Oldenburg: Byrne, *The $1 League,* p. 119.

127   *He confronted New Jersey's owner in a face-to-face encounter:* Ibid., p. 103.

128   *"He had me call my bank":* Kranish and Fisher, *Trump Revealed,* p. 177.

"We got on the phone with [Giants GM] George Young": Reeths, *The United States Football League* p. 159.

130 *"We'll get back to you":* Jack McCallum, "The Man with the Golden Arm," *Sports Illustrated,* March 12, 1984.

On February 19, 1984, Young drove to a small airstrip: Steve Young, with Benedict, *QB,* pp. 90–91.

136 *police in Provo, Utah, once tried to tow the vehicle:* Reeths, *The United States Football League,* p. 198.

137 *"I hope to fix up my car":* "Young Signs Largest Contract," Associated Press, March 6, 1984.

"The funniest part of Steve Young's debut": Jerry Greene, "Young's Big Contract Costs Express Little," *Orlando Sentinel,* March 7, 1984.

Before they parted, Steinberg had one last thing to do: Young, with Benedict, *QB,* p. 97.

138 *"As commissioner, I don't like it":* Bruce Lowitt, "$40 Million Question: Is Young Contract Crazy?" Associated Press, March 7, 1984.

139 *The motive first became apparent to Simmons:* Kranish and Fisher, *Trump Revealed,* p. 173.

140 *"Donald J. Trump, the new owner of":* William N. Wallace, "Trump Would Like to Take On NFL," *New York Times,* Sept. 30, 1983.

"it may be in Don Trump's best interests to pursue": Reeths, *The United States Football League,* p. 161.

141 *"I guarantee you folks in this room":* Kranish and Fisher, *Trump Revealed,* p. 174.

9. Wild Fields

143 *Yet on the morning of August 12, 1982:* "'I Absolutely Won't Leave,'" *Columbus (IN) Republic,* Aug. 13, 1982.

147 *From the very beginning:* Henry Walker, "A-Head," *Nashville Scene,* Feb. 15, 1996.

149 *Ricky Sanders was an obscure:* Jake Russell, "Redskin Legends: Ricky Sanders — Rise to Success," July 9, 2006, *Hog Wire, Washington Redskins* (blog), www.thehogs.net.

156 *Which (league honchos convinced themselves):* Bill Jauss, "Surgery for Blitz Image," *Chicago Tribune,* Feb. 5, 1984.

"I called the Hall of Fame in Canton, Ohio": Reeths, *The United States Football League,* p. 177.

Hoffman had paid a mere $500,000: Ibid., p. 175.

157 *So the Blitz sent Bud Holmes, Payton's agent:* "Payton's Agent: Offers Generous," *Arizona Republic,* Feb. 2, 1984.

During the second week of the preseason: Chuck Garfien, "The Story of the Chicago Blitz and the Craziest Trade in Sports History," csnchicago.com, Feb. 1, 2017.

158 *"They couldn't even buy toilet paper":* Reeths, *The United States Football League,* p. 209.

10. Gerbils and Snowballs

163 *The USFL opened the 1984 season:* Barr, "The Life & Death of the L.A. Express."

164 *"The good news for the Washington Federals":* Ron Cook, "Quarterback Woes," *Pittsburgh Press,* March 17, 1984.

165 *He was released, but spent the ensuing months back near:* "Settlement Reached with Morgantown in Fatal Crash," *Pittsburgh Press,* March 3, 1986.

Gist's death had been big news: Ed Bouchette, "Pass Ball, Spaghetti to Raugh," *Pittsburgh Post-Gazette*, April 5, 1984.

166 *The weather was atrocious — a steady downpour:* "Williams Puts Outlaws Past Maulers, 7-3," *Shreveport Times*, Feb. 27, 1984.

It was his lowest yardage total since being: Ed Bouchette, "Rozier, Mauler Debuts Outlawed," *Pittsburgh Post-Gazette*, Jan. 27, 1984.

167 *He became, according to Russ Franke of the:* Russ Franke, "Stoudt: Pittsburgh Snow Job Helped Birmingham Win," *Pittsburgh Press*, March 18, 1984.

"I tried to commit suicide": Reeths, *The United States Football League*, p. 155.

in 1981, during a visit to a Seattle country music lounge: John Clayton, "Bad Break: Stoudt Sacks Himself for Big Loss," *Pittsburgh Press*, Nov. 10, 1981.

168 *"We were second-class citizens":* Christy Johnson, "Lee Is New Grid Coach at Triway," *Wooster (OH) Daily Record*, March 20, 1009.

169 *Afterward, Pendry walked to his car:* "Sports Notes," *Tyrone Daily Herald*, March 12, 1984.

The team's facility was located inside: Ron Cook, "Travelin' Men," *Pittsburgh Press*, April 1, 1984.

In the early-morning hours of April 13: Ron Cook, "Maulers Withhold Huther's Pay over Injuries from Fight," *Pittsburgh Press*, April 25, 1984.

171 *"I remember the water flowing in":* Gene Frenette, "Where Are They Now: Former Jacksonville Bulls Coach Lindy Infante," *Florida Times-Union*, Aug. 7, 2010.

The Maulers lost 26–2: Gerry Dulac, "Pittsburgh Maulers: Just One of Those Years," *Pittsburgh Post-Gazette*, March 7, 2004.

173 *Raised in poverty by a single mother in nearby Philadelphia, Mississippi:* Bill Reynolds, "Marcus Dupree: Yesterday's News," *Pineview (LA) Town Talk*, August 31, 1986.

"By the time he was an eighth grader": Billy Bowles, "That 'Strange, Bewitching State . . .' of Marcus Dupree," *Detroit Free Press*, Oct. 30, 1983.

"He was the best player on the field": Dillon Bell, "Marcus Dupree: The Best That Never Was," https://prezi.com/dw6gzsitrro9/marcus-dupree-the-best-that-never-was, March 8, 2013.

174 *"If you'd have been in shape":* Austin Murphy, "The Oklahoma Kid," *Sports Illustrated*, Oct. 11, 2004.

He missed the deadline for arriving on campus: Spot McAllen, "Sooners' Marcus Dupree Has Heisman Pressure," *Greenville News*, Sept. 4, 1983.

"I figured," he said, "that I might as well try [to go pro]": Ralph Wiley, "Take the Money and Run," *Sports Illustrated*, March 12, 1984.

a University of Arizona punter named Bob Boris: "Judgment Favors Outlaws' Punter," *Oklahoman*, Feb. 29, 1984.

175 *Dupree wore black boots:* Kevin Spain, "When Marcus Dupree Arrived in New Orleans, It Was with Great Fanfare," *New Orleans Times-Picayune*, Nov. 11, 2010.

"He's got to perform": "Dupree Signs with USFL," *Indianapolis Star*, March 4, 1984.

"He really ran a lot better than we expected he would": Rick Bentley, "Breakers Top Showboats Before 45,269," *Alexandria (LA) Town Talk*, March 12, 1984.

11. Falling

177 *At but 35 years old, George Heddleston:* Jim O'Brien, "49er Publicist in Line for GM Job of USFL Franchise Here," *Pittsburgh Press*, July 5, 1983.

180 *"[Donald Trump] would keep after Pilson":* Byrne, *The $1 League*, p. 137.

"I think he became a Pied Piper": Reeths, *The United States Football League*, p. 251.

181 *"The* New York Times *is supposed to be the most respected"*: Paul Domowitch, "Bandits' Bassett: Switch Story Trumped Up," *Sporting News*, April 30, 1984.

185 *"Well, maybe the guy is worth all the money"*: Chris Dufresne, "Steve Young Leads Express — in Rushing," *Los Angeles Times*, April 2, 1984.

On one memorable afternoon, Gillman looked over a practice field: Young, with Benedict, *QB*, p. 103.

Yet no matter what magic Young pulled out of his hat: Ibid., p. 104.

186 *During a game against the Gold, Oldenburg grabbed:* Ralph Wiley, "This One Was a Game and a Half," *Sports Illustrated*, July 9, 1984.

"IMI apparently is, indeed, kind of magical": G. Christian Hill and Victor F. Zonana, "Loan Broker Is Said to Lure Weak S&Ls into Many Shaky Deals," *Wall Street Journal*, June 8, 1984.

187 *"There have always been people"*: G. Bruce Knecht, "The Unraveling of a 'Billionaire,'" *New York Times*, June 3, 1984.

"It may be that we didn't do our homework": Byrne, *The $1 League*, p. 173.

in the words of Jim Byrne, "arranged for additional financing": Ibid.

188 *Yet upon closer examination, the drop:* William Oscar Johnson, "For Spring Football, the Future Is No Go," *Sports Illustrated*, July 23, 1984.

Putting league before self, Bassett: "Schnellenberger Eyes USFL bid," *Pittsburgh Post-Gazette*, May 9, 1984.

189 *If Bassett required a game to back up his point:* Larry Woody, "Sellout Crowd Watches Showboat Loss," *Tennessean*, June 17, 1984.

190 *"I take him out for a big spare ribs dinner"*: Reeths, *The United States Football League*, p. 217.

194 *Unlike Trump, whose fortune had been inherited from a wealthy father:* Sally A. Downey, "From Synagogues to Sports Teams, 'Builder of Dreams,'" *Philadelphia Inquirer*, Sept. 16, 2012.

195 *"He hit me so hard I nearly lost consciousness"*: Young, with Benedict, *QB*, p. 106.

196 *"Flat on his back and in obvious pain"*: Wiley, "This One Was a Game and a Half."

196 *"I spent four hours in the bath [afterward]"*: Chris Dufresne, "Hadl Thinks Young May Be a Marked Man," *Los Angeles Times*, July 4, 1984.

"Their defense was dog-tired": Walt Jayroe, "Wranglers Outdraw Gamblers, 17–16," *Arizona Republic*, July 2, 1985.

197 *By refusing to press the Gamblers' speedy:* Bob Hurt, "Allen's Defensive Scheme One for the Books," *Arizona Republic*, July 2, 1985.

This was strange enough news, but the weather report for the 12:30 Saturday kickoff: Bud Wilkinson, "ABC, USFL Wilt, Give Wranglers 8:30 p.m. Start," *Arizona Republic*, July 6, 1984.

Afterward, Allen, 62 but spry: Ralph Wiley, "Hot in the Heat, by George," *Sports Illustrated*, July 18, 1984.

198 *"It would have to be a great offer"*: Dufresne, "Hadl Thinks Young May Be a Marked Man."

The USFL had approved a budget of $1.35: Byrne, *The $1 League*, p. 178.

199 *In his initial efforts to have Tampa host the game:* Ibid., p. 181.

"It makes us look bad that somebody's": "3 Wrangler Staffers Arrested," *Philadelphia Daily News*, July 13, 1984.

200 *Wiley, the excellent SI scribe, described Philadelphia's 23–3 romp:* Ralph Wiley, "The Wranglers Were Star Struck," *Sports Illustrated*, July 23, 1984.

12. Movement

205 *That's why, when — in the spring of 1984:* Dave Anderson, "USFL at the Crossroads," *New York Times,* July 1, 1984.

206 *"Patrick was thanked for her efforts":* Byrne, *The $1 League,* p. 197.

207 *"A pretty good size group of writers came":* Reeths, *The United States Football League,* p. 251.

208 *"The vote was eventually called unanimous, but it really wasn't":* Ibid., p. 253.

210 *"The USFL ceased being a league when owners":* Howard Balzer, "Vacuum at the Top," *Sporting News,* Sept. 8, 1984.

211 *"They think they know what they're doing, but they don't":* Walt Jayroe, "As Dream Ends, Bruce Allen Blasts Merger of Wranglers," *Arizona Republic,* Oct. 21, 1984.

212 *The Orioles had recently signed a three-year lease:* "Baltimore Shooting for Stars," *Camden Courier-Post,* Aug. 11, 1984.

213 *The man who submitted the victorious concept:* Brian Schmitz, "Call Them the Renegades," *Orlando Sentinel,* Oct. 19, 1984.

214 *"I find this is an extraordinarily important day":* "Portland Gets Breakers; Domed Stadium Urged," *Statesman Journal* (Salem, OR), Nov. 14, 1984.
his assets were valued to be between $100 million: "Express Find an Owner," United Press International, Oct. 30, 1984.
the film would be commissioned at the same time the team: Dufresne, "Derailing of the Express."

217 *"Doug is a better person than he is a":* Steve Wulf, "Mr. Touchdown Scores Again," *Sports Illustrated,* Feb. 4, 1985.
"I don't know if we'll be lucky enough to sign him": D. J. Arneson, *Doug Flutie,* p. 63.
In all the hubbub, Trump: Byrne, *The $1 League,* p. 257.

218 *If one asked Trump a football question:* Tim Rohan, "Donald Trump and the USFL: A 'Beautiful' Circus," SI.com, July 12, 2016.
Tim Rohan, the excellent Sports Illustrated *football writer:* Ibid.
"When a guy goes out and spends more money": Max J. Rosenthal, "The Trump Files: When Donald Demanded Other People Pay for His Overpriced Quarterback," *Mother Jones,* July 20, 2016.

219 *"He's a perfect college player":* Wulf, "Mr. Touchdown Scores Again."
"It was kind of mutual": Reeths, *The United States Football League,* p. 268.

220 *"It is . . . becoming clearer and clearer":* Michael MacCambridge, *America's Game,* p. 357.
The suit, which aimed to limit the NFL to two networks: "USFL Files $1.32 Billion Lawsuit Against NFL," *Ukiah Daily Journal,* Oct. 18, 1984.

221 *Yes, Roy Cohn — the attorney who discovered initial:* Michelle Dean, "A Mentor in Shamelessness: The Man Who Taught Trump the Power of Publicity," *Guardian,* April 20, 2016.
"a master performing on television": Nicholas von Hoffman, "The Snarling Death of Roy M. Cohn," *Life,* March 1988.
"I can only make analogy to a criminal case": "USFL Claims Secret NFL Group Gunning for It," *Tallahassee Democrat,* Oct. 19, 1984.

222 *Trump, sitting alongside Cohn:* Byrne, *The $1 League,* p. 224.
"The USFL's own missteps": Hal Lancaster, "USFL Facing a Fourth-and-Long As It Staggers into Third Season," *Wall Street Journal,* Feb. 28, 1985.

223 *"The networks have told the league to drop dead":* William Taaffe, "The Networks' Cupboards Were Bare," *Sports Illustrated,* Feb. 25, 1985.

13. Wobbling

224 *"I don't know," Bassett said:* Janice Castro, "On Flutie's Wing, and a Prayer," *Time*, March 4, 1985.

225 *Inside the newsroom of the* Tampa Tribune: Bob McCoy, "The USFL's Own Song," *Sporting News*, June 24, 1985.

226 *Naive to the ways of the Trump universe:* Doug Flutie with Lefko, *Flutie*, p. 48.

227 *"All we heard and read all week was Flutie":* Jerry Rutledge, "Stallions Out-rank Generals," *Anniston Star*, Feb. 25, 1985.

228 *could be spotted in his box, half-asleep:* Walt Jayroe, "Semitough," *Arizona Republic*, Feb. 20, 1985.

229 *The Sipe-Rozier pairing was a hot subject around the league:* "Stars Lose in Opener; Bulls' Sipe Sidelines," *Philadelphia Inquirer*, Feb. 25, 1985.
     Three free agents who stood out were Phil Simms: Pat Dooley, "San Diego QB Signs with Bulls," *Jacksonville Times-Union*, March 4, 1985.

230 *His contract was signed inside the office:* Chris Cobbs, "For $2.6 Million, Luther Can Be Bullish About the USFL," *Los Angeles Times*, March 5, 1985.

232 *During a game between Arizona State and Washington:* Walt Jayroe, "Kush Blames Conspiracy for Ouster," *Arizona Republic*, Oct. 18, 1979.

233 *"I'd kind of had it with him":* Jane Gross, "A Nonconformist, Even in His Exit," *New York Times*, Aug. 20, 1983.
     "If I've got to go through boot camp next month": Walt Jayroe, "Allen-Built Squad Uncertain of Kush," *Arizona Republic*, Dec. 15, 1984.

234 *The team's season-ticket base fell from 18,500:* William N. Wallace, "Outlaws Lacking Fans and Patience," *New York Times*, March 1, 1985.

235 *Houston won 41–20 behind Kelly's three:* "Gamblers Arrest Outlaws, but Lose Kelly to Knee Injury," *Galveston Daily News*, May 27, 1985.

236 *"The black man [Hickman] had a hold of the white man's [Teerlinck] mouth":* Walt Jayroe, "Fight with Player Reportedly Leads to Coach's Firing," *Arizona Republic*, May 28, 1985.

14. Blank Guns

239 *He never returned, and when his family relocated:* Paul Burka, "The Man in the Black Hat," *Texas Monthly*, June 1984 and July 1984.
     His dream was to become a watch repairman: Douglas Martin, "Clinton Manges, Volatile Texas Oilman and Rancher, Dies at 87," *New York Times*, Sept. 28, 2010.

240 *"There is a role for the bull in the pasture":* Roy Bragg, "South Texas Tycoon Manges Dies," *San Antonio Express-News*, Sept. 24, 2010.

241 *"It would choke a horse":* Michael Marks, "So Long, 'Slingers: An Oral History of the San Antonio Gunslingers," *San Antonio Current*, Sept. 29, 2015.
     Or maybe it should have alarmed some in league offices: "Gunslinger Strategy Confined to Courtroom," *Galveston Daily News*, Oct. 3, 1983.

242 *"The sideline is the worst possible place":* Edwin Shrake, "Darrell Should Win as Often," *Sports Illustrated*, Nov. 17, 1975.

244 *Larry Kuharich, the offensive coordinator:* Denne H. Freeman, "Texas Teams Enjoy Rousing USFL Debut," *Longview News-Journal*, May 27, 1984.

246 *last taken a snap* nine years earlier: "Buddy Roberts," *Bryan (TX) Eagle,* Dec. 9, 1976.

247 *"Tonight's winner of the 1984 Dodge Charger is":* Reeths, *The United States Football League,* p. 213.

249 *One notable injury was correctly reported:* Byrne, *The $1 League,* p. 177.

250 *The defense led the league in interceptions:* Tom Labinski, "Something Old, Something New for Guns," *New Braunfels Herald-Zeitung,* Jan. 10, 1985.

252 *San Antonio's punter was Ken Hartley:* "The New England Patriots Are Still Trying to Find a Punter," United Press International, Oct. 7, 1981.

254 *In 1984 the team employed Alvin White:* Richard Dunn, "Alvin White, Millennium Hall of Fame," *Daily Pilot,* Feb. 21, 2000.

257 *Neither Hahn nor Aiken dared say a word:* Brian Schmitz, "Renegades Disarm Gunslingers, 21–20," *Orlando Sentinel,* May 14, 1985.

258 *"I remember my eyes welled up":* Dave Joseph, "Bates Knows No Limits," *Fort Lauderdale Sun-Sentinel,* Nov. 14, 2004.

15. Hookers Have Business Cards?

267 *Ryan later complained to the league:* "USFL Writer Claims Sex Harassment," United Press International, March 14, 1985.

268 *The wayward existence turned even more complicated:* "The City of Philadelphia Is Suing the Baltimore Stars," United Press International, June 4, 1985.

271 *"Well, the L.A. Express really isn't owned":* Jim Murray, "Springtime Is for Lovers, Not Football," *Los Angeles Express,* Feb. 22, 1985.

272 *A month before kickoff, the USFL's other franchises:* Ralph Wiley, "Wish You Were Here," *Sports Illustrated,* March 4, 1985.

Three players were even sold off to (gasp!) the NFL: Michael Janofsky, "Express Find Bonus in Selling to N.F.L.," *New York Times,* Feb. 8, 1985.

"What killed them was the injured-player provision": Barr, "The Life & Death of the L.A. Express."

Rumors swirled that the team would relocate to Hawaii: Chris Dufresne, "Express Beats the Odds . . . Takes On Gamblers," *Los Angeles Times,* Feb. 24, 1985.

One afternoon John Hadl excused himself from a meeting: Young, with Benedict, *QB,* p. 108.

On another morning, a painter charged: Dufresne, "Derailing of the Express."

273 *What those select few in attendance viewed:* Wiley, "Wish You Were Here."

274 *Three days later the franchise dismantled:* Rick Reilly, "Stand Up, Sit Down, Take a Hike," *Los Angeles Times,* Match 4, 1985.

"We've still got the uniforms": Dufresne, "Derailing of the Express."

275 *To celebrate, Wymon Henderson, a cornerback:* Barr, "The Life & Death of the L.A. Express."

The same organization that canned the cheerleaders: Dufresne, "Derailing of the Express."

Young had skipped his college graduation: "Gagliano Leads Gold by Express with Two-Touchdown Performance," *Desert Sun* (Palm Springs, FL), May 31, 1985.

276 *"I was hoping that some pro team":* John Lynch, "A Peek at the Big Time: Instead of a Major Event, USFL Game at Pierce 10 Years Ago Was Minor League," *Los Angeles Times,* August 20, 1995.

277   *En route from the team's Manhattan Beach facility:* Young, with Benedict, *QB*, p. 110.
      Hadl, plopped down in the front seat, removed his checkbook: Tim Hoffarth, "June 15,
      1985: The Express at Pierce College, As I Remember It," www.insidesocal.com, June
      15, 2010.

278   *"I thought maybe the cheerleaders would decorate":* Chris Dufresne, "The Result of
      Express vs. Arizona Game: 8,200," *Los Angeles Times,* June 16, 1985.

279   *"I was the first person he looked at":* Barry Cooper, "Young Goes from Millionaire
      Quarterback to Reserve Fullback," *Orlando Sentinel,* June 22, 1985.

16. Cancer

280   *"They gave me the works":* Jill Lieber, "Rebels with a Good Cause," *Sports Illustrated,*
      May 27, 1985.

281   *"[Donald] was abusive to anybody who didn't":* Reeths, *The United States Football
      League,* p. 250.

282   *"We lost 15,000 season ticket holders":* Argovitz and Miller, *Super Agent,* p. 276.
      "After three years": Gerald Eskenazi, "Bassett Is Adamant on a New League," *New York
      Times,* May 5, 1985.

283   *"Once tall, handsome and athletic":* Byrne, *The $1 League,* p. 286.
      "I believe that Bassett never wanted the USFL": Dick Schneider, "To Bassett, It Was All
      Fun and Games," *Fort Myers News-Press,* May 15, 1986.

284   *"Many want to dump their highest-paid players":* "USFL Owners Find No Clues to
      Solve League's Troubles," *Arizona Republic,* July 15, 1985.

285   *"The current position is that if they want to go":* Paul Zimmerman, "A Heavenly Night
      for the Stars," *Sports Illustrated,* July 22, 1985.
      In a book proposal he later submitted to publishing houses: Usher, with King, book
      proposal for *End Over End.*
      Yet the letter in the game program was one of sunny optimism: Harry Usher, "Letter
      from the Commissioner," 1985 USFL Championship Game program, July 1985.

287   *Approximately one month before the title game:* Chris Thorne, "Meeting at the Mead-
      owlands," 1985 USFL Championship Game program, July 1985.

288   *"If they can drive the length of the field on us":* Zimmerman, "A Heavenly Night for the
      Stars."

290   *He professed to often being lazy:* Larry Weisman, "Jets' Newton Faces Uphill Battle
      Again," *Poughkeepsie Journal,* Aug. 10, 1982.
      Earlier in the game, TV cameras caught Sumner: Greg Bishop, "The End of the USFL,"
      SportsIllustrated.com, 2015.
      Hebert stood silently for 20 minutes: Ibid.

17. Legal Eagles

292   *Other than assuring journalists that:* Dave Anderson, "'L.' Is for Limbo," *New York
      Times,* July 14, 1985.

293   *In July, he declared the USFL's next season:* "Usher Prefers 12-Team USFL," *New York
      Times,* July 3, 1985.
      "I regret that the players are unpaid": Tracy G. Triefler, "Gunslingers Under USFL's
      Inspection," *Longview (TX) News-Journal,* July 24, 1985.

"We're in the habit of changing lemons into lemonades": "Breakers Forced to Release 39 Players," *Statesman Journal* (Salem, OR), Aug. 1, 1985.

"Maybe if we'd had a winning season in Portland": "Canizaro Won't Pay Breakers," *Statesman Journal* (Salem, OR), Nov. 13, 1985.

Joe Cribbs, the star halfback, asked out of his contract: Mike Abbott, "Cribbs Seeks Return to Life in the NFL," *Longview (TX) News-Journal*, July 24, 1985.

294   *"Now all the struggling league has to do is find a network"*: Kevin Mulligan, "TV Talk," *Philadelphia Daily News*, Nov. 15, 1985.

A few days later, the Gold hired: "Denver Team Hold Auction," *Poughkeepsie Journal*, Feb. 25, 1986.

"What I'd really like to do": Rick Telander, "Life with Lord Jim," *Sports Illustrated*, July 21, 1986.

"The Generals' future in the USFL": John Delery, "Generals Happy Their Hibernation's Over," *Morristown (NJ) Daily Record*, May 29, 1986.

"If we played the Chicago Bears": William N. Wallace, "2 Mergers Leave 8 Teams in USFL," *New York Times*, Feb. 20, 1986.

295   *By February, the USFL officially slimmed itself down to eight teams*: "USFL Announces New Alignment," *Hagerstown (MD) Morning Herald*, May 20, 1944.

296   *"The whole time I was in the USFL"*: Howard Balzer, "Leaving the USFL Behind," *Sporting News*, Aug. 15, 1985.

The Vikings also obtained Anthony Carter: "Carter Clears Waivers," *New York Times*, Aug. 24, 1985.

The native Louisianan agreed: "Saints Sign USFL Quarterback Hebert," *Los Angeles Times*, Aug. 8, 1985.

"I'm happy to be in a place where people": Balzer, "Leaving the USFL Behind."

"They arrived in the NFL by means of buy-outs": Rick Telander, "Hanging Around with the Big Boys," *Sports Illustrated*, Sept. 6, 1985.

"We always knew there was talent over there": Ibid.

"We shook hands": Delery, "Generals Happy Their Hibernation's Over."

297   *"He is an incredible self-promoter"*: Lois Romano, "Donald Trump, Holding All the Cards! The Tower! The Team! The Money! The Future!" *Washington Post*, Nov. 15, 1984.

298   *"has come to personify the two sides of modern corporate law practice"*: David Margolick, "Can a Tarnished Star Regain His Luster?" *New York Times*, Feb. 25, 1990.

His rate was $400 per hour, and his salary from: Richard Behar, "A Lawyer's Precipitous Fall from Grace," *Time*, June 24, 2001.

299   *The NFL did commission a paper*: William Nack, "Give the First Round to the USFL," *Sports Illustrated*, July 7, 1986.

in 1984 the NFL did hire a Harvard Business School: Robert B. Fiske Jr., *Prosecutor Defender Counselor*, p. 138. Don Worthington, "Bobby Hebert Signs Contract with Michigan," *Alexandria (LA) Town Talk*, Jan. 23, 1983.

The 46-page document, which the USFL got hold of: Will McDonough, "How NFL Captured the Case," *Boston Globe*, Aug. 17, 1986.

300   *A couple of weeks before the start of the trial*: MacCambridge, *America's Game*, p. 359.

301   *"a melodramatic tour de force"*: Ibid., p. 360.

Before the jury were nine charges against the NFL: Michael Janofsky, "Historic Week for Pro Football?" *New York Times*, July 28, 1986.

"A victory by the NFL or a USFL victory": Ibid.

302 *"merely pathetic, a sour has-been"*: Larry Felser, "$3 Court Case Produced Both Winners, Losers," *Sporting News,* Aug. 18, 1986.

"that the reason the USFL decided to move": Larry Felser, "'Smoking Guns' Could Shoot Down USFL," *Sporting News,* June 23, 1986.

303 *"They had everything their way at the beginning"*: McDonough, "How NFL Captured the Case."

Over the course of five interminable days: Michael Janofsky, "Court Hears of Trump Contact," *New York Times,* May 22, 1986.

304 *Trump's testimony was decidedly different*: Michael Janofsky, "Trump Describes Talks with Rozelle," *New York Times,* June 24, 1986.

Rozelle couldn't believe what he was hearing: Dave Goldberg, "Six Jurors Decide Fate of USFL," Associated Press, July 20, 1986.

"Rothman characterized Trump as the worst kind of snake": Richard Hoffer, "USFL Awarded Only $3 in Antitrust Decision," *Los Angeles Times,* July 30, 1986.

recalled Patricia Sibilia, a juror: Kranish and Fisher, *Trump Revealed,* p. 185.

305 *On July 29, 1986, at exactly 3:55 p.m.*: Michael Janofsky, "USFL Loses in Antitrust Case; Jury Assigns Just $1 in Damages," *New York Times,* July 30, 1986.

306 *Patricia McCabe, a reference clerk for AT&T*: Hoffer, "USFL Awarded Only $3 in Antitrust Decision."

"At one point, Mrs. Sanchez": Ira Berkow, "The Heat in the Jury Room," *New York Times,* July 31, 1986.

Everyone in the courtroom was standing: Fiske Jr., *Prosecutor Defender Counselor,* p. 144.

307 *It had, according to the jury*: Michael Janofsky, "Football Trial: View from Top," *New York Times,* August 3, 1986.

Yes, the jury would clear the NFL: Berkow, "The Heat in the Jury Room."

308 *Rozelle hadn't made it to the courtroom in time for the verdict*: Fiske Jr., *Prosecutor Defender Counselor,* p. 144.

Later on, Sanchez, also one who empathized: Hoffer, "USFL Awarded Only $3 in Antitrust Decision."

18. Suspended

310 *On August 7, 1986, the vast majority of the 530 players*: Walker, *Fallen Generals,* p. 373.

"There are a half dozen USFL players who will be stars": Jill Lieber, "So Long, USFL —Now What?" *Sports Illustrated,* Aug. 18, 1986.

The first expatriate to jump: Vito Stellino, "All but Six Stars Free to Look for Positions in NFL," *Baltimore Sun,* August 12, 1986.

311 *Gibson was beyond giddy*: Butch John, "Gibson Not Moaning over USFL's Demise," *Clarion-Ledger,* August 13, 1986.

This did not sit well with four-time Pro Bowler Tony Dorsett: Rick Telander, "Walker-Dorsett," *Sports Illustrated,* Nov. 17, 1986.

312 *"If this team does not pay me"*: "Walker, Bryant Move to NFL," Associated Press, August 14, 1986.

"I imagine some guys looked at him": Brian Schmitz, "USFL Players Winning Trial on the Field," *Orlando Sentinel,* Sept. 7, 1986.

313 *The Buffalo Bills reached four straight*: R. J. Ochoa, "Newton's Law: Nate Newton Owns #61," insidethestar.com, July 14, 2015.

314 *"playing for the New England Patriots in a preseason game"*: Walt Jayroe, "Express Seeks Greener Pastures," *Arizona Republic*, June 13, 1985.

On January 17, 1988, Harry Usher wrote an open letter: Harry Usher, "From Usher to Rozelle," *New York Times*, Jan. 17, 1988.

316 *Few took the league's demise worse*: Jerry Greene, "Engelberg's Death: A Loss to the Entire Football World," *Orlando Sentinel*, July 31, 1987.

He cursed 100,000 times per minute: Larry Guest, "Engelberg Spent His Life Easing Others' Burdens," *Orlando Sentinel*, July 31, 1987.

Engelberg referred to everyone as "Bubba": Greene, "Engelberg's Death."

When the USFL folded, Engelberg: Gary Taylor, "Ex-'Gades GM Engelberg Dead," *Orlando Sentinel*, July 30, 1987.

317 *"[Katolin] had the look of a fighter about him"*: Ed Meyer, "Browns Won't Lack Effort, If Center Katolin Has His Way," *Akron Beacon Journal*, Oct. 4, 1987.

Clare Farnsworth of the Seattle Post-Intelligencer *later came up*: Ethan Trex, "Scab Story: The 1987 NFL Strike," mentalfloss.com, March 12, 2011.

BIBLIOGRAPHY

Allen, George. *Strategies for Winning: A Top Coach's Game Plan for Victory in Football and Life.* New York: McGraw-Hill, 1990.

Allen, Jennifer. *Fifth Quarter: The Scrimmage of a Football Coach's Daughter.* New York: Random House, 2000.

Argovitz, Jerry, and J. David Miller. *Super Agent: The One Book the NFL and NCAA Don't Want You to Read.* New York: Sports Publishing, 2013.

Arneson, D. J. *Doug Flutie: USFL Superstar.* New York: Modern Publishing, 1985.

Byrne, Jim. *The $1 League: The Rise and Fall of the USFL.* New York: Prentice Hall, 1986.

Damergis, Mike. *USFL: The Rebel League the NFL Didn't Respect but Feared.* New York: Pennylane Productions, 2007.

D'Antonio, Michael. *The Truth About Trump.* New York: Thomas Dunne, 2015.

Dixon, Dave. *The Saints, the Superdome, and the Scandal.* Gretna, LA: Pelican Publishing, 2008.

Fiske Jr., Robert B. *Prosecutor Defender Counselor: The Memoirs of Robert B. Fiske Jr.* Kittery, ME: Smith/Kerr Associates, 2014.

Flutie, Doug, with Perry Lefko. *Flutie.* Toronto: Sports Publishing, 1998.

Garza, Rich, with Greg Singleton. *Available.* San Antonio, TX: Next Big Step, 2014.

Hollander, Zander. *The Complete Handbook of Pro Football.* New York: Penguin, 1990.

Hurt, Harry, III. *Lost Tycoon: The Many Lives of Donald J. Trump.* New York: W. W. Norton, 1993.

Johnston, David Cay. *The Making of Donald Trump.* Brooklyn, NY: Melville House, 2016.

Kranish, Michael, and Marc Fisher. *Trump Revealed: An American Journey of Ambition, Ego, Money, and Power.* New York: Scribner, 2016.

Levy, Marv. *Where Else Would You Rather Be?* New York: Sports Publishing, 2004.

Livsey, Laury. *The Steve Young Story.* Rocklin, CA: Prime Publishing, 1996.

MacCambridge, Michael. *America's Game: The Epic Story of How Pro Football Captured a Nation.* New York: Random House, 2004.

Polian, Bill. *The Game Plan: The Art of Building a Winning Football Team.* Chicago: Triumph Books, 2014.

Reeths, Paul. *The United States Football League, 1982–1986.* Jefferson, NC: McFarland, 2017.

Sporting News Official USFL Guide and Register, 1984. St. Louis, MO: Sporting News, 1984.

Sporting News Official USFL Guide and Register, 1985. St. Louis, MO: Sporting News, 1985.

Spurrier, Steve, with Buddy Martin. *Head Ball Coach: My Life in Football*. New York: Blue Rider, 2016.

Ueberroth, Peter, with Richard Levin and Amy Quinn. *Made in America*. New York: William Morrow, 1985.

Walker, Curtis. *Fallen Generals: The History of the New Jersey Generals, the USFL's Glamour Team (1983–1986)*. San Bernardino, CA: Curtis Walker, 2016.

Walker, Herschel, with Gary Brozek and Charlene Maxfield. *Breaking Free: My Life with Dissociative Identity Disorder*. New York: Simon & Schuster, 2008.

Williams, Doug, with Bryce Hunter. *Quarterblack: Shattering the NFL Myth*. Chicago: Bonus Books, 1990.

Williams, Steve, with Tom Caiazzo. *How Dr. Death Became Dr. Life*. Champaign, IL: Sports Publishing, 2007.

Winters, Bill, and JoAnne V. McGrath. *From the Outhouse to the Penthouse and Somewhere in Between: The Story of One Free Agent's Trip Through the Ranks of Pro Football*. Bloomington, IN: 1stBooks, 2002.

Young, Steve, with Jeff Benedict. *QB: My Life Behind the Spiral*. Boston: Houghton Mifflin Harcourt, 2016.

INDEX